ELECTRONIC ENGINEERING

PROCEEDINGS OF THE 4TH INTERNATIONAL CONFERENCE OF ELECTRONIC ENGINEERING AND INFORMATION SCIENCE (ICEEIS 2017), JANUARY 7–8, 2017, HAIKOU, P.R. CHINA

Electronic Engineering

Editor

Dongxing Wang
Harbin University of Science and Technology, Harbin, China

CRC Press
Taylor & Francis Group
Boca Raton London New York Leiden

CRC Press is an imprint of the
Taylor & Francis Group, an **informa** business

A BALKEMA BOOK

CRC Press/Balkema is an imprint of the Taylor & Francis Group, an informa business

© 2018 Taylor & Francis Group, London, UK

Typeset by MPS Limited, Chennai, India

All rights reserved. No part of this publication or the information contained herein may be reproduced, stored in a retrieval system, or transmitted in any form or by any means, electronic, mechanical, by photocopying, recording or otherwise, without written prior permission from the publishers.

Although all care is taken to ensure integrity and the quality of this publication and the information herein, no responsibility is assumed by the publishers nor the author for any damage to the property or persons as a result of operation or use of this publication and/or the information contained herein.

Published by: CRC Press/Balkema
        Schipholweg 107C, 2316 XC Leiden, The Netherlands
        e-mail: Pub.NL@taylorandfrancis.com
        www.crcpress.com – www.taylorandfrancis.com

ISBN: 978-1-138-60260-1 (Hbk)
ISBN: 978-0-429-46950-3 (eBook)

Electronic Engineering – Wang (ed)
© 2018 Taylor & Francis Group, London, ISBN 978-1-138-60260-1

Table of contents

Preface	IX
Reviewers/scientific committee members	XI

Measurement and selection of infant formula milk powder traceability granularity
W. Wang & X.Y. Gu
1

An algorithm for detecting atomicity bugs in concurrent programs based on
communication invariants
L.Y. Li, J.D. Sun & S.X. Zhu
5

Analysis of the light effect on the ZnPc/Al/ZnPc barrier height of OTFTs
S. Zhao, M. Zhu & Z.J. Cui
9

Robust visual tracking with integrated correlation filters
H. Zhang, L. Zhang, Y.W. Li, Y.R. Wang & L. He
13

A hash function based on iterating the logistic map with scalable positions
J.H. Liu & C.P. Ma
17

The design and implementation of a third-order active analogue high-pass filter
D.D. Han, H.J. Yang & C.Y. Wang
21

Properties of copper phthalocyanine thin film transistors fabricated in vertical structures
M.Z. Yang, D.X. Wang, X.C. Liu & Q.X. Feng
27

Active contour segmentation model based on global and local Gaussian fitting
F.Z. Zhao, H.Y. Liang, X.L. Wu & D.H. Ding
31

Linguistic multi-criteria group decision-making method using incomplete weights information
L.W. Qu, H. Liu, Y.Y. Zuo, S. Zhang, X.S. Chen, V.V. Krasnoproshin, C.Y. Jiang & H. Liu
37

The design of the CAN bus interface controller based on Verilog HDL
H.J. Yang & M.Y. Ren
41

Research on the technology development model of China Mobile Communications
Corporation based on three networks convergence
R. Zhang, Y. An & Y.S. Li
45

Design and implementation of UART interface based on RS232
H. Guo
49

The analysis of the stability of the repetitive controller in an Uninterruptible Power
System (UPS) inverter
J.J. Ma, W.W. Kong, J. Xu, X.W. Zang & Y.H. Qiu
53

Realisation of a remote video monitoring system based on embedded technology
X.Y. Fan, M.X. Song & S.C. Hu
57

Design and implementation of a remote monitoring mobile client in a litchi orchard,
based on an Android platform
G.X. Yu, J.X. Xie, W.X. Wang, H.Z. Lu, X. Xin & Y.H. Wang
61

Calculation and simulation of the impact torsional vibration response of a stator
current for a local fault in a rolling bearing
X.J. Shi, Q.K. Gao, W.T. Li & H. Guo
69

An analysis of the electromagnetic field of the winding inter-turn short-circuit
of an asynchronous motor
S.Y. Ding & Y. Wang
75

V

Continuous max-flow medical image segmentation based on CUDA 81
B. Wang, Y.H. Wu & X. Liu

Vulnerabilities detection in open-source software based on model checking 85
Y. Li, S.B. Huang, X.X. Wang, Y.M. Li & R.H. Chi

Design and realisation of an intelligent agricultural greenhouse control system 89
J.F. Liu, Y.X. Wu, H.L. Zhu & C. Liu

Research on intuitionistic fuzzy measure in decision making 93
S. Zhang, Z.C. Huang & G.F. Kang

Design and research of positioning function of automobile guidance system based on WinCE 97
N. Wang, B.F. Ao, H.P. Tian & F. Liu

Thermal shock resistance of $(ZrB_2 + 3Y\text{-}ZrO_2)/BN$ composites 101
L. Chen, Y.J. Wang, L.X. Zhou & Y. Zhou

Evaluation model for traffic pollution control using multi-attribute group decision-making
based on pure linguistic information 109
H. Liu, S. Zhang, V.V. Krasnoproshin, C.X. Zhang, B. Zhang, B. Yu, H.W. Xuan & Y.D. Jiang

A bid evaluation model of a hydropower project based on the theory of intuitionistic fuzzy sets 113
H. Liu, S. Zhang, C.X. Zhang, Y.P. Liu, B. Yu, B. Zhang, V.V. Krasnoproshin, H.W. Xuan & J.F. Gao

Realisation of a virtual campus roaming system based on the Unity 3D game engine 117
J.H. Dong, J.W. Dong & C. Liu

An efficient high-order masking scheme for the S-Box of AES using composite field arithmetic 121
J.X. Jiang, Y.Y. Zhao, J. Hou, X.X. Feng & H. Huang

Experimental study on the effect of hydroxyl grinding aids on the properties of cement 125
B. Yang, X.F. Wang, X.F. He, C.X. Yang & K.M. Wong

An input delay approach in guaranteed cost sampled-data control 129
L.Y. Fan, J.N. Zhang & J.J. Song

Design of a multilevel cache datapath for coarse-grained reconfigurable array 133
L. Wang, X. Wang, J.L. Zhu, X.G. Guan, H. Huang, H.B. Shen & H. Yuan

Research on path planning for large-scale surface 3D stereo vision measurement 139
Y.J. Qiao & Y.Q. Fan

Research on the application of a chaotic theory for electronic information
security and secrecy management 145
Y.Y. Lai

Research on feature selection algorithms based on random forest 151
D.J. Yao & X.J. Zhan

Mobile-terminal sharing system for B-ultrasonic video 155
Q.H. Shang & F.Y. Liang

A novel area-efficient Rotating S-boxes Masking (RSM) scheme for AES 159
J.X. Jiang, J. Hou, Y.Y. Zhao, X.X. Feng & H. Huang

Lower confidence limits for complex system reliability 163
Q. Wang & G.Z. Zhang

Stiffness analysis of heavy spindle tops 171
B.W. Gao & W.L. Han

Experimental research into fuzzy control strategy of hydraulic robot actuator 175
G.H. Han, C.J. Zhang, Y.N. Liu & Y.C. Shi

Research into DDoS attack detection based on a decision tree constructed in parallel on the
basis of mutual information 181
C. Zhao, H.Q. Wang, F.F. Guo & G.N. Hao

Random addition-chains based DPA countermeasure for AES 185
H. Huang, X.X. Feng, J. Hou, Y.Y. Zhao & J.X. Jiang

Diagonal gait planning and simulation of hydraulic quadruped robot based on ADAMS *J.P. Shao & S.K. Wang*	189
Design and implementation of a mobile intelligent programming learning platform based on Android *C.Z. Ji, B. Yu, Y. Wei & Z.L. Song*	193
Design and implementation of a UDP/IP hardware module based on a field-programmable gate array *Q. Li, G. Luo, G.M. Song, S.Y. Jiang, F. Yuan & C. Wu*	201
A method of building CAD models based on algebraic optimisation of geometric constraints *X.Y. Gao, Y.T. Liu & C.X. Zhang*	207
Use of face similarity to compare the shapes of two CAD models *X.Y. Gao, Y.N. Chen & C.X. Zhang*	211
The research status of direct imaging of exoplanets *Z.J. Han, D. Ren, J.P. Dou, Y.T. Zhu, X. Zhang, J. Guo & C.C. Liu*	215
Time-resolved spectrum research of laser induced Mg plasma emission *Z.Y. Yang, Q. Li, D.Q. Yuan, W.Q. Ni & H.B. Yao*	221
Analysis of the brightness of molten liquid ejection in a millisecond laser interaction with a silicon plate *L. Zhang, X.W. Ni & J. Li*	225
Design of a high linear VCO *M.Y. Ren, M.Y. Qin & B.Z. Song*	229
Design a second-order curvature compensation bandgap reference *M.Y. Ren, B.Z. Song & M.Y. Qin*	233
Study on the corona-resistance property of polyimide/alumina nanocomposite films at elevated temperature *S.T. Minh*	237
Design of CMOS bandgap voltage reference *H. Bao*	243
Amplifier design in an automatic gain control circuit *H. Bao*	245
Design and analysis of CMOS operational amplifier *S. Mallick*	249
Design and simulation of operational amplifier based on CMOS technology *S. Mallick*	251
Author index	253

Electronic Engineering – Wang (ed)
© 2018 Taylor & Francis Group, London, ISBN 978-1-138-60260-1

Preface

The 4th International Conference of Electronic Engineering and Information Science 2017 (ICEEIS2017) was held January 7–8, 2017 in Haikou, P.R. China. This conference was sponsored by the Harbin University of Science and Technology, China. The conference continued the excellent tradition of gathering world-class researchers, engineers and educators engaged in the fields of electronic engineering and information science to meet and present their latest activities.

The main role of ICEEIS 2017 was to bring together innovators from engineering – researchers, scientists, practitioners – to provide a forum to discuss ideas, concepts, and experimental results related to all aspects of electronic engineering and information science. In order to meet the high standards of CRC Press, the organization committee undertook efforts to do the following. Firstly, poor quality papers were refused after a reviewing round by anonymous referee experts. Secondly, periodically, review meetings were held between the reviewers (about six times) for exchanging reviewing feedback. Finally, the conference organization held several preliminary sessions before the conference. Through efforts of different people and departments, the conference was successful and fruitful.

In addition, the conference organizers invite a number of keynote speakers to deliver their speech at the conference. All participants had the chance to discuss with the speakers face to face, which was very helpful for participants.

We hope that you enjoyed the conference and found ICEEIS 2017 rewarding. We are looking forward to seeing more friends at the next conference.

Dongxing Wang

Electronic Engineering – Wang (ed)
© 2018 Taylor & Francis Group, London, ISBN 978-1-138-60260-1

Reviewers/scientific committee members

Zhang Liuyang	Harbin University of Science and Technology, China
Xiong Yanling	Harbin University of Science and Technology, China
Ji Guangju	Harbin University of Science and Technology, China
Zhao Bo	Harbin University of Science and Technology, China
Gao Wei	Harbin University of Science and Technology, China
Yu Xuelian	Harbin University of Science and Technology, China
Yang WenLong	Harbin University of Science and Technology, China
Shen Tao	Harbin University of Science and Technology, China
Song Mingxin	Harbin University of Science and Technology, China
Wei Qi	Harbin University of Science and Technology, China
Wu Haibin	Harbin University of Science and Technology, China
Liu Mingzhu	Harbin University of Science and Technology, China
Shi Yunbo	Harbin University of Science and Technology, China
Ren Mingyuan	Harbin University of Science and Technology, China

Electronic Engineering – Wang (ed)
© 2018 Taylor & Francis Group, London, ISBN 978-1-138-60260-1

Measurement and selection of infant formula milk powder traceability granularity

W. Wang & X.Y. Gu
School of Economics, Harbin University of Science and Technology, Harbin, China

ABSTRACT: Based on the analysis of the infant formula milk powder traceability system's structure, this thesis makes measurements of the traceability granularity levels according to Traceable Units (TUs) and Critical Traceability Points (CTPs). Combined with the production practice of Heilongjiang LD Dairy, this thesis draws the following conclusions: there is a positive correlation between traceability granularity levels, amounts of ingredient and complexity of the processing techniques; there is a negative correlation between traceability granularity levels and batch size of terminal products; and that dairy enterprises can choose different traceability granularity levels according to the consumers' demand for the traceability and quantity of recalled products.

Keywords: infant formula milk powder; traceability granularity; traceability units; critical traceability points

1 INTRODUCTION

Because of the frequent occurrence of infant formula milk powder safety accidents, the Chinese government had enhanced the supervision and regulation of the dairy enterprises. Since May 2014, all the infant formula milk powder production enterprises in China have been compelled to implement the traceability system. A traceability system is a kind of information system, which can realise the trace and track of product information, through the collection and recording of critical information in the production process. Among the related researches, Wang (2013) put forward that it is essential to implement an electronic information management system for infant formula milk powder products in China, and that this system should have the functions of tracing, tracking, regulating, query and warning. In the field of researches on a traceability system's efficiency, the concept and measurement methods of granularity in the information theory had been used. Bollen et al. (2007) pointed out that granularity reflects the sizes of traceable units, while Karlsen et al. (2010) put forward that the size of traceable units cannot reflect the granularity levels accurately. Through the research of an aquatic products traceability system, Karlsen et al. (2011) pointed out that optimal traceability granularity is the condition whereby the enterprises provide the information as sufficient and detailed as possible, as long as the cost is affordable. Qian et al. (2014) created a granularity evaluation model and made an evaluation on the performance of food processing enterprises' traceability systems.

Since the establishment of an infant formula milk powder traceability system in China is still at an initial stage, the above researches on granularity model can be references for the system design and performance evaluation of a Chinese infant formula milk powder traceability system. Therefore, based on the researches at home and abroad, this thesis introduces the concept of traceability granularity into the design and application of an infant formula milk powder traceability system. Combined with the production practice of Heilongjiang LD Dairy, the thesis studies the measurement and selection of traceability granularity from the perspective of dairy enterprises.

2 STRUCTURE OF TRACEABILITY SYSTEM

The infant formula milk powder's traceability system has three main parts, namely the collection of information, the design of product identifiers and the construction of a central database. The hierarchical structure of an infant formula milk powder's traceability system is shown in Figure 1.

(1) Information collection. Establishing an information recording system can realise the information collection of every segment in the infant formula milk powder's supply chain. The information recording can be by real time online collection. Alternatively, it can be by regular offline collection from the various sensors, portable devices or paper recording, followed by manual data input

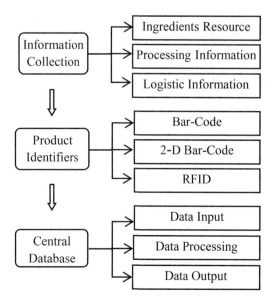

Figure 1. Hierarchical structure of infant formula milk powder's traceability system.

3 MEASUREMENT OF TRACEABILITY GRANULARITY

Because of the mandatory implementation policy, by October 2015, 102 enterprises in mainland China which produce infant formula milk powder had all established the traceability system. Although these traceability systems are of similar structure, the application effects differ greatly. In order to evaluate the construction and implementation conditions of traceability systems, this thesis introduces the concept of granularity. In information theory, granularity represents the degree of information attached on the research object. Granularity in the traceability system depends on the sizes and accuracy degree of traceability information. In order to describe the granularity clearly, Traceable Units (TUs) and Critical Traceability Points (CTPs) are used. TUs are the smallest descriptive information unit. The quantity of TUs can reflect the amount of information attached on the traceability granularity. CTPs are the nodes of information transfer in the traceability chain. The quantity of CTPs and the accuracy of coding will directly influence the accuracy of granularity information. Therefore, combined with the production practice of Heilongjiang LD Dairy, this thesis will discuss how to measure the granularity in a traceability system.

3.1 TUs

Since the production lines of infant formula milk powder work every day, a great amount of traceability information has also been produced with the flow of raw materials and terminal products. In order to realise the trace and track of production information, trace objects should be identified. TUs are the smallest descriptive information unit. Figure 2 shows the TUs in the production process of infant formula milk powder of Heilongjiang LD Dairy enterprises. It is found that every procedure in the processing can be regarded as being a TU, from the ingredients, to the mixture of ingredients, and finally to the end products with packing.

3.2 CTPs

According to the production practice of Heilongjiang LD Dairy, it is found that the production process of infant formula milk powder has several links, such as the reception of raw materials, the manufacturing of milk powder and packing. When the TUs are changed from one link to another, the information attached is vulnerable to being lost and damaged. For example, certain raw material is purchased and stored in bags in the warehouse. After a period of time, it will be mixed with other material, and be homogenised and dried together. These processes will gradually change the physical and chemical features of the material. So once the treatment and conversion have taken place, the partial information of traceability granularity may be lost and damaged.

into the computer. It includes the basic information of infant formula milk powder production enterprises, the product's processing information, the ingredients' resource information and the logistic information of the terminal products. The enterprises may gather different kinds of information according to the traceability requirements.

(2) Product identifiers. The Chinese infant formula milk powder market currently has mainly three kinds of identifying technologies, which are the bar-code, the two-dimension bar-code (2D bar-code) and Radio Frequency Identification (RFID). Bar-code is low in data storage capacity and space utilisation efficiency, and it is hard to repair the information once a bar-code is damaged. The data storage capacity of 2D bar-code is larger than that of bar-code, and the data can be encrypted. Compared with the bar-code and 2D bar-code, RFID has the following features: it is easy to read, it can be read from a long distance, the information stored is of high accuracy, and it cannot be affected easily by brutal and dirty environments. When choosing the product identifiers, enterprises will consider the packing, product value, and the traceability requirements of infant formula milk powders of different brands. Balancing the cost and the efficiency, 2D bar-codes are now widely used.

(3) Central database. In order to realise the trace and track of product information along the supply chain of infant formula milk powder, dairy enterprises need to establish the central database. The data gathered from the production, processing, distribution and sales processes, is integrated and stored in the database.

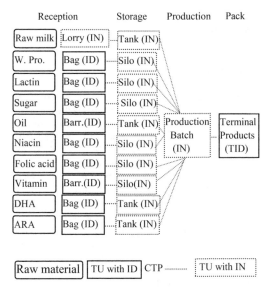

Figure 2. Identification of TUs and CTPs.

In order to safeguard the accuracy of traceability information, it is essential to identify the nodes of these changing links. The best way to identify the nodes is by coding. Every batch of material has unique numbers, which are made by the material suppliers, and similarly with the personal IDs. Before entering the production line, these materials will be stored in certain containers. These materials will be given new coding according to the reference numbers of the containers, which are the Internal Numbers (INs).

CTPs are the points at which the coding number of the materials or products changes, such as where the coding of the materials is changed from ID to IN. From Figure 2, it is found that there are three kinds of CTPs during the production of infant formula milk powder, which are the CTPs between the material IDs and container INs, the CTPs between the container INs and production batch INs, and the CTPs between the production batch INs and terminal products IDs. The terminal products IDs can also be called the Traceability IDs (TIDs), which will appear on the package of terminal products. The TIDs may be in the form of a bar-code, 2D bar-code or RFID. Consumers can make inquiries and trace the products' information through these TIDs.

3.3 Selection range of traceability granularity

The traceability granularity of infant formula milk powder can be represented by the number of TUs and CTPs, while in the practice of dairy enterprises, TUs and CTPs are related to the kinds of raw materials, complexity of processing techniques and the batch size of terminal products. Generally speaking, the more are the kinds of the raw materials, the larger are the numbers of the TUs, and then the traceability granularity is much finer. The more complex the producing techniques, the larger are the number of the CTPs, the more is the change of the coding, and then the traceability granularity is much finer. Since kinds of raw material and complexity of process techniques are constant when the formula of infant formula milk powder is fixed, the relationship between traceability granularity and terminal product batches will be discussed, based on the case of Heilongjiang LD Dairy.

According to production practice of Heilongjiang LD Dairy, the output of infant formula milk powder X is 3.6 tonnes for each working day. If 900 g of infant formula milk powder are packed in a can, LD Dairy will produce 400 cans of milk powder per day. If the production lines operate five days a week, or 52 weeks a year, LD Dairy will produce 20,000 cans each week, and 1,040,000 cans a year. When one can is chosen as batch size, every can of infant formula milk powder has a unique TID. When one day is chosen as batch size of the terminal products, the 400 cans of products share the same TID, and the enterprise will need 260 TIDs in a year. Similarly, if one week is chosen, then there will be 52 TIDs in a year; if one year is chosen, 1,040,000 cans of infant formula milk powder will share the same TID. If the information provided by the traceability system is comprehensive and accurate, it is called 'fine granularity'; conversely, if the information is general and vague, it is called 'coarse granularity'. So, traceability granularity is related to the batch size of terminal products. The bigger the batch size is, the coarser the traceability granularity, and vice versa.

4 SELECTION OF TRACEABILITY GRANULARITY

When selecting the traceability granularity, dairy enterprises will consider the information demand of consumers and the quantity of recalled products.

4.1 Consumers' demand on traceability information

Although the granularity level of taking one can as the batch size is obviously higher than that of one year, if the information consumers' need is about origin of place, then the granularity level of one year can satisfy the need. If consumers want to know the information about the raw materials' supply and quality inspection, then the granularity level of one can or one day will be more appropriate. For dairy enterprises, the finer the traceability granularity that is selected, the higher the cost that will be spent on the construction and application of the traceability system. Therefore, dairy enterprises should pay close attention to the changes in consumers' demand on traceability information, and adjust the granularity level according to the information types and information amount of the consumers' demand.

Table 1. Relationship between granularity level and quantity of recalled products.

R.M.	A	B	C	D	E
Raw milk	10,000	4th–5th	2,500	75.69	4,444
W.Pro.	2,000	4th–5th	475	14.42	4,666
Lactin	500	2nd–3rd	67	2.02	8,291
Sugar	500	2nd–3rd	65	1.96	8,546
Oil	1,000	2nd–3rd	126	3.81	8,817
DHA	100	3rd–4th	20	0.60	5,555
ARA	100	3rd–4th	20	0.60	5,555
Vitamin	100	3rd	15	0.45	7,406
Niacin	100	2nd	10	0.30	11,110
Folic acid	50	2nd	5	0.15	11,110

A: Purchase quantity of the raw material each time (kg)
B: Purchase frequency
C: Quantity of raw material used for one tonne (kg)
D: Meddling percent of a specific in a mixture (%)
E: Quantity recalled if raw material contaminated (kg)

4.2 Quantity of recalled products

Except function of information tracing of products, the traceability system can also be used in the recall of tainted products. It is necessary to discuss the relationship between the granularity level and the quantity of recalled products. Recently, contamination of raw materials has been one of the main factors causing the poor quality and safety accidents involving infant formula milk powder. Therefore, the influence of granularity level on the quantity of recalled products is studied, based on the production practices of the Heilongjiang LD Dairy.

Table 1 shows the detailed production data of infant formula milk powder X of Heilongjiang LD Dairy. It is known that ten kinds of raw materials are needed in the production, and these raw materials come from seven suppliers. Take raw milk as an example, where 2,500 kg of raw milk are used to produce 1 tonne of infant formula milk powder. If 900 g of milk powder is packed in a can, 1,111 cans of milk powder can be produced. If 10,000 kg of raw milk are purchased at one time, then 4,444 cans of milk powder can be produced using this batch of raw milk. In case this batch of raw milk is contaminated, if the granularity level is one can, then 4,444 cans of products will be recalled; if the granularity level is one day, then 8,000 cans of products will be recalled; if the granularity level is one week, then 40,000 cans of products will be recalled. In conclusion, granularity levels can affect the quantity of recalled products. The finer traceability granularity means a smaller quantity of recalled products. Besides, the quantity of recalled products can also be decreased through reducing the amount of single purchases and increasing the frequency of purchases. Take folic acid as another example of an ingredient. If the purchase amount is reduced from 50 kg to 25 kg, then the number of recalled products can decrease from 11,110 to 5,555 cans.

5 CONCLUSIONS

Combined with the production practice of Heilongjiang LD Dairy, this paper discusses the measurement and selection of traceability granularity from the aspect of enterprises. It is found that the traceability granularity can be described by TUs and CTPs. The granularity levels positively relate to the amount of ingredients and the complexity of processing techniques, and negatively relate to the batch size of terminal products. When choosing the granularity levels, dairy enterprises should consider the information of consumers' demand and the amount of recalled products. Infant formula milk powder enterprises should pay close attention to the changes of consumers' demands on traceability information, especially the information types and information amount. Since the finer the granularity levels are, the smaller is the quantity of recalled products, and enterprises should select the finest granularity level at a bearable cost.

ACKNOWLEDGEMENT

The authors gratefully acknowledge the support of the National Social Science Fund (grant No. 15BJY108).

REFERENCES

Bollen, A.F., Riden, C.P. & Cox, N.R. (2007). Agricultural supply system traceability, Part I: Role of packing procedures and effects of fruit mixing. *Biosystems Engineering*, 98(4), 391–400.

Karlsen, K.M., Donnelly, K.A.-M. & Olsen, P. (2011). Granularity and its importance for traceability in a farmed salmon supply chain. *Journal of Food Engineering*, 102(1), 1–8.

Karlsen, K.M., Olsen, P. & Donnelly, K.A.-M. (2010). Implementing traceability: Practical challenges at a mineral water bottling plant. *British Food Journal*, 112(2), 187–197.

Kim, H.M., Fox, M.S. & Gruninger, M. (1995). An ontology for enterprise modeling. In *Proceedings of the Fourth Workshop on Enabling Technologies: Infrastructure for Collaborative Enterprises WETICE 95* (pp. 105–116). Los Alamitos, USA.

Qian, J.P., Liu, X.X., Yang, X., Xing, B. & Ji, Z. (2014). Construction of index system for traceability granularity evaluation of traceability system. *Transactions of the Chinese Society of Agricultural Engineering (Transactions of the CSAE)*, 30(1), 98–104.

Qian, J.P., & Yang, X.T., et al. (2015). Model for traceability granularity evaluation of traceability system in agricultural products. *Systems Engineering*, 35(1), 1–6.

Wang, R.L. (2013). Regulation and traceability of food safety: The case of infant formula milk powder. *Information China*, 2013(Z1), 86–88.

Electronic Engineering – Wang (ed)
© 2018 Taylor & Francis Group, London, ISBN 978-1-138-60260-1

An algorithm for detecting atomicity bugs in concurrent programs based on communication invariants

L.Y. Li, J.D. Sun & S.X. Zhu
Harbin University of Science and Technology, Harbin, China

ABSTRACT: Communication between threads may occur when the multi-core program is running, and the uncertainty of execution order may lead to concurrency bugs. The atomicity characteristic is extremely important for the research into concurrency bug detection. To address the problem, an algorithm based on communication invariants was proposed in this paper to detect atomicity bugs. It can be proved by experiment that atomicity bugs can be efficiently detected by this algorithm.

Keywords: concurrency bug detection; atomicity violation; invariant; concurrent program

1 INTRODUCTION

At present, the parallel architecture has become a typical configuration. The multi-thread technique can improve the efficiency of a program. However, it can cause a concurrency bug that is nondeterministic. An example of this is the 2003 Northeast blackout, which caused a loss of more than $30,000,000,000 (Poulsen, 2004). How to detect the concurrency bugs has been the key to reducing the loss.

Concurrency bugs can be divided into three types: data races, atomicity violations and ordering violations. Some researchers think that the program atomicity characteristic is extremely significant for the research of concurrency bug detection. The atomicity violations often appear in some famous software, like Apache and MySQL (Fox et al., 2009). Therefore, a detecting algorithm for atomicity bugs is proposed in this paper.

Generally, the research of concurrency bug detection can be divided into three categories: symptom-based detection, invariant-based detection, and dynamic bug avoidance (Alam et al., 2008). Among them, the invariant-based detection is widely researched, and some novel detection methods have been proposed, such as AI, Buggaboo, DefUse, and DIDUCE (Hangal & Lam, 2002; Lucia, 2008; Shi et al., 2010; Zhang et al., 2014). If any invariant is violated, it is highly probable that a concurrency bug occurs.

In summary, an algorithm is designed in this paper to extract the communication invariants. It uses a hash table to divide traces into groups, and then checks the relationship between invariants to determine whether there is an atomicity bug. Context is added into the detection step to improve the bug recognition ratio.

2 RELATED WORK

In this paper, we select the Read-After-Write (RAW) dependency as the communication invariant (Lee & Tucker, 2012). The invariants are represented by ordered pair, <W, R>. Obviously, the atomicity bugs will occur when programs access the shared variable in different thread.

The communication invariants are <LcWr, LcRd> and <RmWr, LcRd>. <LcWr, LcRd> means a local thread read data from the shared variable which is written by a local thread. <RmWr, LcRd> means a local thread read data from the shared variable which is written by a removed thread (Lucia & Ceze, 2009). Whether the write instruction is remote or local is relative to the read instruction.

The program named Bank, which is shown in Figure 1, demonstrates an instance which has atomicity bugs. It may be run in the following manner. When the main thread runs to pthread_create, the program is switched to thread1. The main thread will not be executed until thread1 ends. Because the variant num is modified by thread1, the output is 20 rather than 30. Note that the code in line 10 writes data to the num, and the code in line 15 reads data from it. The invariant, <LcWr, LcRd>, has occurred and if thread1 modifies the num, then another invariant, <RmWr, LcRd>, is engendered. These invariants both intersect each other, which leads to the atomicity bugs.

Atomicity violations do not necessarily lead to bugs. Figure 2 shows a program segment with benign atomicity violations. The J1 and J2 become an invariant—<LcWr, LcRd>. Also, the I1 and J2 become a <RmWr, LcRd>. They intersect each other. However, they do not lead to the bug, and also meet the expectation of programmers.

```
1  pthread_t t;
2  void* thread1(int *num)
3  {
4      *num = 0;
5      return;
6  }
7
8  int main()
9  {
10     int num = 10;
11     int sum = 0;
12     void *ret1;
13     int base = 20;
14     pthread_create(&t,NULL,thread1,&num);
15     sum = base + num;
16     pthread_join(t,&ret1);
17     printf("%d\n",sum);
18     return 0;
19 }
```

Figure 1. A program with atomicity bugs.

```
Thread 1        Thread 2
                J1 flag = 0;
                J2 while(flag != 1)
                J3 {
I1 flag = 1     J4  ...
                J5 }
```

Figure 2. A program segment with benign atomicity violations.

3 ALGORITHM DESIGN

The system for detecting atomicity bugs can be divided into three modules: trace extraction module, trace analysis module, and bug detection module.

3.1 Trace extraction

Pin is used to extract program execution traces. Pin, which is developed by Intel(R), is widely used for the instrumentation of programs (Intel, 2016). In this module, Thread ID (TID), Instruction Counter (IC), Instruction Pointer (IP), read or write operation (W/R), and the variable address are extracted. Figure 3 shows the trace extraction process.

This Figure shows that when traces are extracted, the following operations should be done. If the repeated write instructions happened and were in the same thread, only one write instruction should be recorded. If repeated read instructions happened and were in the same thread, only one read instruction should be recorded as well.

3.2 Invariant extraction

Firstly, the traces are excessive. Each of the original traces collected by Pin only records one assembly instruction without contexts. Therefore, it is essential

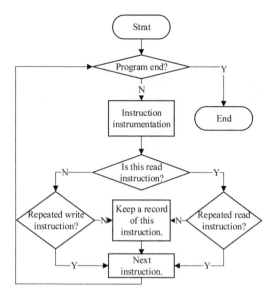

Figure 3. The trace extraction process.

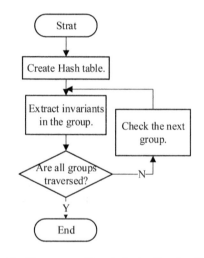

Figure 4. The process of invariant extraction algorithm.

to keep a relative order of read and write operations when the invariants are extracted.

Then, it is possible to lead to concurrency bugs only when the program accesses the shared variable. So, the unshared variable is not essential to be recorded.

The process of the invariant extraction algorithm is shown in Figure 4. The expatiation is as follows.

Firstly, we use a hash table to divide this data and remove redundant data.

Then, if the group is NULL, it means that the invariants in this group are extracted completely. The invariants should be extracted from the next group. Otherwise, extract invariants from this group continuously.

In each group, before the read operation, if there is a write operation which is in the same thread as the write operation which is pointed to by the pointer, then

```
    Thread 1                Thread 2
 I1 p = malloc(sizeof(int));
                         J1 if(p != NULL)
    I2 free p;
                         J2 *p = 10;
```

Figure 5. A program segment with atomicity bug.

```
    Thread 1    Thread 2
 I1 X = 1;
               J1 while(X != 0)
                   {…}
    I2 X = 0;
```

Figure 6. A program segment with no bugs.

the pointer points to the next write operation. Otherwise, the invariant is matched successfully. Repeat this operation until all invariants are extracted completely from this group. Then check the next group.

Finally, if all invariants are extracted from all groups, then the algorithm ends.

3.3 Atomicity bug detection

Another bug will be proposed in this section, in addition to the atomicity bug mentioned in Section 2. The bug is shown in Figure 5.

Both invariants, <I1, J1> and <I2, J2>, would occur in the program segment. Due to the execution of I2, the atomicity of Thread 2 is destroyed. As a result, due to the wrong execution, an atomicity bug occurs.

By contrast, it may be the benign atomicity violations. Figure 6 shows a program segment that violates the atomicity, but there is no bug in it.

Regarding the above described problems, the bug detection algorithm shown in Figure 7 is proposed.

Firstly, sort the invariants by the IC of the read operation, and remove the redundant data.

Then, in order to increase the correctness ratio of detection, the context is added. If all invariants traversal is complete, the detection ends. Otherwise, detect five consecutive invariants. If two of them meet the following relations, the atomicity bug occurs: (1) Their read operations have the same IC, and the latter's read and write operations are in different threads. In addition, their write operations are in different threads; and (2) They access the same shared variable, but IPs of their read operations are different. Their read operations are in the same thread, but their write operations are in different threads.

Finally, check the next five invariants until all invariants are traversed.

4 EXPERIMENT

In this paper, four programs, FFT, LU, RADIX, and Bank, are detected respectively by the atomicity bug detection algorithm.

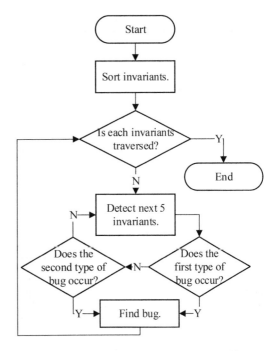

Figure 7. The process of detection.

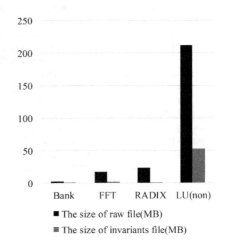

Figure 8. The comparison of the quantity of invariants before and after extraction.

The comparison of the files is shown in Figure 8. The size of original trace file is large before extraction, but after extraction, a large amount of redundant data is removed.

The experimental results are shown in Table 1 and Table 2, respectively before and after the bugs are injected.

The results show that the algorithm for detecting atomicity bugs in concurrent programs based on communication invariants can detect the atomicity bugs efficiently.

Table 1. Experimental results before injecting bugs.

Program	Code amount	Thread amount	Running time s	Number of bugs
Bank	32	2	0.12	0
FFT	898	4	5.19	11
LU(non)	677	4	117	80
RADIX	832	2	3.54	12

Table 2. Experimental results after injecting bugs.

Program	Code amount	Thread amount	Running time s	Number of bugs inject	Number of bugs bugs
Bank	32	2	0.13	1	1
FFT	898	4	5.99	1	12
LU(non)	677	4	141.10	1	88
RADIX	832	2	3.44	1	13

5 CONCLUSIONS

Concurrency bugs may occur randomly due to the uncertainty of the execution order of the concurrent program at run-time. Concurrency bugs have become the limit. It is vital to know how to detect concurrency bugs. In order to address this problem, an algorithm for detecting the atomicity bugs is proposed in this paper.

This algorithm detects bugs offline, and it can reduce the impact of the test program executions. In the invariant extraction step, a hash table based invariant extraction algorithm improves extraction efficiency. The experiment results show that this algorithm can extract communication invariants by analysing program traces, and it can detect atomicity bugs efficiently after feature matching.

ACKNOWLEDGEMENTS

This research has been supported by National Natural Science Foundation of China (61502123), Heilongjiang Provincial Youth Science Foundation (QC2015084) and China Postdoctoral Science Foundation (2015M571429).

REFERENCES

Alam, M.U., Begam, R., Rahman, S. & Muzahid, A. (2008). Concurrency bug detection and avoidance through continuous learning of invariants using neural networks in hardware.

Fox, A., Griffith, R., Joseph, A., Katz, R., Konwinski, A., Lee, G. & Stoica, I. (2009). Above the clouds: A Berkeley view of cloud computing. *Dept. Electrical Eng. and Comput. Sciences, University of California, Berkeley, Rep. UCB/EECS*, 28(13).

Hangal, S. & Lam, M.S. (2002). Tracking down software bugs using automatic anomaly detection. In *Proceedings of the 24th international conference on Software engineering* (pp. 291–301). IEEE.

Intel. (2016). *Pin 3.0 user guide.*

Lee, J. & Tucker, K.N. (2012). Learning? An examination of data based concurrency bug detection.

Lucia, B. (2008). Buggaboo-A machine learning method for determining the bugginess of concurrent program executions (or: CSE573 Final Project Report).

Lucia, B. & Ceze, L. (2009). Finding concurrency bugs with context-aware communication graphs. In *Proceedings of the 42nd Annual IEEE/ACM International Symposium on Microarchitecture* (pp. 553–563).

Poulsen, K. (2004). Software bug contributed to blackout. Security Focus.

Shi, Y., Park, S., Yin, Z., Lu, S., Zhou, Y., Chen, W. & Zheng, W. (2010). Do I use the wrong definition?: DeFuse: Definition-use invariants for detecting concurrency and sequential bugs. *ACM Sigplan Notices*, 45(10) 160–174.

Zhang, M., Wu, Y., Lu, S., Qi, S., Ren, J. & Zheng, W. (2014). AI: A lightweight system for tolerating concurrency bugs. In *Proceedings of the 22nd ACM SIGSOFT International Symposium on Foundations of Software Engineering*, (pp. 330–340).

Electronic Engineering – Wang (ed)
© 2018 Taylor & Francis Group, London, ISBN 978-1-138-60260-1

Analysis of the light effect on the ZnPc/Al/ZnPc barrier height of OTFTs

S. Zhao, M. Zhu & Z.J. Cui

Key Laboratory of Engineering Dielectrics and its Application, Department of Electronic Science and Technology, College of Applied Science, Harbin University of Science and Technology, Heilongjiang Harbin, China

ABSTRACT: The device structure of the organic photoelectric thin-film transistors is ITO/ZnPc/Al/ZnPc/Cu. These test results showed that the height of the barrier was reduced under light conditions, which can increase the current. The light amplification factor of the current between the emitter and the collector was analysed. It can be shown that the amplification range is 2.3–3.2 times when the voltage between the emitter and the collector is 2 V, that the range is 1.8–2.8 times when the voltage between the emitter and the collector is 3 V, and that it is bigger when the voltage between the base and the collector is 1 V, as compared to when the voltage is 0 V. It can be concluded that the device by illumination can improve the current, and with the increase of the base voltage and the decrease of the collector voltage, the effect of the light amplification improves.

Keywords: zinc phthalocyanine; the current-voltage characteristics; the Schottky barrier; the light amplification factor

1 INTRODUCTION

In recent years, the research and application of organic materials has been increasing day by day. Optoelectronic devices using organic materials can be used in low-cost, large-area and flexible devices, especially LCDs and OLEDs. An organic thin-film transistor with a vertical structure has high stability and high luminous efficiency in display, and it has successful achieved of the application of the flexible devices in organic display drivers (Kudo & Chiba, 2006; Cho et al., 2006; Uemura et al., 2012). The transistor has a very short conductive channel, so it will reduce the driver voltage and improve the working current effectively (Zorba & Gao, 2005).

Phthalocyanine materials have been widely used in the field of optoelectronics, for example in photovoltaic devices, optoelectronic sensors and image sensors. In order to better realise the photosensitive characteristics of optoelectronic devices, this paper used ZnPc as the active layer of the organic thin-film transistor. Zinc phthalocyanine is also called ZnPc. It is a transport type material of hole, which has good photoelectric properties (Schon & Kloc, 2007; Eder et al., 2004; Heremans, 2014).

2 PREPARATION OF THE DEVICES AND ANALYSIS OF THE STRUCTURE

The substrate material of the photoelectric transistor is glass. Each substrate can be formed from four ZnPc photoelectric TFTs, and the structure of each transistor is ITO/ZnPc/Al/ZnPc/Cu.

Figure 1 is the structure schematic diagram and the characteristic test circuit of a single OTFT. Figure 1(a) shows the three-dimensional structure of an OTFT, and Figure 1(b) shows the characteristic test circuit of an OTFT.

The process of preparation is as follows. First, the ITO collector on the glass substrate is prepared by RF magnetron sputtering; then, the first ZnPc film is prepared by vacuum evaporation. Second, the Al gate electrode on the first ZnPc layer is prepared by DC magnetron sputtering; then, the second ZnPc film is prepared. Third, the Cu emitter is prepared by DC magnetron sputtering. Finally, lead out all electrodes by pressing indium particles. The purity of Cu and Al are 99.9%, the temperature of evaporated ZnPc films is 350°C, and the substrate temperature is room temperature. All measurements are carried out at room temperature.

Figure 2 is the ideal energy band diagram of OTFTs. According to the contact theory of metal-semiconductors, the Al electrode and ZnPc forms Schottky contact; ZnPc and other materials (ITO and Cu) both form ohmic contact. The thickness of Al film is directly proportional to the size of the Schottky barrier. Therefore, the thickness of Al film has a great influence on the performance of the device. So, the thickness of Al film should be optimised to a desired thickness. To avoid the increase of leakage current, Al film thickness is not too thin. To avoid carriers not through the film normally, Al film thickness is not too thick.

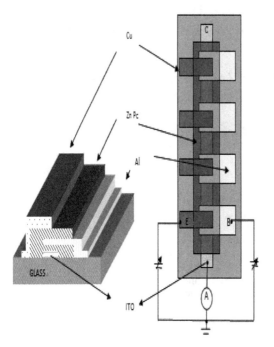

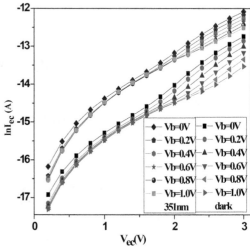

Figure 3. The characteristic curve of V_{ec} and $\ln I_{ec}$, both in the dark and under 351 nm illumination of ZnPc OTFTs. Relationship between $\ln I_{ec}$ and V_{ec} when V_b is from 0.2 V to 1.0 V step 0.2 V, and V_{ec} is from 0 V to 3.0 V step 0.2 V.

Figure 1. The structure schematic diagram (on the left) and the characteristic test circuit of OTFT (on the right). ITO is the collector of OTFTs, Cu is the emitter, Al is the base electrode, and ZnPc as the active layer of OTFTs. The collector is connected to ground. The emitter and the base electrode are connected to different power sources.

When V_{bc} is certain, the height of the barrier inside the device is certain, with the increase of V_{ec}, have more carriers through the barrier, thus the current increasing. When V_{ec} is certain, the number of the free carriers is certain, with the increase of V_{bc}, the height of the barrier inside the device increasing, so carriers through the barrier becomes very difficult, thus the current decreasing.

Equation 1 is the relationship between I_{ec} and the bias voltage.

$$I_{ec} = I_0 \exp(\frac{qV^*}{kT}) \quad (1)$$

where V^* is a function with V_{ec} and V_{bc}; T is 350 K; q/kT is 33.1126; I_0 is as shown in Equation 2.

$$I_0 = SA^*T^2 \exp(-\frac{q\varphi_b}{kT}) \quad (2)$$

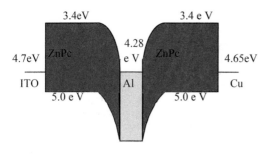

Figure 2. The ideal energy band diagram of OTFTs. The work function of Al is 4.28 eV, the HOMO and LUMO energy level of ZnPc are 5.0 eV and 3.4 eV respectively, and the work function of ITO and Cu are 4.7 eV and 4.65 eV respectively.

where S is effective area of the device, with a value of 0.04 cm^2; A^* is 31.5 A/CM^{2}K^2.

When $V_{bc} = 0$ V, Equation 3 compares the current of the two states of light and dark.

$$I_{ec1} - I_{ec2} = SA^*T^2 \exp[\frac{qV_{ec}}{kT}(\frac{1}{b_1} - \frac{1}{b_2})] \quad (3)$$

3 CHARACTERISTIC TEST AND RESULT ANALYSIS OF THE DEVICES

Figure 3 shows the output characteristics of ZnPc OTFTs, both in the dark and under light conditions. It can be seen from Figure 3 that I_{ec} was significantly increased under the light conditions. Under the same conditions, when V_{bc} is certain, with the increase of V_{ec}, I_{ec} was increased; when the V_{ec} is certain, with the increase of V_{bc}, I_{ec} was reduced. The phenomenon occurs upon many factors.

where $b_x = \exp(q\varphi_{bx}/kT)$; I_{ec1} and b_1 as the current and the parameter under light situation respectively; I_{ec2} and b_2 as the current and the parameter in the dark respectively.

Equation 3 was calculated. Equation 4 gives:

$$\ln(I_{ec1} - I_{ec2}) = \ln(SA^*T^2) + \frac{qV_{ec}}{kT} + \ln(\frac{1}{b_1} - \frac{1}{b_2}) \quad (4)$$

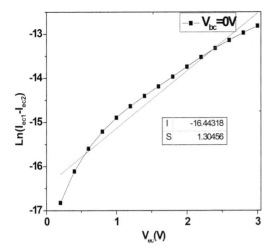

Figure 4. The curve of the relationship between V_{ec} and $\ln(I_{ec1}-I_{ec2})$ when V_{ec} is from 0 V to 3.0 V step 0.2 V, and V_{bc} is 0 V. The intercept is -16.44318, and the slope is 1.30456.

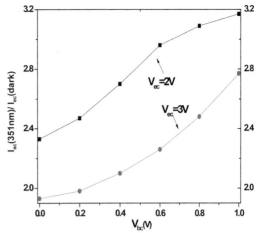

Figure 5. The curve of the relationship between I_{ec} (351 nm)/I_{ec} (dark) and V_{bc}; also called the relationship between β_L and V_{bc}, when V_{ec} is 2 V and V_{ec} is 3 V, and V_{bc} is from 0 V to 1.0 V step 0.2 V.

Figure 4 shows the fitting analysis of Equation 4.

In Figure 4, intercept $S = \ln(SA^*T^2) + \ln(1/b_1 - 1/b_2) = 1.30456$, $1/b_1 - 1/b_2 = 2.254 \times 10^{-5}$ were obtained. Therefore $b_1 < b_2$, so it can be concluded that the light can reduce the height of the barrier inside the device.

Then, the relationship between the amplification factor and the base voltage was analysed in the dark and under 351 nm illumination. We only analysed two cases, where V_{ec} is 2 V and V_{ec} is 3 V respectively The relationship is shown in Figure 5. The amplification factor of I_{ec} is also called β_L. β_L can be expressed by Equation 5.

$$\beta_L = \frac{I_L}{I_{dark}} \qquad (5)$$

where I_L is the current under 351 nm illumination; I_{dark} is the current in the dark.

In Figure 5, the range of β_L is 2.3–3.2 times when V_{ec} is 2 V; β_L is largest when V_{bc} is 1 V, and β_L is smallest when V_{bc} is 0 V. The range of β_L is 1.8–2.8 times when V_{ec} is 3 V; β_L is largest when V_{bc} is 1 V, and β_L is smallest when V_{bc} is 0 V. β_L is bigger when V_{ec} is 2 V than when V_{ec} is 3 V. In conclusion, the device by illumination can improve the current of OTFTs, and the device has a great light amplification effect. With the increase of V_{bc} and the decrease of V_{ec}, the light amplification effect improves.

4 CONCLUSIONS

In this paper, a transistor with organic material ZnPc as active layer was successfully prepared. The transistor has a vertical structure of ITO/ZnPc/Al/ZnPc/Cu.

After the optical and electrical tests are carried out, the results show that the transistor has unsaturated characteristics, and because ZnPc is a light-sensitive material, so the transistor has a great current amplification in the light conditions. The output current is related to the applied voltage and illumination. Illumination can reduce the height of the barrier inside the device, so that more carriers can through the Schottky barrier, which can increase the current. The light amplification factor of I_{ec} of the device was analysed. It can be concluded that illumination can improve the current of the device. With the increase of V_{bc} and the decrease of V_{ec}, the light amplification factor of I_{ec} becomes higher. β_L is highest when V_{ec} is 2 V and V_{bc} is 1 V, the highest amplification value being about 3.2 times.

REFERENCES

Cho, S.M., Han, S.H., Kim, J.H. & Jang, J. (2006). Photoleakage currents in organic thin-film transistors. *Appl. Phys. Lett.*, 071106–071109.

Eder, F., Klauk, H., Halik, M., Zschieschang, U., Schmid, G. & Dehm, C. (2004). Organic electronics on paper. *Appl. Phys. Lett.*, 2673–2676.

Heremans, P. (2014). Electronics on plastic foil, for applications in flexible OLED displays, sensor arrays and circuits. *Active-Matrix Flatpanel Displays and Devices*.

Kudo, K. & Chiba, I. (2006). Recent progress on organic thin film transistors and flexible display applications. *Nanotechnology Materials and Devices Conference*, 290.

Schon, J.H. & Kloc, C. (2001). Organic metal-semiconductor field-effect phototransistors. *Appl. Phys. Lett.*, 3538–3541.

Uemura, T., Nakayama, K. & Hirose, Y. (2012). Band-like transport in solution-crystallized organic transistors. *Current Applied Physics*, 87.

Zorba, S. & Gao, Y. (2005). Feasibility of static induction transistor with organic semiconductors. *Appl. Phys. Lett.*, 193508-193510.

Electronic Engineering – Wang (ed)
© 2018 Taylor & Francis Group, London, ISBN 978-1-138-60260-1

Robust visual tracking with integrated correlation filters

H. Zhang, L. Zhang, Y.W. Li & Y.R. Wang
School of Astronautics, Beihang University, Beijing, China

L. He
Luoyang electro-optical equipment research institute, Luoyang, China

ABSTRACT: Visual tracking is an important research area in computer vision. Various trackers have been proposed recently. Among these trackers, the correlation filter-based trackers have achieved compelling results in different benchmarks, proving the great strengths in efficiency and robustness. However, conventional correlation filter methods mainly utilize a fixed-sized window for tracking, which restricts target search region and limits the robustness of dealing with scale variations and fast motions. In this paper, we propose a tracking method with integrated correlation filters. The task is decomposed into translation and scale estimation. Two correlation filters are introduced in translation estimation. One correlation filter is used for tracking, and the other one is for re-detection. Then a correlation filter trained on a scale pyramid representation is applied to estimate scales of the target. Experimental results on a large benchmark dataset indicate that the proposed algorithm performs favorably against recent state-of-the-art trackers regarding accuracy and robustness.

Keywords: Visual tracking; integrated trackers model; correlation filter

1 INTRODUCTION

Visual tracking is an active research topic in computer vision, owing to its wide applications such as human-computer interfaces, robotics, and surveillance. Its objective is to estimate the location of an initialized visual target in each frame of an image sequence. However, visual tracking is a difficult problem due to the factors such as occlusion, deformation, fast motion, illumination variations, background clutter, and scale variations.

In the past decade, various tracking algorithms have been proposed, which can be roughly categorized as generative and discriminative (Chen et al. 2015). Generative methods handle tracking by searching for the region's most similar to the target model, and discriminative algorithms intend to differentiate the target from backgrounds. Different from generative methods, discriminative algorithms utilize both target and background information for tracking. Recently, discriminative methods have been demonstrated to be more competing (Wu et al. 2013). Particularly, correlation filter-based discriminative trackers are efficient for visual tracking by utilizing circulant matrix and fast Fourier transform (Chen et al. 2015). However, conventional correlation filter-based methods, such as MOSSE (Bolme et al. 2010) and KCF (Henriques et al. 2015), mainly employ fixed-sized windows for tracking, which are not able to handle the problem of scale variations and limits the performance for fast motion

due to restricted target search region. In addition, their methods lack the mechanism of re-detection.

In this paper, we integrate three correlation filters for visual tracking. Among them, two correlation filters are utilized for locating the tracked target. The correlation filter for tracking can provide favorable performance in general cases, and the filter for re-detection has different adaptive rate and window size with the other one to against fast motion and occlusion. After locating the target, the filter for scale estimation is applied on the appearance pyramid.

The proposed method achieves a very appealing performance in accuracy and robustness against state-of-the-art trackers in a comprehensive evaluation on a large-scale benchmark with 50 challenging image sequences.

The remainder of this paper is organized as follows. Section 2 introduces the correlation filter tracker and our method. Section 3 presents the experimental results and discussions. We draw the conclusions in Section 4.

2 THE TRACKER

In our work, the tracking work is decomposed into translation and scale estimation. First, we train two discriminative correlation filters to estimate translation. Then we estimate the scale by learning a discriminative correlation filter on a scale pyramid representation.

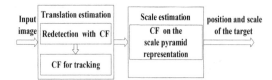

Figure 1. Flowchart of the proposed tracking method. CF is short for correlation filter. Three CFs are integrated. We estimate the target position with re-detection strategy. Then a CF on the scale pyramid representation is applied to estimate the size of the target.

2.1 Correlation filters

Correlation filters are considered as classifiers recently by lots of correlation filter-based methods (Danelljan et al. 2014). The classifier is trained on an image path of size $M \times N$, which is centered around the target. All cyclic shifts $x_{m,n}, (m, n) \in \{0, 1, \ldots, M\} \times \{0, 1, \ldots, N\}$, are used as training samples. $y_{m,n}$, which is the label of $x_{m,n}$, is usually a Gaussian function.

The problem can be treated as minimizing the objective function of ridge regression (Wu et al. 2013)

$$\min_{w} \sum_{m,n} |\langle \varphi(x_{m,n}), w \rangle - y(m,n)|^2 + \lambda \|w\|^2 \quad (1)$$

where λ is a regularization parameter ($\lambda \geq 0$) and φ denotes the mapping to a kernel space by the kernel κ. The inner product of x and x' is computed as $\langle \varphi(x), \varphi(x') \rangle = \kappa(x, x')$. The filter can be trained in the dual space, and $w = \sum_{m,n} \alpha(m,n) \varphi(x_{m,n})$, the coefficient α can be learnt by

$$F(\alpha) = \frac{F(y)}{F(\mathbf{k}^x) + \lambda} \quad (2)$$

where F denotes the discrete Fourier operator. $\mathbf{k}^x = \kappa(x_{m,n}, x)$. The patch z in the next frame with the same size of x is used to compute the confidence map as

$$\dot{y} = F^{-1}(F(\mathbf{k}^z) \odot F(\alpha)) \quad (3)$$

where F^{-1} presents the inverse discrete Fourier transform. $\mathbf{k}^z = \kappa(z_{m,n}, \dot{x})$. $\dot{x}$ denotes the learned target appearance. $\odot$ is the element-wise product. The location of the maximum value of $\dot{y}$ is considered as the new position of the tracked target.

2.2 Integrated correlation filters

Since correlation filter based trackers mainly utilize a fixed-sized window for tracking, the trackers easily fail to track if the target moves fast and runs out from the tracking window. Additionally, their methods lack the re-detection strategy to deal with occlusions. Figure 1 illustrates the flowchart of the proposed method.

Our tracker integrates three correlation filters for different purposes. Two filters are applied in translation estimation, and the other is for scale estimation.

In translation estimation, the proposed tracker first estimates the translation according to a common correlation filter. If the confidence value of the filter results drops below a threshold T, the correlation filter for re-detection will work. If the re-detection filter has better results (a larger confidence value) than the common one, its tracking result will replace the result of the common one. The correlation filter for re-detection has a large window to introduce more samples and extend the searching region. In addition, the re-detection filter learns from the reliably tracked target, and a threshold T_r is used to block unwanted updates. The re-detection filter updates when the confidence value of the re-detection filter is larger than T_r.

We use the linear kernel $\kappa(x, x') = x^T x'$ in our filters, and the correlation filters is extended using multi-dimensional image features. With the help of circulant matrix, $F(\alpha)$ can be expressed by

$$F(\alpha) = \frac{F(y)}{\sum_d F(\dot{x}_d)^* \odot F(\dot{x}_d) + \lambda} \quad (4)$$

$$\dot{y} = F^{-1}((\sum_d F(\dot{x}_d)^* \odot F(z_d)) \odot F(\alpha)) \quad (5)$$

where d is the dimension number of the features, $*$ is the complex-conjugate.

In order to adapt to appearance changes, the target model and the transformed classifier coefficients $F(\alpha)$ need to be updated over time. They are updated with a learning rate η as

$$F(\dot{x}^t) = (1 - \eta) F(\dot{x}^{t-1}) + \eta F(x^t) \quad (6)$$

$$F(\bar{\alpha}^t) = (1 - \eta) F(\bar{\alpha}^{t-1}) + \eta F(\alpha^t) \quad (7)$$

where t presents the index of the current frame, $\bar{\alpha}$ denotes the learned coefficient model.

We use raw pixels, Histogram of Gradient (HOG) (Felzenszwalb et al. 2010) and Color Name (CN) (Van De Weijer et al. 2009) for image presentation. HOG features are extracted using a cell size of 4×4. CN is an 11-dimensional color representation that is mapped from the RGB values. Due to the 11 dimensional features sum up to 1, thus we use 10-dimensional features instead. All these features are multiplied by a cosine window to eliminate discontinuities at the image boundaries and also make the target center nearby have larger weight.

After locating the target, we take a series of samples of different sizes at the target position, and resize them to the same size and construct a scale pyramid representation. Similar to DSST (Danelljan et al. 2014), the set of scale factors is $S = \{a^n | n = [(1 - N)/2], \ldots, [(N + 1)/2]\}, N$ presents the number of samples and a is the scale increment factor. Then, a correlation filter with a searching strategy is trained on the scale pyramid representation. The scale is obtained by finding the maximum value of the correlation response map.

3 EXPERIMENTS

In this section, we evaluate the proposed tracking algorithm on a benchmark dataset (Wu et al. 2013) that has 50 challenging image sequences. Both quantitative and qualitative comparisons with state-of-the-art-trackers are provided in the experiments.

3.1 Experimental setup

The regularization parameter is set to $\lambda = 10^{-4}$. For computational efficiency, we do not use the re-detection filter to scan the entire frame. The window size of the re-detection filter is set to 4 times of the target size, which is sufficient for most cases. We use $N = 33$ number of scales with the scale increment factor $a = 1.02$. The threshold for re-detection is set to $T = 0.15$. The update threshold of the re-detection filter is set to $T_r = 0.45$. The learning rate is set to $\eta = 0.01$. We use the same parameter values for all the sequences. The proposed tracker is implemented in Matlab, and all the experiments are performed on an Intel i5-4460 CPU (3.2 GHz) PC with 6 GB memory.

To evaluate the performance of our method, distance precision and overlap success rate are used for evaluation (Wu et al. 2013). Distance precision is calculated as the percentage of frames whose estimated location is within the given threshold distance of the ground truth. Overlap success rate is the percentage of frames where the bounding box overlap oversteps a threshold. In addition, we also supply the speed in the frames per second (FPS).

3.2 Experimental results

We compare our method on the benchmark with comparisons to 4 state-of-the-art methods, including SAMF (Li & Zhu 2014), RPT (Li et al. 2015), KCF (Henriques et al. 2015), and DSST (Danelljan et al. 2014).

3.2.1 Overall performance

Figure 2 presents the precision plots and success plots over all the 50 sequences. The two legends show the mean distance precision at 20 pixels and the area under the curve (AUC), respectively. According to the presented results, our method performs favorably compared to other trackers in both precision and success plots. Particularly, our method significantly outperforms RPT by 5.3 % in success plots.

The benchmark dataset is annotated with 11 attributes, such as scale variation, occlusion, and so forth. In Table 1, we supply the mean distance precision corresponding to 11 attributes at 10 pixels and FPS. It can be found that our method achieves the best performance in the vast majority of attributes. Regarding mean distance precision of all sequences, the proposed tracker provides a gain of 4.8 % over SAMF. Particularly, in the out-of-plane rotation evaluation, our tracker achieves the best score 0.718, which is 6.2% higher than the score of SAMF. For scale variations,

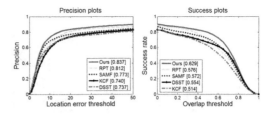

Figure 2. Precision and success plots over all the 50 sequences. The proposed tracker outperforms other trackers.

Table 1. Attribute-based mean distance precision at 10 pixels. The best result is shown in bold font.

	Ours	DSST	KCF	SAMF	RPT
LR*	0.518	0.479	0.323	0.498	0.356
BC*	0.693	0.553	0.624	0.617	0.662
OV*	0.532	0.389	0.527	0.524	0.440
IPR*	0.665	0.610	0.563	0.590	0.629
FM*	0.530	0.372	0.453	0.460	0.527
MB*	0.568	0.429	0.508	0.444	0.581
DEF*	0.744	0.524	0.616	0.715	0.579
OCC*	0.758	0.580	0.577	0.715	0.552
SV*	0.738	0.643	0.526	0.656	0.636
OPR*	0.718	0.592	0.572	0.656	0.609
IV*	0.645	0.600	0.525	0.582	0.577
ALL	0.734	0.614	0.592	0.686	0.645
FPS	12.1	12.0	251.3	14.8	3.9

*LR: Low Resolution, BC: Background Clutters, OV: Out-of-View, IPR: In-Plane Rotation, FM: Fast Motion, MB: Motion Blur, DEF: Deformation, OCC: Occlusion, SV: Scale Variation, OPR: Out-of-Plane Rotation, IV: Illumination Variation.

our method achieves the score of 0.738 and significantly outperforms SAMF by 8.2%. In the cases of fast motion, the proposed tracker achieves a gain of 7% compared to SAMF. For occlusions, our tracker also obtains a gain of 4.3% compared to SAMF. Due to the motion information is utilized in RPT, it also performs considerably well. KCF runs fastest, but the speed of our tracker is also very close to DSST and SAMF. RPT runs at 3.9 FPS that is the slowest. Although our tracker is not specially designed for illumination variation, in-plane rotations and deformations, the proposed method also achieves appealing performances on these challenging image sequences.

3.2.2 Qualitative evaluation

We also provide a qualitative comparison of the proposed tracker with the four state-of-the-art methods on 12 benchmark sequences. The results are shown in Figure 3. KCF performs well in terms of deformation and out-of-plane rotation (Bolt and Liquor). However, its performance on scale variations (Doll, Dog, and Singer1) is limited by the fixed-sized window. DSST has the capability of estimating scale changes but does not perform well on occlusions and out-of-plane rotation (Girl and Jogging1). SAMF performs well

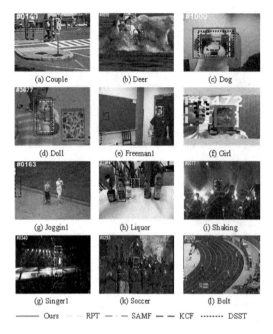

Figure 3. Screenshots of tracking results on 12 benchmark sequences. The target is initiated in the first frame of the sequence.

on scale variations and out-of-plane rotation (Dog, Jogging1, Bolt, and Girl), but fails to handle fast motions (Soccer, Deer, and Couple). RPT performs well on fast motions and motion blurs (Soccer and Deer) but does not perform well on deformations and occlusions (Bolt, Girl, and Jogging1). Overall, our tracker achieves outstanding performance on the 12 sequences. The proposed tracker performs well on scale variations, fast motions, and deformations (Doll, Jogging1, and Bolt), and is less sensitive to occlusions, motion blur and out-of-plane rotations (Liquor, Soccer, and Freeman1).

4 CONCLUSIONS

In this paper, we integrate three correlation filters for visual tracking. A common correlation filter works for usual tracking. A re-detection strategy is introduced by utilizing the correlation filter with different adaptive rates and window size. To efficiently estimate the scale of the target, a scale pyramid representation is applied, and then a correlation filter is trained on the scale pyramid representation. In addition, an online updating scheme is utilized to adapt to appearance changes. The method has a vast improvement in fast motions, scale variations, and occlusions. Experiments on 50 challenging benchmark sequences show that the proposed tracker performs favorably compared to the state-of-the-art methods regarding accuracy and robustness.

ACKNOWLEDGMENTS

This work is supported by the National Natural Science Foundation of China Grant No. 61571026. Lei He is the corresponding author. Thanks for the co-authors' guidance and help.

REFERENCES

Bolme, D. S. et al. (2010). Visual object tracking using adaptive correlation filters. *Computer Vision and Pattern Recognition (CVPR)*: 2544–2550.
Chen, Z. et al. (2015). An Experimental Survey on Correlation Filter-based Tracking. *arXiv preprint arXiv*: 1509.05520.
Danelljan, M. et al. (2014). Accurate scale estimation for robust visual tracking. *British Machine Vision Conference, Nottingham. BMVA Press.*
Felzenszwalb, P. F. et al. (2010). Object detection with discriminatively trained part-based models. *IEEE transactions on pattern analysis and machine intelligence,* 32(9): 1627–1645.
Henriques, J. F. et al. (2015). High-speed tracking with kernelized correlation filters. *IEEE Transactions on Pattern Analysis and Machine Intelligence,* 37(3): 583–596.
Li, Y. & Zhu, J. (2014). A scale adaptive kernel correlation filter tracker with feature integration. *European Conference on Computer Vision:* 254–265.
Li, Y. et al. (2015). Reliable patch trackers: Robust visual tracking by exploiting reliable patches. *Proceedings of the IEEE Conference on Computer Vision and Pattern Recognition*: 353–361.
Van, D. & Weijer, J. et al. (2009). Learning color names for real-world applications. *IEEE Transactions on Image Processing,* 18(7): 1512–1523.
Wu, Y. et al. (2013). Online object tracking: A benchmark. *Proceedings of the IEEE conference on computer vision and pattern recognition*: 2411–2418.

A hash function based on iterating the logistic map with scalable positions

J.H. Liu & C.P. Ma
College of Computer Science and Technology, Harbin University of Science and Technology, Harbin, China

ABSTRACT: An algorithm for a hash function is proposed using a chaotic map. The chaotic map is built, based on the logistic map with scalable positions. The hash value is constructed by iterating the chaotic map step by step. The iterating value of the chaotic map is dynamically dependent on previous iterations of chaos. In addition, the corresponding information is converted into different positions of the logistic map. The algorithm of the hash function is measured as for changes to the original message condition, collision analysis and statistical analysis. The experimental results indicate that the probabilities of mean change are close enough to 50%. Computational simulation and theoretical analysis imply that the proposed algorithm can resist birthday attack, statistical attack and meet-in-the-middle attack. It can be applied to information security.

Keywords: logistic map; scalable position; hash function

1 INTRODUCTION

In recent years, the hash function has been widely used with chaotic maps in the fields of information security and data encryption. Since a chaotic map possesses many dynamic properties, for instance, extreme sensitivity to the initial parameters and control parameters (Pecora & Carroll, 1990), it is a very effective random sequence generator.

The study of chaos theory in information security has attracted great attention. Chaotic systems have been widely applied in digital cryptography (Matthews, 1989), multichannel spread-spectrum communication (Xiao et al., 1996), random number generators, and secure communication systems (Boutayeb et al., 2002).

Moreover, the synchronising hyperchaotic systems with a scalar transmitted signal are reported (Peng et al., 1996). Chaotic systems are suitable for use in the information security field (Zhou & Ling, 1997; Baptista, 1998).

The new property of the chaotic system is reported (Liu et al., 2014). For example, the property is introduced for invariable inherited genetic positions in real orbits of the logistic map (Liu, 2015).

Therefore, chaotic maps are been known in the field of information security (Chenaghlu et al., 2016). The proposed algorithm of the hash function is based on the chaotic system.

2 LOGISTIC MAP

Consider the logistic map that is defined as the chaotic system:

$$x_{n+1} = a * x_n (1 - x_n) \quad (1)$$

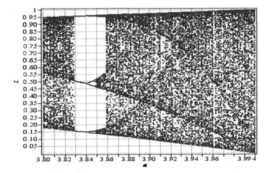

Figure 1. The chaotic area plots steady parts.

Here, a is a floating-point number between 3 and 4, representing the control value; x_0 is the initial value of x_n and denotes the specified value; n plays a role in the nth iteration. Therefore, the input information is inserted into the logistic mapping, based on the scalable position set.

Figure 1 plots the chaotic series of the logistic map. There is a stable periodic orbit when the value of a ranges from 3.83 to 3.86. However, the other parts are fairly characteristic, and it can be made use of as a random number generator.

3 HASH FUNCTION WITH LOGISTIC MAP

The hash function is widely utilised in information security, such as for hashing text messages and user passwords.

A basic procedure of hash functions is to compress a random length of input information into a confirmed output message of 128-bit length. In this section, an

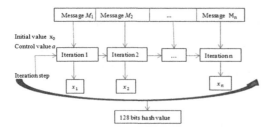

Figure 2. The whole structure of the algorithm is plotted.

improved hash function algorithm is presented, which depends on the logistic map.

3.1 Converting the primitive message into a block

Our scheme can deal with input messages of any length. Every value of iterating the logistic map is obtained enough digital positions.

The original message can be divided into equal-sized sub-blocks, except for the last part. Usually, it is necessary to append some message bits at the end of the original message, in order to fill the final sub-block.

Figure 2 shows the structure of the proposed algorithm. Here, M is separated into n partitions, and each part has n bits. $M = (M_1, M_2, \ldots, M_n), M_i$ ($i = 1, 2, \ldots, n$). N is the length of the sub-block. Finally, the hash value of 128 bits will be gained on completion of the logistic map iteration of Equation 1.

3.2 Algorithm of hash function

The algorithm can be designed as follows.

(i) Assume that the original message is P with length Ln. Each letter of P will be converted into the format of its corresponding ASCII value. Then the corresponding ASCII value is transformed into the binary B.

$$L = \text{Binary}\,(l_1, l_2, \ldots, ln) \quad (2)$$

where l is each letter of P; L is an array that stores transforming data.

(ii) Whether appending some bits message or not, according to Expression 3 0 and 1 series are appended.

$$L_{append} = \begin{cases} 0 & if\,(\text{mod}(n,128)) = 0; \\ 128 - \text{mod}(n,128) & otherwise; \end{cases} \quad (3)$$

(iii) The iteration process of n rounds is as follows:

Step 1: input control parameter $a = 3.9$ splits joint $l_0 \ldots l_3$, the initial parameter $x_0 = 0.7$ splits joint $l_4 \ldots l_7$. The logistic map will be iterated at least 30 times in order to obtain better random sequences.

Step 2: when $i = 2$, computing $x_{31} = (x_{30}$ XOR $(l_{4i} \ldots l_{4i+3}))$, where i is the number of iterations. Repeat the procedure until all bits are processed.

Table 1. NIST test result of a chaotic random sequence.

Test Method	p-value	Result
Mono-Bit	0.70325	Success
Frequency Block	0.29387	Success
Runs	0.20220	Success
Longest Run	0.45445	Success
Binary Matrix Rank	0.16748	Success
Discrete Fourier Transform	0.46170	Success
Non-Overlapping TM	0.21227	Success
Overlapping TM	0.51310	Success
Universal Statistical	0.42622	Success
Linear Complexity	0.19374	Success
Serial	0.73250	Success
Approximate Entropy	0.86335	Success
Cumulative Sums	0.52510	Success
Random Excursions Variant	0.86470	Success
Random Excursions	0.60941	Success

(iv) The seed is chosen randomly by using the following function:

$$\text{Seed} = \text{Rand}\,(N_{arrGroup}) \quad (4)$$

Positions of thirtieth iteration result will be more than twenty thousand digital positions. Therefore, at the last iteration result, the front 2,048 positions are selected in the group. The total number of groups is equal to 32. In other words, $N_{arrGroup} = 32$. The first seed is chosen randomly. Then the seed of the second group is selected according to the first group seed. The final result comprises of 128 hash values.

4 PERFORMANCE AND SECURITY ANALYSIS

This section will analyse the performance of the proposed algorithm.

4.1 NIST test

The National Institute of Standards and Technology (NIST) test suite is used to test the randomness as follows. Table 1 shows the test result of a chaotic random sequence, generated by using the NIST suite. In the NIST suite the template matching is denoted by TM.

The National Institute of Standards and Technology publishes the NIST test suite. It is a well-known means to assess the randomness of binary sequences.

The NIST suite is used to test at most 1,000,000 bits, generated from the present chaotic random sequences. The NIST test includes 15 kinds of test functions. For each test, a p-value larger than a given value of 0.01 represents that the test has passed.

It implies that the chaotic sequence has wonderful randomness. Table 1 summarises that the generated sequence can pass the 15 different tests. It is obvious that the presented algorithm has an extremely random characterisation.

4.2 Sensitivity of the message and initial conditions

This section shows the high sensitivity of the Section 3.2 algorithm for the hash function. The test includes small changes of the input information. The five change conditions are as follows:

C1: The original information is: 'Sensitivity of hash value to the message and initial conditions.';
C2: The character 'S' in the word 'Sensitivity' is replaced by 's' to become 'sensitivity';
C3: Transform the word 'conditions' in the original information into 'condition';
C4: The character 'v' in the word 'value' is replaced by 'n' into 'nalue';
C5: Eliminate the character '.' at the end of the sentence.

The corresponding hexadecimal sequence of hash values is as follows:

Condition1:
6CA2C96D258D2E376E6C28CA40C2DC20
Condition2:
73420E5A4FCF697A94FFCA0A7B5672C1
Condition3:
290A5C9FE83310881B4566E6AAB28790
Condition4:
5DE61CB182B74C5C2C3E54B2A188E3B9
Condition5:
16ADF0E7E224809A1B5EF7C57438F328

The results imply that an arbitrary little change can lead to about 50% changing possibility in the final hash values.

4.3 Statistical analysis of diffusion and confusion

Generally speaking, diffusion and confusion are recognised as being two kinds of commendable encryption algorithms. The generated hash value is in binary format that contains only 1 or 0.

If one bit of the original information is changed, then a new hash value will be generated. It is the desired diffusion action that any little changes in the original message will result in the 50% changing possibility. As we know, four statistical experiments are tested as follows:

Mean changed bit number:

$$B_{mean} = \frac{1}{N}\sum_{1}^{N} B_i \quad (5)$$

Mean changed probability:

$$P = (B_{mean}/n) \times 100\% \quad (6)$$

Standard variance of the changed bit number:

$$\Delta B = \sqrt{\frac{1}{N-1}\sum_{i=1}^{N}(B_i - B_{mean})^2} \quad (7)$$

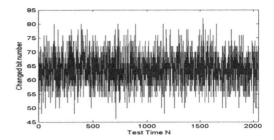

Figure 3. Distribution of changed bit number.

Table 2. The statistical performance of the proposed algorithm.

N	B_{mean}	$P(\%)$	ΔB	$\Delta P(\%)$
256	64.37	49.85	5.48	4.21
512	64.21	49.87	5.65	4.29
1,024	64.09	50.06	5.49	4.36
2,048	64.58	50.26	5.59	4.38

Standard variance:

$$\Delta P = \sqrt{\frac{1}{N-1}\sum_{i=1}^{N}(\frac{B_i}{n} - P)^2} \times 100\% \quad (8)$$

In the above Equations 5–8, N represents the total number of test times, $n = 128$ bits, and B_i is the number of changed bits based on the ith test.

The diffusion and confusion test is carried out as follows. The hash values are generated from the input message. When randomly choosing a bit in the original message, a new hash value is generated. The result comparing the two kinds of hash values is shown in Figure 3.

This kind of the test is executed N times. Figure 3 shows the corresponding distribution of changed bit number, where here $N = 2,048$.

Table 2 shows the test data with $N = 256, 512, 1,024$ and $2,048$. According to this data we can safely deduce that the values arrived at for the mean changed bit number B, and the mean changed percentage of P, are respectively the given 64-bit value and 50%.

In all tests, ΔB and ΔP are very small. The experimental result shows that statistical performance of confusion and diffusion of the proposed hash function is able to arrive at stable conditions.

5 SECURITY ANALYSIS

This section describes analysis of collision resistance and birthday attack resistance.

5.1 Text analysis of collision resistance and meet-in-the-middle attack resistance

In the scheme, the intermediate results are x_{30}. First of all, the 30 times of iteration proceed.

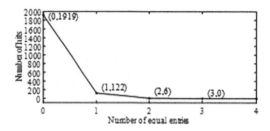

Figure 4. Same value and position distribution of ASCII characters at the hash value.

With the original message P, iterations of chaotic map and the changing bits test can be considered as being the hash round function. Assume that the y_i is a meet-in-the-middle collision of y_j.

According to Section 3.2 in the proposed algorithm, we know that if there is only one seed difference from all $N_{arrGroup}$, then there exists 2^{32} kinds of different values. So y_j is not a collision. In other words, it is infeasible to find a meet-in-the-middle collision in the floating-point domain. Hence, the presented algorithm can resist meet-in-the-middle attack.

5.2 Collision test

Many experiments on collision resistance and their analysis are described in the following test.

First of all, when randomly selecting a piece of input data, the new message will be constructed and kept in the memory array. It can select one bit to be changed, then the new hash value will be built. In comparing with the two times hash value, the result of the hash value with the same position is computed. The Equation can be denoted as:

$$d = \sum_{i}^{N} |t(m_i) - t(m'_i)| \qquad (9)$$

Here m_i and m'_i are the ith new sequence of ASCII value for the obtained message. The experiment executes repeatedly for collision, such as 2,048 times. The test value includes: the minimum, maximum and mean values. The value is 2, 197, 672, and 1,448 respectively. The total ASCII values' distribution at the identical position is shown as Figure 4. We can see clearly that the maximum value of the same character is 2. Therefore, the probability of collision is very small.

5.3 Birthday attack

The test for collision resistance and birthday attack is relative. The hash value is very sensitive for each bit changed value. It is close to the property of chaos. In other words, the small bit shift can obtain the uniform distribution of hash value.

The security of the hash function is dependent on the length of hash value for this kind of the test. When the hash value is equal to 128-bit, there exists the 2^{64} possibility of attack. It implies that the possibility of occurrence is very low.

5.4 Flexibility

With the progress of technology development, algorithms need a longer and stronger hash function. In order to gain the longer hash value, our scheme can be modified as follows: let $N_{arrGroup} = 64$, and each group is to gain one seed. So, we can get the 256-bit hash value sequence.

Compare this with the traditional hash function, such as MD5, which is fixed in length at 128 bits. The hash function is suitable for the actual need of information security.

6 CONCLUSION

The hash function using logistic map with scalable positions is proposed. The initial parameters will join a user's input message. The substantial changes can be found in the final result, even though one bit is changed in the message. Due to making full use of the properties of chaos, the scheme can meet the requirements for information security.

ACKNOWLEDGEMENTS

This work is supported by Natural Science Foundation of Heilongjiang Province of China under Grant No. F201304.

REFERENCES

Baptista, M.S. (1998). Cryptography with chaos. *Physics Letters A, 240*(1–2), 50–54.

Boutayeb, M., Darouach, M. & Rafaralahy, H. (2002). Generalized state-space observers for chaotic synchronization and secure communication. *IEEE Trans. on CAS-I, 49*(3), 345–349.

Chenaghlu, M.A., Jamali, S. & Khasmakhi, N.N. (2016). A novel keyed parallel hashing scheme based on a new chaotic system. *Chaos, Solitons and Fractals, 87,* 216–225.

Liu, J. (2015). The property of invariable inherited genetic positions in real trajectories of logistic map. *Cybernetics and Physics, 4*(3), 82–89.

Liu, J., Zhang, H. & Song, D. (2014). The property of chaotic orbits with lower positions of numerical solutions in the logistic map. *Entropy, 16*(11), 5618–5632.

Matthews, R. (1989). On the derivation of a 'chaotic, encryption algorithm. *Cryptologia, 13*(1), 29–42.

Pecora, L. M. & Carroll, T.L. (1990). Synchronization in chaotic systems. *Phys. Rev. Lett., 64*(8), 821–824.

Peng, J.H., Ding, E.J., Ding, M. & Yang, W. (1996). Synchronizing hyperchaos with a scalar transmitted signal. *Phys. Rev. Lett., 76*(6), 904–907.

Xiao, J.H., Hu, G. & Qu, Z.L. (1996). Synchronization of spatiotemporal chaos and its application to multichannel spread-spectrum communication. *Phys. Rev. Lett., 77*(20), 4162–4165.

Zhou, H. & Ling, X.T. (1997). Problems with the chaotic inverse system encryption approach. *IEEE Transactions on Circuits and Systems I, 44*(3), 268–271.

Electronic Engineering – Wang (ed)
© 2018 Taylor & Francis Group, London, ISBN 978-1-138-60260-1

The design and implementation of a third-order active analogue high-pass filter

D.D. Han, H.J. Yang & C.Y. Wang
School of Software, Harbin University of Science and Technology, Harbin, China

ABSTRACT: Currently, high-order active filters are widely used in various electronic devices and communication systems. A third-order active analogue high-pass filter with folded cascode op-amp as the core circuit and a cut-off frequency of 1 kHz was designed in this paper. The third-order active high-pass filter can be seen as consisting of a first-order active high-pass filter and a second-order active high-pass filter. The circuit schematic is built using the OrCAD® Capture CIS (Component Information System). The simulation result is verified by Hspice. The result shows that the designed specifications meet the requirements and the filtering performance is good.

Keywords: CMOS; high-pass filter; folded Cascode op-amp

1   INTRODUCTION

The filter is widely used in modern electronic equipment and various types of electrical control system. Since the op-amp is added to the structure of an active filter, it allows the filter to have gain. The higher the filter order, the better the filtering effect. Therefore, the study of the high-order active filter has a very important significance. Usually, an active high-pass filter is composed of an op-amp and a Resistor-Capacitor (RC) circuit. Through the study of the filter principle and its different structure, a third-order active analogue high-pass filter was designed in this paper, with a folded cascode op-amp as the core circuit and a cut-off frequency of 1 kHz. This third-order active high-pass filter can be seen as consisting of a first-order active high-pass filter and a second-order active high-pass filter. Moreover, the design method of a third-order active high-pass filter is also suitable for a higher-order filter. The performance indicators of the designed two-stage amplifier are optimised, and finally, the simulation results in line with design specifications are obtained. Then, the frequency domain and time domain characteristics of the third-order active analogue high-pass filter are simulated and analysed.

2   DESIGN AND SIMULATION OF OPERATIONAL AMPLIFIER

As the op-amp in this paper works in the analogue filter, so it generally does not use a high-speed operational amplifier. The folded cascode op-amp is selected by considering characteristics such as gain, swing, power and noise. The design of this folded cascode op-amp is based on a 0.5 um Complementary Metal-Oxide-Semiconductor (CMOS) digital-analogue mixed-technology library of CMSC.

Design indicators are as follows:

DC Gain	$>100\,dB$
Unit Gain Bandwidth	$>4\,MHz$
CMRR	$>80\,dB$
PSRR	$>80\,dB$
Phase Margin	$\approx 60°$
Power Consumption	$<300\,mW$
Input Offset Voltage	$<0.5\,mV$
Slew Rate	$>2\,V/uS$
Load Capacitance	$=5\,pF$

2.1   Design of amplifier

The folded cascode amplifier is used in the circuit structure. This two-stage op-amp can be seen as a cascade of folded cascode circuit and a simple amplifier. It uses Miller compensation capacitors. The design requires a gain of 100 dB or more, so it is assigned to the two-stage amplifier, as shown in Equation 1 and Equation 2. The circuit structure is shown in Figure 1.

$$A_1 = g_{m1}[(g_{m12}r_{o12}r_{o10}) \,\|\, (g_{m8}r_{o8}r_{o5})] = 70dB \qquad (1)$$

$$A_2 = g_{m13}(r_{o6} \,\|\, r_{o13}) = 30dB \qquad (2)$$

In this paper, a circuit block, which itself generates a fixed bias voltage, is used as an operational amplifier to provide a bias voltage. V_{BIAS1}, V_{BIAS2}, V_{BIAS3}, and V_{BIAS4} of Figure 1 are given by a voltage bias circuit. The bias circuit is shown in Figure 2. I_{BIAS1} and

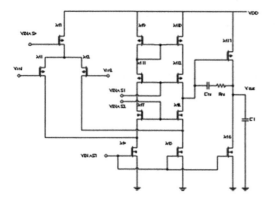

Figure 1. Circuit structure of folded cascode op-amp.

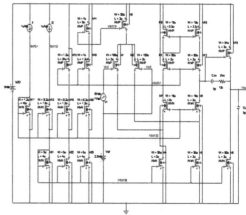

Figure 3. Circuit schematic.

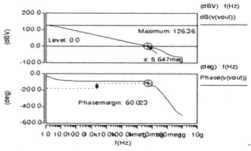

Figure 4. AC simulation result.

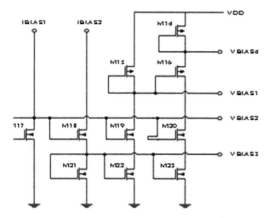

Figure 2. Bias circuit.

I_{BIAS2} are the reference currents introduced from the outside, and the width-to-length ratio of M16–M21 is the same as the previous design method. M15 and M17 are gate-to-drain short-circuits, so gate voltage can be generated.

The width-to-length ratios of all Metal-Oxide-Semiconductor (MOS) transistors are calculated after analysis. Taking into account all factors, through continuous debugging, the folded cascade two-stage amplifier parameters of each device are determined. The circuit schematic of this amplifier is built by OrCAD® Capture CIS (Component Information System), and is shown in Figure 3.

2.2 Simulation and analysis of amplifier

Gain is the most important measure of the op-amp; in addition, the phase margin is also a very important indicator of circuit system design. Typically, the phase margin is at least 45°, and preferably about 60°. The Alternating Current (AC) simulation result of the folded cascode operational amplifier is shown in Figure 4. The gain of the bandgap operational amplifier is 126.36 dB, which is greater than 100 dB and in line with indicators. The unity gain bandwidth is

```
**** voltage sources

subckt
element  0:v_vdd    0:v_vref   0:v_vl
volts    5.0000     2.5000     0.
current  -54.3972u  0.         0.
power    271.9862u  0.         0.

total voltage source power dissipation=  271.9862u  watts
```

Figure 5. Power consumption.

5.647 MHz, which is greater than 4 MHz and is in line with indicators. The phase margin is equal to 60°.

Under the premise of ensuring normal operation of the circuit, the current levels are as small as possible, thereby reducing power consumption. The power consumption of the op-amp is verified by the simulation report that is generated by Hspice. As can be seen from Figure 5, the power consumption of the folded cascode op-amp is only 271.9 uW. It is far less than the 300 uW in the design specification, so the purpose of the low-power design is achieved.

The frequency characteristic waveform of the CMRR generated by the CosmosScope under AC simulation is shown in Figure 6. As can be seen from the Figure, the CMRR is 103.08 dB, which is greater than 80 dB and in line with design requirements.

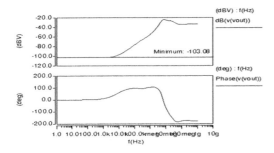

Figure 6. Frequency characteristic waveform of CMRR.

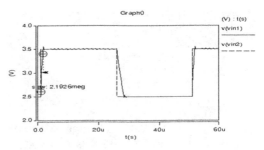

Figure 9. Slew rate.

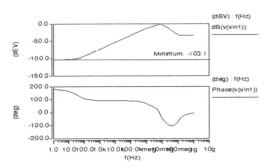

Figure 7. Frequency characteristic waveform of PSRR.

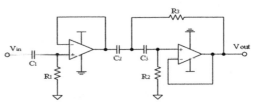

Figure 10. Circuit structure.

Table 1. Bessel filter parameters.

Order number (n)	Filter number	a_1	b_1
3	1	0.756	0
	2	0.9996	0.4772

```
node      =voltage    node      =voltage    node      =voltage
+0:ibias1 =  1.7336   0:ibias2 =  8.5482    0:n00690 =  5.0000
+0:n00704 =  3.4588   0:n00754 =  3.9629    0:n00758 =  3.9702
+0:n00762 =  3.9151   0:n00766 =  3.8944    0:n00770 =  668.1285m
+0:n00774 =  668.1766m 0:n016791 = 2.5000   0:n04256 =  3.4829
+0:n21086 =  846.1366m 0:n21099 = 847.9175m 0:vbias1 =  2.8155
+0:vbias3 =  880.1645m 0:vbias4 =  3.9711   0:vin1   =  2.5000
+0:vin2   =  2.5000
```

Figure 8. Input offset voltage.

The frequency characteristic waveform of the PSRR generated by the CosmosScope under AC simulation is shown in Figure 7. As can be seen from the Figure, the PSRR is 103.1 dB, which is greater than 80 dB and in line with design requirements.

The input offset voltage of the op-amp is shown in Figure 8. As can be seen from the Figure, vout = vin1 = vin2, so the input offset voltage is less than 0.5 mV, in line with design requirements.

The slew rate of the op-amp is shown in Figure 9. As can be seen from the Figure, the slew rate of the op-amp is 2.19 V/us, in line with design requirements.

3 DESIGN AND SIMULATION OF THIRD-ORDER ACTIVE ANALOGUE HIGH-PASS FILTER

After the op-amp used in the filter is designed, the design and simulation of the third-order active high-pass filter can continue. The specific design process is determined by the circuit structure and device parameters of the third-order active high-pass filter.

3.1 Structural design

The structure of the third-order active high-pass filter is shown in Figure 10. As can be seen from the Figure, this third-order active high-pass filter can be seen as consisting of a first-order active high-pass filter and a second-order active high-pass filter. The negative inputs of both op-amps are directly connected to the op-amp's output, instead of adding resistors. The passband gain A of the filter is equal to 1. This is also a major feature of this design method, which is convenient to design, but also reduces the overall area of the filter.

3.2 Parameter design

The design requires that cut-off frequency is equal to 1 kHz, passband gain A is equal to 1, and the characteristics of the Bessel filter are required. It has been determined from Figure 10 that the third-order filter consists of two stages. The Bessel filter parameters are shown in Table 1.

According to the coefficients (a_1, b_1) of each filter, capacitor parameters are firstly set, and then according to the first-order, second-order formula, the resistance of the resistance was obtained.

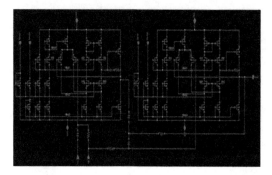

Figure 11. Circuit schematic.

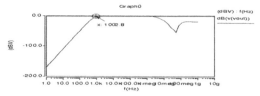

Figure 12. Frequency domain characteristic.

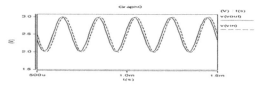

Figure 13. Time domain characteristic (1).

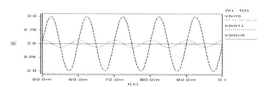

Figure 14. Time domain characteristic (2).

Filter 1: Since the design requires that gain is 1, the gain of each filter is 1. $C_1 = 10$ nF can be assumed, by the formula $R_1 = \dfrac{1}{2\pi f_c a_1 C_1}$. The value of resistor R_1 is obtained, and is shown in Equation 3.

$$R_1 = \frac{1}{2\pi f_c a_1 C_1} = \frac{1}{2\pi \times 1kHz \times 0.756 \times 10nF} \approx 21.1 k\Omega \quad (3)$$

Filter 2: The second stage filter is a second-order high-pass filter. The Sallen-Key type can be used, and it is also assumed that $C_2 = C_3 = 10\,nF$. From the Equations $R_1 = \dfrac{1}{\pi f_c C a_1}$ and $R_2 = \dfrac{a_1}{4\pi f_c C b_1}$, resistance values of R_2 and R_3 are obtained, as shown in Equation 4 and Equation 5. At this point, all circuit parameters of the third-order active high-pass filter have been identified.

$$R_2 = \frac{1}{\pi f_c C a_1} = \frac{1}{\pi \times 1kHz \times 0.9996 \times 10nF} \approx 31.8 k\Omega \quad (4)$$

$$R_3 = \frac{a_1}{4\pi f_c C b_1} = \frac{0.9996}{4\pi \times 1kHz \times 0.4772 \times 10nF} \approx 16.7 k\Omega \quad (5)$$

3.3 Overall design

The circuit schematic of the third-order active high-pass filter is shown in Figure 11 using Cadence.

3.4 AC simulation and results analysis

After drawing the circuit schematic of the filter, the netlist file is generated, and the AC simulation of the netlist file is completed using Hspice. Then the simulation waveform of its frequency domain characteristic is viewed using CosmosScope.

The frequency domain characteristic of the third-order active high-pass filter, based on the previous design parameters, are shown in Figure 12. As can be seen from the Figure, the filter achieves a cut-off frequency of 1,002.8 Hz at a gain drop of 3 dB. The maximum gain is in a smooth line and is 0 dB, so the effect that the input signal is outputted as it is realised.

3.5 Transient simulation and results analysis

An input signal of 5 kHz is added. Since 5 kHz is greater than the cut-off frequency of the filter, the output waveform of the filter should be an intact output. The simulation results show that the output waveform of the filter is basically the same as that of the input signal. The simulation results are shown in Figure 13.

The frequency of the input signal is then changed to 100 Hz. The simulation results are shown in Figure 14. As can be seen from the Figure, the output waveform is close to a straight line, indicating that the filter has a good filtering effect.

4 CONCLUSIONS

A third-order active analogue high-pass filter is designed in this paper. The design process begins with the design of the folded cascode op-amp, and then circuit structure and parameters of the third-order active analogue high-pass filter are designed. The folded cascode op-amp is simulated by Hspice. According to the simulation results, the designed circuit is optimised, and ultimately, a more satisfactory op-amp performance is achieved. The whole filter is then designed and simulated, and its frequency domain and time domain characteristics are analysed.

ACKNOWLEDGEMENTS

This paper is sponsored by Natural Science Foundation of Heilongjiang Province (Grant No. QC2016089 and F2015042).

REFERENCES

Alcaso, A.N. & Cardoso, A.J.M. (2005). Power supply harmonic filter behaviour in a twelve-pulse LCI drive system under power converter faults. In *Proceedings of the Power Electronics Specialists Conference, PESC05 IEEE 36th, Recife* (pp. 2890–2901).

Han, B. (2013a). Matrix splitting with symmetry and symmetric tight framelet filter banks with two high-pass filters. *Appl. Comput. Harmon. Anal.*, 35(2), 200–227.

Han, B. (2013b). Symmetric tight framelet filter banks with three high-pass filters. *Appl. Comput. Harmon. Anal.* Retrieved from http://dx.doi.0rg/l0.1016/j.acha.2013.11.001

Palaskas, Y., Tsividis, Y., Prodanov, V. & Boccuzzi, V. (2004). A 'divide and conquer' technique for implementing wide dynamic range continuous-time filters. *IEEE Journal of Solid-State Circuits*, 39(2), 279–307.

Electronic Engineering – Wang (ed)
© 2018 Taylor & Francis Group, London, ISBN 978-1-138-60260-1

Properties of copper phthalocyanine thin film transistors fabricated in vertical structures

M.Z. Yang, D.X. Wang, X.C. Liu & Q.X. Feng
Key Laboratory of Engineering Dielectrics and its Application, Department of Electronic Science and Technology, College of Applied Science, Harbin University of Science and Technology, Heilongjiang Harbin, China

ABSTRACT: Copper Phthalocyanine Thin Film Transistors (CuPc TFTs) in five layers were fabricated in this paper. The submicron conductive channel improves the characteristics of the TFTs. The relationship between current and voltage is unsaturated in the transistors. This is because the complex current transmission mechanism formed by the vertical structure and the nature of copper phthalocyanine when V_{GS} is 0.8 V and V_{DS} is 3 V, g_m is 4.90×10^{-4} S, r_d is 1.02×10^4 Ω and μ is 4.99. The dynamic characteristics are that the phase shift of I_{DS} is changed from 0° to 90°, and the cut-off frequency is 2.6 kHz.

Keywords: CuPc; thin film transistors; submicron conductive channel; cut-off frequency

1 INTRODUCTION

There are the properties of abundance, simply adjusted molecular structure, diversification of the film forming process and the low price of organic semiconductor materials, so that the organic electronic device has a certain advantage in development. Phthalocyanine dye is one of the most promising materials that has so far been prepared for optoelectronic devices; it can be fabricated into optical recording devices, gas sensors and solar cells with good optoelectronic properties, excellent film growth characteristics and chemical stability. Phthalocyanine dye is purified simply, and it is used in preparing high purity film without decomposition. We mainly study the static characteristics and current transmission mechanism of CuPc thin film transistors in vertical structures in this paper (Jiti et al., 2010).

2 DEVICE FABRICATION

The glass substrate is maintained at room temperature during the preparation. The evaporation temperature of copper phthalocyanine is 400°C, the evaporation rate is about 0.5 Å/s and the vacuum is 4×10^{-5} torr.

Firstly, there was a copper electrode deposited on the glass substrate by DC magnetron sputtering. Secondly, a copper phthalocyanine thin film was prepared by vacuum evaporation. Thirdly, a layer of semi-conducting Al thin film was prepared as the gate. Fourthly, a copper phthalocyanine thin film was fabricated. Finally, the other copper electrode was prepared.

The structure of the CuPc/Al/CuPc in the CuPc TFT is vertical. There was Schottky contact between the Al and CuPc, then two Schottky diodes formed back to back, and the favourable Ohmic contact was formed between the CuPc and Cu thin film. The effective area of the VOTFT is 0.025 cm².

3 DEVICE TESTING

The characteristics of the copper phthalocyanine thin film transistor are tested by Keithley 4200, Keithley 428, 33220A and Tektronix 3021B. All of the electrical characteristics are measured at room temperature in the atmosphere. The source is grounded through an ammeter, V_{DS} is the voltage between the drain and source in a range of 0 V to 3 V. V_{GS} varies from 0.6 V to 1.6 V and the step is 0.2 V. The structures of this vertical thin film transistor and the circuit of static characteristic testing are shown in Figure 1. Its dynamic characteristics are tested in the case of the V_{GS}, which is applied to the sine wave AC signal.

4 RESULTS AND ANALYSIS

In an ideal CuPc TFT, holes in the CuPc flow into the Al thin film when they come into contact. The surface of the Al thin film is then positively charged, and the surface of the CuPc is negatively charged, the amount of changes is equivalent. V_{DG} is applied forward bias when $V_{DS} > V_{GS} > 0$ V, the Schottky barrier reduces. V_{GS} is applied back bias voltage, and the Schottky barrier increases, the sharp of barrier region between gate

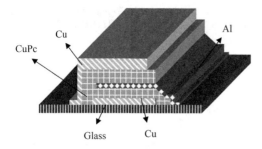

(a) Structure of CuPc thin film transistor.

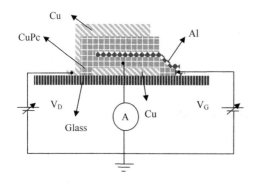

(b) Circuit of static characteristic testing in the transistor.

Figure 1. Structure and testing circuit of CuPc TFT.

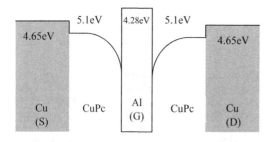

Figure 2. The band diagram of the thin film transistor.

and source is similar to triangle. The Schottky barrier between the gate and the source is higher, the barrier span is relatively smaller, and the main transmission mechanism is tunnelling through the barrier. The band diagram of the TFT is shown in Figure 2.

The increase of the V_{DS} makes the depletion layer near the gate narrower, the Schottky barrier decreases, the barrier potential reduces, that carriers transported over. More carriers are injected into the organic semiconductor layer from the drain, tunnelling through the gate region with CuPc/Al/CuPc Schottky barriers, and eventually reaching the source. This is the working current in the device (Hoshino et al., 2002).

The static characteristic of the CuPc TFT is that CuPc is prepared as an active layer and this is shown in Figure 3.

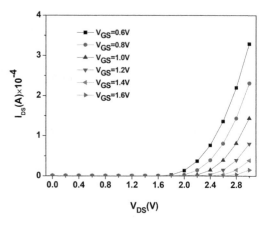

Figure 3. The static characteristic of the CuPc TFT.

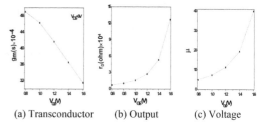

(a) Transconductor (b) Output (c) Voltage

Figure 4. The electrical properties in CuPc TFT.

When V_{DS} varies from 0 V to 3 V, I_{DS} increases with the increase of V_{DS}. When the voltage is low, the characteristic I-V changes exponentially, and it deviates exponentially as the voltage becomes higher. I_{DS} decreases with the increase of V_{GS} when the V_{DS} is constant. There is a submicron conductive channel in CuPc TFT, and the transistor can be used in a lower operating voltage of $V_{GS} = 0.8$ V, $V_{DS} = 3$ V, and can obtain a large current density. The relationship between I and V presents the unsaturated characteristic, and the reason for this is that a complex current transmission mechanism is formed because of the vertical structure and the nature of CuPc (Kwak et al., 2006).

The characteristics of the TFT, such as g_m, r_d and μ, are illustrated in Figure 4. When V_{GS} is 0.8 V and V_{DS} is 3 V, $g_m = 4.90 \times 10^{-4}$ S, $r_d = 1.02 \times 10^4 \Omega$ and $\mu = 4.99$.

In Figure 4(a), the transconductance decreases when V_{GS} varies in the range of 0.6~1.6 V. g_m is greater and V_{GS} could control I_{DS} better.

The relationship between r_d and V_{GS} is shown in Figure 4(b). The output resistance of the thin film transistor increases with the increase of V_{GS}.

The voltage amplification factor is shown in Figure 4(c). μ is related to V_{DS} and V_{GS}, and it has nothing to do with I_{DS}, so the expression is $\mu = g_m \times r_d$. By maintaining the I_{DS} at a certain value, keeping the required V_{DS} in the same current and a different V_{GS} by the linear interpolation method between adjacent

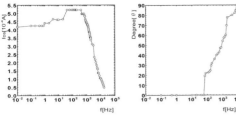

(a) The amplitude frequency characteristic (b) The phase frequency characteristic

Figure 5. The amplitude frequency characteristic and phase frequency characteristic curve of CuPc TFT.

tested data, then the voltage amplification factor μ is obtained. The voltage amplification factor increases with the increase of V_{GS} (Boming et al., 2010).

The amplitude frequency characteristic and phase frequency characteristic are indicated in Figure 5. A small sinusoidal AC signal is applied on the gate/source electrode, and the offset is 0.6 V. I_{DS} starts to fall when f is greater than 2.6 kHz. The phase shift of I_{DS} is varied significantly from 0° to 90°. The cut-off frequency of the transistor is 2.6 kHz. However, in situations involving large amounts of power, the vertical CuPc thin film transistor has a fast transient response and low distortion because of the submicron conductive channel, and the working speed of the transistor is improved. The shortcomings of the low carrier mobility and high resistivity of organic semiconductor materials are made up (Park et al., 2009).

5 CONCLUSION

The vertical copper phthalocyanine thin film transistors prepared in this paper have the submicron conductive channel, which can be used at a lower operating voltage and obtain a larger current. The relationship between the current and the voltage in the transistor is unsaturated. This is because the vertical structure and the nature of CuPc contribute to the complex current transmission mechanism. The static characteristics are that $g_m = 4.90 \times 10^{-4}$ S, $r_d = 1.02 \times 10^4$ Ω and μ = 4.99, when V_{GS} is 0.8 V and V_{DS} is 3 V. The dynamic characteristic is that I_{DS} begins to fall if f is greater than 2.6 kHz. The phase shift of I_{DS} is changed from 0° to 90°. The cut-off frequency is 2.6 kHz.

The copper phthalocyanine thin film transistor can improve the working characteristics and remedy the low carrier mobility and high resistivity of organic semiconductor materials by the short conductive channel length.

REFERENCES

Hoshino, S., Kamata, T., et al. (2002). Effect of active layer thickness on device properties of organic thin-film transistors based on Cu(II) phthalocyanine. *Journal of Applied Physics*, 6028–6032.

Kwak, T. H., Kang, H. S., et al. (2006). *Copper-phthalocyanine based organic thin film transistor* (pp. 630–631). IEEE Conference Publications.

Nukeaw, J. & Tunhoo, B. (2010). *Characterization of copper phthalocyanine thin film photovoltaic on PET substrate prepared by electron-beam evaporation technique* (pp. 660–661). IEEE Conference Publications.

Park, J., Royer, J. E., et al. (2009). Ambient induced degradation and chemically activated recovery in copper phthalocyanine thin film transistors. *Journal of Applied Physics*, 106(3).

Wu, B., Wang, D., et al. (2013). *Device operation of organic semiconductor copper phthalocyanine thin film transistor* (pp. 206–208). IEEE Conference Publications.

Electronic Engineering – Wang (ed)
© 2018 Taylor & Francis Group, London, ISBN 978-1-138-60260-1

Active contour segmentation model based on global and local Gaussian fitting

F.Z. Zhao, H.Y. Liang & X.L. Wu
College of Computer Science & Information Engineering, Hezhou University, Guangxi, China

D.H. Ding
College of Computer Science & Information Engineering, Hunan University of Arts and Science, Changde, China

ABSTRACT: This paper presents a novel active contour segmentation model for image segmentation. In the proposed model, the image intensities are described by the global and local Gaussian distributions with different means and variances, respectively. In this combination, we increase a weighting coefficient by which the proposed model can adjust the ratio between the global and local region fitting energies. In addition, to improve the accuracy of fitting information, the Heaviside function is improved. The proposed model is able to distinguish regions with similar intensity means but different variances. By adaptively updating the weighting coefficient, this algorithm can speed up the convergence rate. Comparative experiment results demonstrate that the proposed model is effective with application to both the synthetic and medical images.

Keywords: active contour model; image segmentation; Gaussian distributions; Heaviside function

1 INTRODUCTION

Image segmentation is always one of the fundamental problems in image processing and computer vision. Classically, it is the process of dividing images into meaningful subsets with approximately similar properties. In the past thirty years, many researchers have been continuously proposed a wide variety of methods for image segmentation (Xiao et al. 2013).

Up to now, the best known and most influential algorithm is active contour model, which is based on the theory of curve and surface evolutions and geometric flows have been studied extensively and used in the field of image segmentation successfully (Kass et al. 1998). Generally speaking, active contour models can be categorized into two different classes: edge-based models and region-based models (Caselles et al. 1997). The edge-based active contour models mainly use the edge information such as image gradient to drive the active contour toward the object boundaries and stop the contour evolution (Xu and Prince 1998). However, the defects of the edge-based active contours models are that they are not only very sensitive to noise but also difficult to detect weak edges. Moreover, the segmentation results are highly dependent on the placement of the initial curve. Compared with the edge-based models, the region-based active contour models utilize the image statistical information to detect objects, which have many advantages. Firstly, the region-based models can effectively segment the images with weak boundaries. Secondly, they are less

sensitive to the noise. Moreover, the region-based models are usually less dependent on the initialization since they exploit the global region information the global region information of the image statistics (Wang et al. 2015). Therefore, in this paper, we mainly focus on the region-based models.

Region-based active contour models use a certain region descriptor to guide the motion of the active contour. Therefore, they are less sensitive to initial contours and have better performance for images with weak object boundaries. A large variety of region-based models have been proposed over the past thirty years. The well-known Chan-Vese (CV) models are based on the assumption that image intensities are statistically homogeneous in each region, and therefore they fail to segment images with intensity inhomogeneity (Chan and Vese 2001). Recently, Li et al. proposed a local binary fitting (LBF) model to overcome the difficulty in segmentation caused by intensity inhomogeneity. The LBF model draws upon spatially varying local region information and thus is able to deal with intensity inhomogeneity. By using local region information, the LBF model is able to provide desirable segmentation results even in the presence of intensity inhomogeneity. However, the LBF model based on local information, the drawback is the high dependence on initial position of the contour. To improve the robustness to initialization, Wang et al. defined an energy function (Local and Global Intensity Fitting, LGIF) with a local intensity fitting term used in LBF model and an auxiliary global intensity fitting term used in C-V model. However, the LGIF model also has several deficiencies as follows. First, the non-convexity of the LGIF energy functional may cause local minima. Second, for different images, it is necessary to choose appropriate weight values to control the local and global intensity fitting force. Third, the segmentation result is still slightly dependent on the location of the initial contour, because the weight between the local term and the global term is a fixed positive constant.

These aforementioned models draw upon intensity means, which enables them to cope with intensity inhomogeneity. However, when the intensity means of the image is consistent, the aforementioned models will be invalid. To overcome this problem, Wang et al. proposed a novel active contour model which is driven by local Gaussian distribution fitting (LGDF) energy with local means and variances as variables. The local intensity means and variances are strictly derived from a variational principle. However, LGDF model only utilize the local intensity information to establish the energy function, so the convergence speed is slow. In addition, it utilizes the traditional Heaviside function to fitting image information, which has some shortcomings in accuracy.

In this paper, we proposed a novel active contour model for image segmentation. Note that our method is different from the LGDF model. First, in the proposed model, the Heaviside function is improved, which can improve the fitting accuracy of intensity information. Second, the global and local region fitting energies are described by a combination of the global and local Gaussian distributions with different means and variances, respectively. In this combination of the proposed model, we utilized a weighting coefficient by which we can dynamically adjust the ratio between the global and local region fitting energies and the weighting coefficient can be adaptively updated with the contour evolution. The proposed algorithm is effective and fits to the image with intensity inhomogeneity. In addition, it is insensitive to the initial contour and can be specially used to detect the desired objects. Experiments on both the synthetic and medical images demonstrate desirable performances of the proposed method.

2 BACKGROUND

2.1 The CV model

Chan and Vese proposed CV model, which assumed that the original image is a piecewise constant function. For an image I on the image domain Ω, they proposed to minimize the following energy:

$$E^{CV}(c_1, c_2, C) = \lambda_1 \int_{inside(C)} |I(x) - c_1|^2 dx$$
$$+ \lambda_2 \int_{outside(C)} |I(x) - c_2|^2 dx + \nu |C| \quad (1)$$

where $inside(C)$ and $outside(C)$ represent the region inside and outside of the contour C, respectively. c_1

and c_2 are two constants that approximate the image in $inside(C)$ and $outside(C)$, respectively. λ_1, λ_2 and ν are nonnegative constant.

The CV model has good performance in image segmentation due to its ability of obtaining a larger convergence range and being less sensitive to the initialization. However, if the intensities with inside C or outside C are not homogeneous, the constants and will not be accurate. As a consequence, the CV model generally fails to segment images with intensity inhomogeneity. Similarly, more general piecewise constant models in a multiphase level set framework are not good at such images either.

2.2 The LBF model

To overcome the difficulty caused by intensity inhomogeneity, Li et al. proposed the LBF model, which can segment images with intensity inhomogeneity, using the local intensity information efficiently, and has achieved promising results. The data fitting term is defined in the form of kernel function. They proposed to minimize the local binary fitting energy:

$$E^{LBF}(f_1(x), f_2(x), \phi(x)) = \lambda_1 \int \left[\int K_\sigma(x - y) |I(y) - f_1(x)|^2 dy \right] dx$$
$$+ \lambda_2 \int \left[\int K_\sigma(x - y) |I(y) - f_1(x)|^2 dy \right] dx + \nu L(\phi(x)) + \mu P(\phi(x)) \quad (2)$$

where $\lambda_1, \lambda_2, \nu$ and μ are weighting positive constants. K_σ is a Gaussian kernel function, and σ is a constant to control the local region size. The third term is the length term to smooth the contour and the forth term is the level set regularization term to penalize the deviation of the level set function ϕ from a signed distance function.

Because of using local region information, specifically local intensity mean, the LBF model is able to provide desirable segmentation results even in the presence of intensity inhomogeneity. However, the LBF model is sensitive to initialization to some extent and is easy to fall into local minimum, which limits its practical applications.

2.3 The LGDF model

Wang et al. proposed the LGDF model by utilizing a finite Gaussian distribution. The energy function of the LGIF model is defined as follows:

$$E^{LGDF} =$$
$$\int_\Omega \left(\sum_{i=1}^N \int_{\Omega_i} -\omega(x - y) \left[\log(\sigma_i(x)) + \frac{(I(y) - u_i(x))^2}{2\sigma_i^2(x)} \right] dy \right) dx$$
$$+ \nu L(\phi(x)) + \mu P(\phi(x)) \quad (3)$$

where $\omega(x - y)$ is a discrete Gaussian kernel function. ν and μ are nonnegative constant. μ_i $(i = 1, 2)$ are called intensity means. σ_i $(i = 1, 2)$ are called standard deviations. $L(\phi)$ is length term and $P(\phi)$ is regularization term.

Because of the LGDF model only utilize the local intensity information to establish the energy function, so the convergence speed is slow. In addition, it utilizes the traditional Heaviside function to fitting image information, which has some shortcomings in accuracy.

3 THE PROPOSED MODEL

In image segmentation, the accuracy of segmentation largely depends on the local intensity information. Meanwhile, to enhance the robustness to noise and reduce the possibility of getting stuck in local minima, the global intensity information plays an important role. Particularly, the global intensity information is crucial to decrease the sensitivity to the location of initial contour.

Motivated by the contributions in, and the LBF model, we improved Heaviside function, and proposed an active contour segmentation model based on global and local Gaussian fitting, which is LGGDF (Local and Global Gaussian Distribution Fitting, LGGDF) model. In the proposed model, the fitting energy term is described by a combination of the global and local Gaussian distributions with different means and variances, respectively. In this combination of the proposed model, we utilize a weighting coefficient by which we can dynamically adjust the ratio between the global and local region fitting energies and the weighting coefficient can be adaptively updated with the contour evolution. The energy function is defined as follows:

$$E(C, u_1(x), u_2(x), \sigma_1^2(x), \sigma_2^2(x), u_3, u_4, \sigma_3^2, \sigma_4^2)$$

$$= (1 - W(x)) \left\{ \int \left[\int_{\Omega_1} K(x-y) \log p_{1,x}(I(y), u_1(x), \sigma_1^2(x)) dy \right] dx \right.$$

$$+ \int \left[\int_{\Omega_2} K(x-y) \log p_{2,x}(I(y), u_2(x), \sigma_2^2(x)) dy \right] dx \right\}$$

$$- W(x) \left\{ \int_{\Omega_1} \log p_3(I(x), u_3, \sigma_3^2) + \int_{\Omega_2} \log p_4(I(x), u_4, \sigma_4^2) \right\} dx$$

$$+ vL(\phi(x)) + \mu P(\phi(x)) \tag{4}$$

where the first term is the local Gaussian fitting energy. The second term is the global Gaussian fitting energy. The third term is the length term. The fourth term is the regularization term. ϕ is the level set function. v and μ are nonnegative constant. μ_i $(i=1,2)$ are called local means, and μ_i $(i=3,4)$ called global means. σ_i $(i=1,2)$ are called local standard deviations, and similarly σ_i $(i=3,4)$ are called global standard deviations. $K(x-y)$ is the window function. $W(x)$ is the weighting coefficient.

The selection of window function is various. We can utilize a discrete Gaussian kernel template after clipping or a typical constant function template. The window function of this paper is as follows:

$$K(x,r) = \begin{cases} 1, & |x-y| \le r \\ 0, & |x-y| > r \end{cases} \tag{5}$$

The weighting coefficient $W(x)$ is as follows:

$$W(x) = \rho \cdot average(C_N(x))(1 - C_N(x)) \tag{6}$$

where ρ is a nonnegative constant. $average(C_N(x))$ is the mean which is $C_N(x)$ of the image and it reflects the intensity contrast. If the original image has higher contrast, the global energy fitting term will be enhanced adaptively by the weight coefficient. In contrast, the global energy fitting term is enhanced will be enhanced.

$C_N(x)$ is defined as follows:

$$C_N(x) = \frac{M_{max} - M_{min}}{M_g} \tag{7}$$

where N is the size of the local window and it is 11×11 in this paper. M_{max} and M_{min} are the maximum and minimum of gray value in the local window, respectively. M_g is the maximum of gray value in the whole image domain. For gray image, the value of M_g is usually 255. $C_N(x)$ fluctuates between 0 and 1. In general, $C_N(x)$ is smaller in the smoothing area of the image and it is larger at the edge of the target.

Using the standard gradient descent algorithm to minimize the LGGDF model, we derive the level set evolution equation as follows:

$$\frac{\partial \phi}{\partial t} = -\delta_\varepsilon(\phi) \left[(1 - W(x))(e_1 - e_2) + W(x)(e_3 - e_4) \right]$$

$$+ v\delta_\varepsilon(\phi) div(\frac{\nabla \phi}{|\nabla \phi|}) + \mu div(d_p(|\nabla \phi|) \nabla \phi) \tag{8}$$

where the first term is the data force term. The second term is the length term. The third term is the regularization term.

In Eq. (8), e_1, e_2, e_3 and e_4 are respectively as follows:

$$e_i = \int K(y-x)(\log \sigma_i(y) + \frac{(I(x) - u_i(y))^2}{2\sigma_i^2(y)}) dy, \quad i=1,2 \tag{9}$$

$$e_j = \log \sigma_j + \frac{(I(x) - u_j^2)}{2\sigma_j^2}) dy, \quad j=3,4 \tag{10}$$

where $u_1(x), u_2(x), u_3(x), u_4(x)$ and $\sigma_1(x), \sigma_2(x), \sigma_3(x), \sigma_4(x)$ are respectively as follows:

$$\begin{cases} u_1(x) = \dfrac{\int_\Omega K(y-x) \cdot H(\phi(y)) dy}{\int_\Omega K(y-x) H(\phi(y)) dy} \\[4mm] u_2(x) = \dfrac{\int_\Omega K(y-x) \cdot [1 - H(\phi(y))] dy}{\int_\Omega K(y-x)[1 - H(\phi(y))] dy} \end{cases} \tag{11}$$

$$\begin{cases} u_3(x) = \dfrac{\int_\Omega H(\phi(y)) dy}{\int_\Omega H(\phi(y)) dy} \\[4mm] u_4(x) = \dfrac{\int_\Omega (1 - H(\phi(y))) dy}{\int_\Omega (1 - H(\phi(y))) dy} \end{cases} \tag{12}$$

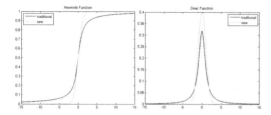

Figure 1. The diagrams of Heaviside function and Dirac function.

$$\begin{cases} \sigma_1^2(x) = \dfrac{\int_\Omega K(y-x)\cdot(u_1(x)-I(y))^2 \cdot H(\phi(y))dy}{\int_\Omega K(y-x)\cdot H(\phi(y))dy} \\ \sigma_2^2(x) = \dfrac{\int_\Omega K(y-x)\cdot(u_2(x)-I(y))^2 \cdot [1-H(\phi(y))]dy}{\int_\Omega K(y-x)[1-H(\phi(y))]dy} \end{cases} \quad (13)$$

$$\begin{cases} \sigma_3^2(x) = \dfrac{\int_\Omega (u_3(x)-I(y))^2 \cdot H(\phi(y))dy}{\int_\Omega H(\phi(y))dy} \\ \sigma_4^2(x) = \dfrac{\int_\Omega (u_4(x)-I(y))^2 \cdot [1-H(\phi(y))]dy}{\int_\Omega [1-H(\phi(y))]dy} \end{cases} \quad (14)$$

In the LGDF model, $H(\phi)$ is the conventional Heaviside function and its expression is as follows:

$$H_\varepsilon(x) = \frac{1}{2}\left\{1+\frac{2}{\pi}\left[\arctan(\frac{x}{\varepsilon})\right]\right\} \quad (15)$$

The corresponding Dirac function is as follows:

$$\delta_\varepsilon(x) = \frac{1}{\pi}\frac{\varepsilon}{\varepsilon^2+x^2} \quad (16)$$

In this paper, we improve the Heaviside function and the new Heaviside function is as follows:

$$H_{new}(x) = \frac{1}{2}+\frac{1}{\sqrt{\pi}}\int_0^{\frac{t}{\sqrt{2}\sigma}} e^{-x^2} dx \quad (17)$$

In Eq. (17), the second term is an error function. To derivate the Eq. (17) and we will get the corresponding improved Dirac function which is as follows:

$$\delta_{new}(x) = H_{new}'(x) = \frac{1}{\sqrt{2\pi}\sigma} e^{-\frac{x^2}{2\sigma^2}} \quad (18)$$

The corresponding Heaviside function and Dirac function are shown in figure 1, where (a) is the Heaviside function and (b) is the Dirac function. The red curve shows the traditional Heaviside function and the corresponding Dirac function and the green curve shows the improved Heaviside function and the corresponding Dirac function.

Can be seen from the Fig. 1, the performance of the improved Heaviside function and Dirac function is superior to the traditional Heaviside function and Dirac function obviously.

The implementation of our method is straightforward, and is presented below:

Step 1: Initialize a level set function ϕ.
Step 2: Update $u_1(x)$ and $u_2(x)$ using Eq. (11).
Step 3: Update $u_3(x)$ and $u_4(x)$ using Eq. (12).
Step 4: Update $\sigma_1(x)$ and $\sigma_2(x)$ using Eq. (13).
Step 5: Update $\sigma_3(x)$ and $\sigma_4(x)$ using Eq. (14).
Step 6: Update $e_1(x)$ and $e_2(x)$ using Eq. (9).
Step 7: Update $e_3(x)$ and $e_4(x)$ using Eq. (10).
Step 8: Update the level set function using Eq. (8).
Step 9: Regularize the level set function with the regularization term.
Step 10: Return to step 2 until the convergence criteria is met.

4 EXPERIMENTAL RESULTS AND ANALYSIS

This section validates the performance of the proposed model with various synthetic and real images from different modalities. We compared our method with CV model, LBF model, LGDF model. The energy fitting functions of the CV model and the LBF model utilize the image intensity means to model. The energy fitting function of the LGDF model utilizes the local intensity means and variance to model, respectively.

All experiments are performed on a PC with Pentium Dual-Core CPU 2.0 GHz and 2.0 GHz RAM, using MATLAB2012a. We tested the proposed method with the following parameters: $\Delta t = 0.1$, $\upsilon = 0.0008 \times 255^2$, $\mu = 1$, $\sigma = 5$. In order to speed up the contour evolution, the initial contour is set to a binary function whose value is 2. In this paper, the initial contours and the final contours are plotted as green contours and red contours, respectively.

4.1 Synthetic images

In order to demonstrate the capability of our method in dealing with intensity inhomogeneity and speed up the convergence, we first compare our method with CV, LBF and LGDF to segment a synthetic image with intensity inhomogeneity. The size of the image is 127×96. The characteristics of this image are the target and background without noise, while behind the T type exist artifact. The segmentation results with the same initial contours are illustrated in Fig. 2.

As can be seen from Fig. 2, with the same initial contours, the CV and LBF model fail to segment the image. The LGDF model Shows a certain ability to segment the image, while the convergence speeding is lower than our method. By contrast, our method successfully segments it, as shown in the last column of Fig. 2.

In order to illustrate LGGDF model can segment heterogeneous image with the same intensity means and different variances, our model is applied to another

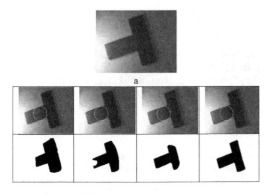

Figure 2. Segmentation results of CV, LBF, LGDF and our method with intensity inhomogeneity (a) original image. Column 1: the results of CV model. Column 2: the results of LBF model. Column 3: the results of LGDF model. Column 4: the results of our model.

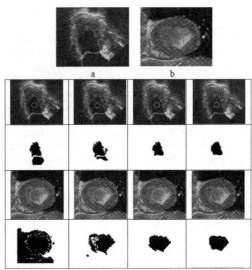

Figure 4. Segmentation results of medical images. (a) and (b) are original images; Column 1: the results of CV model. Column 2: the results of LBF model. Column 3: the results of LGDF model. Column 4: the results of our model.

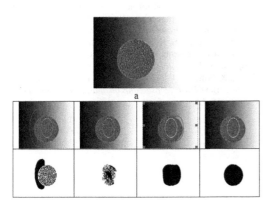

Figure 3. Segmentation results of CV, LBF, LGDF and our method for intensity inhomogeneity with the same means and different variances (a) original image. Column 1: the results of CV model. Column 2: the results of LBF model. Column 3: the results of LGIF model. Column 4: the results of our model.

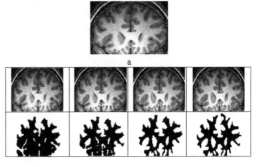

Figure 5. Segmentation results of a brain image. (a) original image; Column 1: the results of CV model. Column 2: the results of LBF model. Column 3: the results of LGDF model. Column 4: the results of our model.

synthetic image. The size of the image is 238×124. The intensity means with background and target are the same, which are 114. The variances are 0 and 1276, respectively. In order to highlight the performance of LGGDF model, our model compared with the CV model, LBF model and LGDF model. The segmentation results are illustrated in Fig. 3.

As can be seen from Fig. 3, the CV and LBF models fail to segment the image with the same means and different variances. Because of they only utilizing the intensity means information, without involving the variances information, they are unable to deal with the mage with the same means and different variances. The LGDF model, utilizing the means and variances information for statistical modeling, shows a certain ability to segment the image. However, the LGDF model without global region information and its segmentation results are not very desired. By contrast, our method can segment the image successfully by utilizing local and global region information and improving the Heaviside function.

4.2 Medical images

In order to evaluate the performance of the LGGDF model, our model is applied to the real medical image. Fig. 4 shows two medical images with intensity inhomogeneity. (a) shows the left ventricle in a tagged MR image and (b) shows an ultrasound image of left ventricle. It is clearly seen that the two images are corrupted by noise, severe intensity inhomogeneity and weak boundaries. Compare our method with CV, LBF and LGDF to segment them. Fig. 4 shows the segmentation results.

As can be seen from Fig. 4, the CV and LBF models fail to segment the medical images. The LGDF model shows a certain ability to segment the image, while the segmentation results are not quite ideal. In contrast, our model can accurately recover the object shapes.

In order to further illustrate the capability of our method in dealing with intensity inhomogeneity, we test our method for brain MR image. Consequently, the CV and the LBF models fail to extract white matter from the image. The LGDF model shows the intermediate results. It can be seen that our method achieves satisfactory results for the image. This experiment further shows the robustness of our method to intensity inhomogeneity.

5 CONCLUSIONS

In this paper, we have presented a region-based active contour based on global and local Gaussian fitting. The main contribution of this work lies in that we show the robustness of our method to intensity inhomogeneity. The segmentation accuracy is enhanced by improving Heaviside function. Comparative experiments on image segmentation show that our method can achieve accurate segmentation results with the same initial contours. However, we don't estimate the bias fields of the original images. It will be the content of the research in the next step of work about this paper.

ACKNOWLEDGMENTS

This research is supported by the Programs, which are Science and Technology Research Key Project (No. ZD2014129) of Guangxi Province Colleges, and the Fundamental Research Funds for the Hezhou University (No. 2016ZZZK11).

REFERENCES

Caselles, V., Kimmel, R., & Sapiro, G. (1997). Geodesic active contours. *International journal of computer vision*, *22*(1), 61–79.

Chan, T. F., & Vese, L. A. (2001). Active contours without edges. *IEEE Transactions on image processing*, *10*(2), 266–277.

Kass, M., Witkin, A., & Terzopoulos, D. (1988). Snakes: Active contour models. *International journal of computer vision*, *1*(4), 321–331.

Wang, X. F., Min, H., Zou, L., & Zhang, Y. G. (2015). A novel level set method for image segmentation by incorporating local statistical analysis and global similarity measurement. *Pattern Recognition*, *48*(1), 189–204.

Xiao, C., Gan, J., & Hu, X. (2013). Fast level set image and video segmentation using new evolution indicator operators. *The Visual Computer*, *29*(1), 27–39.

Xu, C., & Prince, J. L. (1998). Snakes, shapes, and gradient vector flow. *IEEE Transactions on image processing*, *7*(3), 359–369.

Electronic Engineering – Wang (ed)
© 2018 Taylor & Francis Group, London, ISBN 978-1-138-60260-1

Linguistic multi-criteria group decision-making method using incomplete weights information

L.W. Qu
School of Software, Harbin University of Science and Technology, Harbin, China

H. Liu
School of Software, Harbin University of Science and Technology, Harbin, China
Postdoctoral Research Center of Computer Science and Technology,
Harbin University of Science and Technology, Harbin, China

Y.Y. Zuo
School of Software, Harbin University of Science and Technology, Harbin, China

S. Zhang
School of Computer Science and Technology, Harbin Engineering University, Harbin, China
School of Informatics and Technology, Heilongjiang University, Harbin, China

X.S. Chen
School of Software, Harbin University of Science and Technology, Harbin, China

V.V. Krasnoproshin
Faculty of Applied Mathematics and Computer Science, Belarusian State University, Minsk, Belarus

C.Y. Jiang & H. Liu
School of Software, Harbin University of Science and Technology, Harbin, China

ABSTRACT: To solve the group decision-making problems of uncertain linguistic multiple criteria and incomplete weights data, a decision analysis method was proposed. Describing the uncertain linguistic variables, calculating steps were given to solve the group decision-making problems. The core was combining uncertain linguistic decision matrices provided by experts to a group decision matrix with informative possibilities. Then, a quadratic programming model was settled through calculating the difference values between each alternative and positive/negative ideal point to determine the criteria weights, thus giving the relative proximity of each alternative so as to rank all of the alternatives. A case study was undertaken to show how to practice the method and the analysis showed that the method is feasible and is simple to perform.

Keywords: multiple criteria group decision-making; uncertain linguistic variable; incomplete information; alternatives ranking

1 INTRODUCTION

For practical multi-criteria group Decision-Making (DM) problems, the decision makers often use more convenient and natural language to give evaluation information, due to the difficulty and fuzziness of objective things. In recent years, the Multiple Criteria Group Decision-Making (MCGDM) problem, based on linguistic evaluation data, has been the subject of attention from scholars (Chiclana et al., 2013; Vahdani et al., 2013; Wang et al., 2016; Xia et al., 2013; Zhang et al., 2013). However, because of the complexity of objective things, and the uncertainty and fuzziness of

the human mind, experts in the Linguistic Multiple Criteria Group Decision-Making (LMCGDM) analysis of the evaluation of the property is often given in the form of uncertainty in the form of language.

Wei applied the Uncertain Linguistic Ordered Weighted Averaging (ULOWA) operator (Wei, 2009) and Liu applied the Interval-Valued Uncertain Linguistic Ordered Weighted Averaging (IULOWA) operator (Liu et al., 2016) respectively, showing the MCGDM method based on the non-deterministic linguistic data. It is noticed that the above decision method only applies to cases with known criteria evaluation values and a criteria weight value;

and it cannot solve the multi-criteria group decision problem where the criteria weight information is unknown or incomplete.

In this paper, the uncertain decision matrix from each expert is transformed into the group decision matrix with possibility degree information, for the MCGDM problem with unknown criteria weights and criteria value formed by Uncertain Linguistic Variables (ULV). Then, a MCGDM analysis method is given, which is consistent with the ideal point method (Höhle & Rodabaugh, 2012; Vahdani et al., 2013; Xu, 2015).

2 PROBLEM DESCRIPTION

In order to facilitate the description and analysis of MCGDM with uncertainties, this paper firstly gives the following concepts and their properties.

$S = \{s_i, |i = 0, 1, \cdots, T\}$ is proposed as a linguistic variables set consisting of $T + 1$ linguistic variables. As an example, a linguistic variables set consisting of seven linguistic variables can be presented as:

$$S = \begin{cases} s_0 = \text{verypoor(vp)}, s_1 = \text{poor(p)}, s_2 = \text{mediumpoor(mp)}, \\ s_3 = \text{fair(f)}, s_4 = \text{mediumgood(mg)} s_5 = \text{good(g)}, \\ s_6 = \text{verygood(vg)} \end{cases}.$$

where S has the following properties:

1) Orderliness: if $i > j$, then s_i ">" s_j and ">" represents 'is better than'. And if "<" represents 'is worse than', ">" represents 'is equal to' correspondingly.
2) Inverse operator 'Neg': if $j = T - i$, then $Neg(s_i) = s_j$, and $T + 1$ represents the number of elements of set S.
3) Maximisation and minimisation operators: if s_i ">" s_j, then $Max\{s_i, s_j\} = s_i$, and $Min\{s_i, s_j\} = s_j$.

$\tilde{S} = \{\tilde{s}_i, |i = 0, 1, \cdots, n\}$ is proposed as a ULV set, where $\tilde{s}_i = [s_\alpha, s_\beta]$, s_α, s_β represents the upper and lower bounds of ULVs $\tilde{s}_i$ and $s_\alpha, s_\beta \in S$, s_α, s_β ">" $s_{\frac{T}{2}}$ or s_α, s_β "≤" $s_{\frac{T}{2}}$.

In the ULV set, taking linguistic variable $s_{\frac{T}{2}}$ as the bound, the linguistic variables on both sides of the bound are opposite in semantic meaning.

In order to avoid confusion, and to ensure the consistency of the logic, the upper and lower bounds of the ULV $\tilde{s}_i$ should be on the left or right of $s_{\frac{T}{2}}$.

A group decision-making problem is formalised as below with multiple uncertain linguistic criteria based on incomplete criteria of weight information. $X = \{X_1, X_2, \ldots, X_m\}$ represents a set of m alternatives, where $m \geq 2$, and P_j represents the jth criteria. $w = (w_1, w_2, \ldots, w_n)^T \in W$ represents the unknown criteria of the weight vector, representing the weight of criterion P_j, and $\sum_{j=1}^{n} w_j = 1$, $w_j \geq 0$ ($j = 1, 2, \ldots, n$). W represents a set of mathematical expressions with

incomplete information for the criteria weight, specifically including the following five cases (Wei, 2015):

1) $w_i - w_j \geq \varepsilon_i$;
2) $w_i \geq \alpha_i w_j$;
3) $w_i - w_j \geq w_k - w_l, i \neq j \neq k \neq l$;
4) $\beta_i^- \leq w_i \leq \beta_i^+$;
5) $\rho_i^- w_j \leq w_i \leq \rho_i^+ w_j$ or $\rho_i^- \leq w_i / w_j \leq \rho_i^+, w_j \neq 0$.

where $\forall i, \varepsilon_i, \beta_i^-, \beta_i^+ \in [0, 1]$; $\alpha_i, \rho_i^-, \rho_i^+$ are positive real numbers; $e = \{e_1, e_2, \ldots, e_m\}$ represents a set of l evaluation experts ($l \geq 2$), e_k represents the kth expert; $u = \{u_1, u_2, \ldots, u_n\}^T$ represents the expert weight vector, where u_k represents the weight of expert e_k, $\sum_{k=1}^{k} u_k = 1$, and $u_k \geq 0$ ($k = 1, 2, \ldots, l$); $\tilde{A}^k = [\tilde{a}_{ij}^k]_{m \times n}$ are Decision Matrices with Uncertain Linguistic Evaluation Information (DMULEI), $\tilde{a}_{ij}^k$ represents an evaluation result of the criterion P_j on alternative X_i given by the decision maker e_k, and $\tilde{a}_{ij}^k \in S$. Then the problem is transferred into the following.

Given DMULEI $\tilde{A}^k$, the expert weight vector u and the unknown criteria weight vector w, how to choose the optimal alternative of set X.

3 DECISION-MAKING METHOD

According to the elementary impression of the ideal point method, the analysis method for solving MCGDM problems with incomplete criteria weight information is given below:

1) Normalise $\tilde{A}^k$ as $\tilde{B}^k = [\tilde{b}_{ij}^k]_{m \times n}$.

$$\tilde{b}_{ij}^k = \begin{cases} \tilde{a}_{ij}^k, p_j \text{ is benefit type attribute;} \\ Neg(\tilde{a}_{ij}^k), p_j \text{ is cost type attribute.} \end{cases} \quad (1)$$

where $\tilde{a}_{ij}^k \in \tilde{S}$. And for $\tilde{a}_{ij}^k = [s_{\alpha(ij)}^k, s_{\beta(ij)}^k]$, $Neg(\tilde{a}_{ij}^k) = [s_{\alpha'(ij)}^k, s_{\beta'(ij)}^k]$ is given, where $\alpha' = T - \alpha, \beta' = T - \beta$ and $\tilde{b}_{ij}^k \in \tilde{S}$.

2) Integrate uncertain linguistic DM matrix $\tilde{B}^k = [\tilde{b}_{ij}^k]_{m \times n}$ into $\bar{C} = [\bar{c}_{ij}]_{m \times n}$, which is a DM matrix with possibility degree information.

$$\bar{c}_{ij} = (c_{0(ij)}, \cdots, c_{T(ij)}) = \left(\sum_{k=1}^{l} u_k I(s_{0(ij)}^k), \cdots, \sum_{k=1}^{l} u_k I(s_{g(ij)}^k) \right) \quad (2)$$

where $g = 0, 1, \ldots, T$, $I(s_{g(ij)}^k)$ is an indicative function.

$$I(s_{g(ij)}^k) = \begin{cases} 1, s_{g(ij)}^k \in \tilde{b}_{ij}^k; \\ 0, s_{g(ij)}^k \in \tilde{b}_{ij}^k. \end{cases} \quad (3)$$

3) Build Weighted Group Decision Matrix (WGDM) $\bar{R} = [\bar{r}_{ij}]_{m \times n}$, on the basis of the group DM matrix $\bar{C}$.

$$\bar{r}_{ij} = (r_{0(ij)}, \cdots, r_{T(ij)}) = w_j(c_{0(ij)}, \cdots, c_{T(ij)}) \quad (4)$$

4) Construct Positive Ideal Point (PIP) $\bar{v}^+ = (\bar{v}_1^+, \bar{v}_2^+, \ldots, \bar{v}_n^+)$ and Negative Ideal Point (NIP) $\bar{v}^- = (\bar{v}_1^-, \bar{v}_2^-, \ldots, \bar{v}_n^-)$, on the basis of WGDM $\bar{R}$, where $j = 1, 2, \ldots, n$.

$$\bar{v}_j^+ = (\bar{v}_{0(j)}^+, \cdots, \bar{v}_{T(j)}^+) =$$
$$(\max_i r_{0(ij)}, \cdots, \max_i r_{T(ij)}) = \quad (5)$$
$$(w_j \max_i c_{0(ij)}, \cdots, w_j \max_i c_{T(ij)});$$

$$\bar{v}_j^- = (\bar{v}_{0(j)}^-, \cdots, \bar{v}_{T(j)}^-) =$$
$$(\min_i r_{0(ij)}, \cdots, \min_i r_{T(ij)}) = \quad (6)$$
$$(w_j \min_i c_{0(ij)}, \cdots, w_j \min_i c_{T(ij)});$$

5) Based on $\bar{R}$, $\bar{v}_j^+$ and $\bar{v}_j^-$, the differences between each scheme and the PIP and NIP have been determined. Given the calculation formulas of d_i^+ and d_i^-, which are the difference values of alternative X_i with PIP and NIP:

$$d_i^+ = \sum_{j=1}^n d(\bar{r}_{ij}, \bar{v}_j^+) = \sum_{j=1}^n [w_j^2(c_{0(ij)} - \max_i c_{0(ij)})^2 + \quad (7)$$
$$\cdots + w_j^2(c_{T(ij)} - \max_i c_{T(ij)})^2]$$

$$d_i^- = \sum_{j=1}^n d(\bar{r}_{ij}, \bar{v}_j^-) = \sum_{j=1}^n [w_j^2(c_{0(ij)} -, \min_i c_{0(ij)})^2 + \quad (8)$$
$$\cdots + w_j^2(c_{T(ij)} - \min_i c_{T(ij)})^2]$$

where $d(\bar{r}_{ij}, \bar{v}_j^+)$ represents the distance between $\bar{r}_{ij}$ and $\bar{v}_j^+$. For receiving d_i^+ and d_i^-, w_j needs to be calculated. Therefore, a Multiple Objective Optimisation (MOO) model is established as below:

$$\min D_i^+ = \sum_{j=1}^n d(\bar{r}_{ij}, \bar{v}_j^+), \quad (9)$$

$$\max D_i^- = \sum_{j=1}^n d(\bar{r}_{ij}, \bar{v}_j^-), \quad (10)$$

$$s.t. \quad w \in W, \quad (11)$$

$$\sum_{j=1}^n w_j = 1, \quad (12)$$

$$w_j \geq 0. \quad (13)$$

6) The objective functions represented by Equations 9 and 10 can be synthesised. Consequently, the MOO problem can be converted to a single-objective optimisation problem:

$$\min D = \sum_{i=1}^m \sum_{j=1}^n [d(\bar{r}_{ij}, \bar{v}_j^+) - d(\bar{r}_{ij}, \bar{v}_j^-)] = \quad (14)$$
$$\sum_{i=1}^m \sum_{j=1}^n w_j^2 h_{ij},$$

$$s.t. \quad w \in W, \quad (15)$$

$$\sum_{j=1}^n w_j = 1, \quad (16)$$

$$w_j \geq 0. \quad (17)$$

where $h_{ij} = [(c_{0(ij)} - \max_i c_{0(ij)})^2 + \cdots + (c_{T(ij)} - \max_i c_{T(ij)})^2] - [(c_{0(ij)} - \min_i c_{0(ij)})^2 + \cdots + (c_{T(ij)} - \min_i c_{T(ij)})^2]$. So, the single-objective optimisation problem in Equation 14 is essentially a quadratic programming problem that can be solved using the MATLAB.

7) Compute the relative proximity of each alternative. The relative proximity is z_i^*, and the calculation formula is:

$$z_i^* = d_i^+ \Big/ (d_i^- + d_i^+). \quad (18)$$

8) According to the relative proximity z_i^*, rank the priority of all of the alternatives. The smaller z_i^* is, the better the alternative X_i is.

4 CASE STUDY

An enterprise needs to choose the best e-commerce platform from four alternatives. The decision maker imitates three experts to evaluate these four alternatives, which are X_1, X_2, X_3 and X_4. Four criteria are also considered, which are P_1, P_2, P_3 and P_4. Using Matlab 6.5 and Oracle 11 g, the decision-making proceeds as follows.

The decision maker gave the expert weight information as $v = (\frac{1}{2}, \frac{1}{2}, \frac{1}{3})^T$; however, the criteria weight information given by the decision maker is incomplete, including $0.7w_1 \leq w_2 \leq 0.8\,w_1$, $w_3 - w_2 \leq 0.2$, $0.3 \leq w_4 \leq 0.4$. The uncertain linguistic decision matrix is given by three experts as $\tilde{A}^k = [\tilde{a}_{ij}^k]_{4 \times 4}$ ($k = 1, 2, 3$) in the following:

$$\tilde{A}^1 = \begin{bmatrix} [s_1, s_2] & [s_5, s_6] & [s_2, s_3] & [s_5, s_6] \\ [s_0, s_2] & [s_3, s_4] & [s_1, s_3] & [s_3, s_4] \\ [s_0, s_1] & [s_3, s_5] & [s_4, s_6] & [s_2, s_3] \\ [s_0, s_2] & [s_3, s_5] & [s_4, s_5] & [s_5, s_6] \end{bmatrix},$$

$$\tilde{A}^2 = \begin{bmatrix} [s_2, s_3] & [s_5, s_6] & [s_3, s_4] & [s_4, s_5] \\ [s_1, s_2] & [s_5, s_6] & [s_3, s_4] & [s_2, s_4] \\ [s_2, s_3] & [s_4, s_5] & [s_5, s_6] & [s_3, s_5] \\ [s_1, s_3] & [s_4, s_5] & [s_5, s_6] & [s_5, s_6] \end{bmatrix},$$

$$\tilde{A}^3 = \begin{bmatrix} [s_3, s_4] & [s_4, s_5] & [s_4, s_5] & [s_5, s_6] \\ [s_1, s_2] & [s_5, s_6] & [s_3, s_4] & [s_4, s_6] \\ [s_2, s_3] & [s_4, s_5] & [s_5, s_6] & [s_4, s_5] \\ [s_1, s_2] & [s_4, s_6] & [s_5, s_6] & [s_4, s_5] \end{bmatrix}.$$

In order to solve this decision problem, it is possible to carry out the calculation steps described above.

Using Equation 1, the normalised decision matrix $\tilde{B}^k = [\tilde{b}_{ij}^k]$ ($k = 1,2,3$) was received as:

$$\tilde{B}^1 = \begin{bmatrix} [s_4, s_5] & [s_5, s_6] & [s_2, s_3] & [s_5, s_6] \\ [s_4, s_6] & [s_3, s_4] & [s_1, s_3] & [s_3, s_4] \\ [s_5, s_6] & [s_3, s_5] & [s_4, s_6] & [s_2, s_3] \\ [s_4, s_6] & [s_3, s_5] & [s_4, s_5] & [s_5, s_6] \end{bmatrix},$$

$$\tilde{B}^2 = \begin{bmatrix} [s_3, s_4] & [s_5, s_6] & [s_3, s_4] & [s_4, s_5] \\ [s_4, s_5] & [s_5, s_6] & [s_3, s_4] & [s_2, s_4] \\ [s_3, s_4] & [s_4, s_5] & [s_5, s_6] & [s_3, s_5] \\ [s_3, s_5] & [s_4, s_5] & [s_5, s_6] & [s_5, s_6] \end{bmatrix},$$

$$\tilde{B}^3 = \begin{bmatrix} [s_2, s_3] & [s_4, s_5] & [s_4, s_5] & [s_5, s_6] \\ [s_4, s_5] & [s_5, s_6] & [s_3, s_4] & [s_4, s_6] \\ [s_3, s_4] & [s_4, s_5] & [s_5, s_6] & [s_4, s_5] \\ [s_4, s_5] & [s_4, s_6] & [s_5, s_6] & [s_4, s_5] \end{bmatrix}.$$

According to Equations 2–17, the criteria weight vector is $w = (0.154, 0.123, 0.323, 0.4)^T$. The difference values between each alternative and PIP were $d_1^+ = 0.176$, $d_2^+ = 0.07$, $d_3^+ = 0.497$ and $d_4^+ = 0.356$. The difference values between each alternative and NIP were $d_1^- = 0.384$, $d_2^- = 0.24$, $d_3^- = 1.884$ and $d_4^- = 3.134$. By using Equation 18, the relative proximity of each alternative was calculated as $z_1^* = 0.314$, $z_2^* = 0.226$, $z_3^* = 0.209$ and $z_4^* = 0.102$. Consequently, the ranking result of the four alternatives was listed as $X_4 > X_3 > X_2 > X_1$.

5 CONCLUSION

A decision analysis method is proposed for the uncertain LMCGDM problem with incomplete criteria weight information. In this method, the uncertain linguistic matrices of each expert were assembled into a group decision matrix with possibility degree information. Then, the optimisation model is established and solved by approaching the ideal point, and the weight value of the criteria were obtained, in order to choose the optimal alternative or rank alternatives. This method provides a new idea for solving this kind of LMCGDM problem.

ACKNOWLEDGEMENTS

This work is sponsored by the University Nursing Program for Young Scholars with Creative Talents in Heilongjiang Province (UNPYSCT-2016037), Science and Technology Research Project of the Education Department of Heilongjiang Province (12541150), Harbin Science and Technology Innovation Talent Research Special Funds (2016RAQXJ039), Higher Education Research Project in "The thirteenth Five-Year Plan" of Heilongjiang Higher Education Association (16Q081), Education and Teaching Research Project of Harbin University of Science and Technology (220160014) and Entrepreneurship Training Program of Harbin University of Science and Technology in 2016 (No. 33). Also, this project was supported by the Hei Long Jiang Postdoctoral Foundation.

REFERENCES

Chiclana, F., García, J. T., del Moral, M. J. & Herrera-Viedma, E. (2013). A statistical comparative study of different similarity measures of consensus in group decision making. *Information Sciences*, *221*, 110–123.

Höhle, U. & Rodabaugh, S. E. (2012). *Mathematics of fuzzy sets: Logic, topology, and measure theory* (Vol. 3). Springer Science & Business Media.

Liu, X., Ju, Y. & Yang, S. (2016). Some generalized interval-valued hesitant uncertain linguistic aggregation operators and their applications to multiple attribute group decision making. *Soft Computing*, *20*(2), 495–510.

Vahdani, B., Tavakkoli-Moghaddam, R., Mousavi, S. M. & Ghodratnama, A. (2013). Soft computing based on new interval-valued fuzzy modified multi-criteria decision-making method. *Applied Soft Computing*, *13*(1), 165–172.

Wang, J., Wang, J. Q., Zhang, H. Y. & Chen, X. H. (2016). Multi-criteria group decision-making approach based on 2-tuple linguistic aggregation operators with multi-hesitant fuzzy linguistic information. *International Journal of Fuzzy Systems*, *18*(1), 81–97.

Wei, G. W. (2009). Uncertain linguistic hybrid geometric mean operator and its application to group decision making under uncertain linguistic environment. *International Journal of Uncertainty, Fuzziness and Knowledge-Based Systems*, *17*(02), 251–267.

Wei, G. (2015). Approaches to interval intuitionistic trapezoidal fuzzy multiple attribute decision making with incomplete weight information. *International Journal of Fuzzy Systems*, *17*(3), 484–489.

Xia, M., Xu, Z. & Chen, N. (2013). Some hesitant fuzzy aggregation operators with their application in group decision making. *Group Decision and Negotiation*, *22*(2), 259–279.

Xu, Z. (2015). *Uncertain multi-attribute decision making: Methods and applications*. Springer.

Zhang, X., Jin, F. & Liu, P. (2013). A grey relational projection method for multi-attribute decision making based on intuitionistic trapezoidal fuzzy number. *Applied Mathematical Modelling*, *37*(5), 3467–3477.

The design of the CAN bus interface controller based on Verilog HDL

H.J. Yang & M.Y. Ren
School of Software, Harbin University of Science and Technology, Harbin, China

ABSTRACT: This paper mainly introduces the transmission protocols and principles of CAN bus. Based on this, the CAN bus interface controller was designed. The method of generating the modules of the CAN bus interface controller is given, and the related module is designed by Verilog HDL. It is then compiled and simulated by the Quartus II development environment; the waveform file proved that the design was correct, which laid the foundation for the application of the CAN bus interface circuit.

Keywords: CAN bus protocol; packets; data frame; error frame

1 INTRODUCTION

CAN is an abbreviation of Controller Area Network. It is an ISO international standardised serial communication protocol. It is a type of field bus with high communication speeds, easy implementation and high cost performance. It has been widely used throughout the world. The main work of this paper is to design and implement a CAN2.0 protocol standard bus controller using Verilog HDL.

2 THE CAN BUS PROTOCOL

The signal transmission on the CAN bus is carried out with two differential voltage signals, CAN_H and CAN_L. On the bus, '0' and '1' are represented by two complementary logics, 'dominant' and 'recessive'. When all of the nodes are '1', the bus medium is in the 'recessive' state, namely '1'; as long as there is a node to send the '0' logic, the bus will appear in the 'dominant' state.

There are two main types of CRC, the non-standard CRC, which is defined by the user of the generation of CRC polynomials, and the standard CRC set by the international organisation for standardisation to generate the polynomial. The second type of CRC are widely used; the international organisation for the standardisation of several major common CRC is shown in Table 1. In the actual application process, in order to ensure the reliability of the message transmission, we will split the original data into a certain length of the data unit, usually called the data transmission unit for the 'frame'. A frame generally includes synchronisation signals (the start and end of the frame), error control signals (various error detection codes or error correction codes), flow control signals (co-ordinating the rate of the sender and the receiver), control information, data information,

addressing (when the channel sharing, need to ensure that each frame can correctly reach the destination, the recipient will have to know the source of information), and so on. There are four different frame types in CAN bus message transmission: data frame, remote frame, error frame, and overload frame. The data frame and the remote frame have two kinds of frame formats, namely the standard frame format and the extended frame format; they are separated from the previous frame by the frame interval.

2.1 Data frame

The data frame consists of seven parts: frame start, arbitration field, control field, data field, CRC field, response field and end of frame. The length of the data field may be zero. The data frame format is shown in Figure 1.

The SOF indicates the start of a data frame or a remote frame, and consists of a single dominant bit. The station is allowed to start sending messages only when the bus is idle. All of the nodes must be synchronised with the beginning of the frame start bit of the node that first started sending the message. The arbitration phase refers to the priority of the data frame.

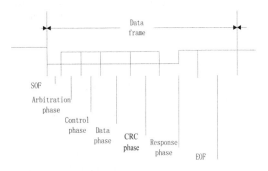

Figure 1. Data frame formats.

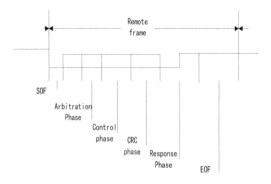

Figure 2. Remote frame formats.

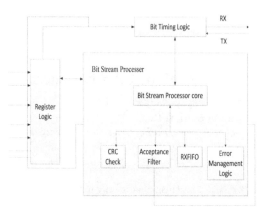

Figure 3. The architecture of the CAN bus interface controller.

The arbitration field for standard frames and extended frames is not the same. The arbitration phase in the standard frame format consists of a 1-bit identifier (ID) and a Remote Transmit Request (RTR) bit. In the extended frame format, the arbitration phase consists of a 29-bit identifier (ID), SRR bit, an Identifier Extension (IDE) bit, and a remote transmit request bit (RTR).

The extended frame identifier consists of a basic ID of 11 bits and an extended ID of 18 bits. The control phase indicates the byte length of the data field in the data frame, consisting of 6 bits of the reserved bit and the data length code. However, the control phase structure is the same in both the standard frame and the extended frame, but the location is not the same.

The data phase represents the data to be transmitted in the data frame. It contains 0 to 8 bytes, one byte for each 8 bits, the first in the high, the low in the post. The CRC phase includes the CRC sequence and the following CRC delimiter. The response phase contains an acknowledge bit and an acknowledge delimiter, which are two bits in length. EOF consists of 7 recessive bits, each of which is defined by the end of the frame.

2.2 Remote frame

The remote frame acts as a data receiving node and can transmit data by sending a remote frame to start its resource node. There are also standard and extended formats for remote frames. In the remote frame there are frame start, arbitration field, control field, CRC field, response field and end of frame. The remote frame structure is shown in Figure 2.

2.3 Error frame

An error frame consists of two fields. The first field is a superposition of error flags from different nodes. The second field indicates the error delimiter.

2.4 Overload frame

Overload frame function: This provides an additional delay for the data frame or remote frame transmission. The overload frame consists of two phases: the overload flag and the overload delimiter.

3 THE DESIGN AND IMPLEMENTATION OF THE CAN BUS INTERFACE CONTROLLER

Within the whole design, we split the CAN bus interface controller into three modules: Register logic module, Bit Timing Logic Module, and Bit stream processor module. The Bit stream processor is one of the most important parts of the CAN bus interface controller. It is composed of five parts: a Bit stream processor core module, a CRC check module, an Acceptance filter module, a RXFIFO module, and an Error Management Logic module. The architecture of the CAN bus interface controller is shown in Figure 3.

3.1 Register logic module

The Register logic module is used to save the frame information, status and commands. At the same time in the send, receive information or error, by rewriting the status register and interrupt register to reflect the current working conditions of the controller. The most important function of the register module is to write or read the data of the corresponding register according to the address. After we define the bit definition of each register, we can design it to write or read the data according to the address. The flow chart of the Register logic module is shown in Figure 4.

3.2 Bit Timing Logic Module

The Bit Timing Logic Module is the base of the controller. It monitors the CAN bus bit stream and processes the bit timing. The design of the Bit Timing Logic Module includes three parts: timing design, sampling point design and bit synchronisation design. The module was designed using a state

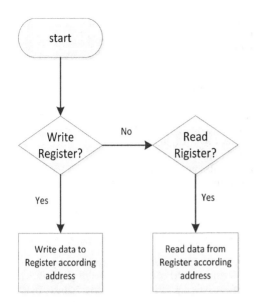

Figure 4. The flow chart of the Register logic module.

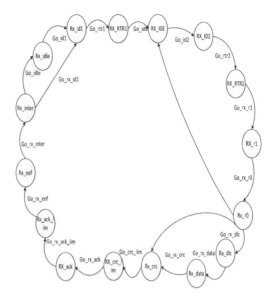

Figure 6. The state machine of data transmission.

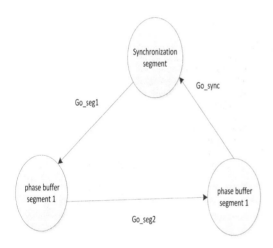

Figure 5. Timing design state machines.

machine with three states. The state machine is shown in Figure 5.

Define a counter to count the time span Tq. When the count value reaches the TSEG1, TSEG2 defined in the register and the length of the synchronisation segment defined in the design, the system will generate the corresponding transition conditions: go_seg1, go_seg2, go_sync. After judging these transition conditions, it cycles between the three states above.

3.3 *The Bit stream processor module*

The Bit stream processor module is the core module of the controller. It is used to achieve the following four functions: (1) Send and receive messages, including the transmission of the buffer, and fill the receiving bit of the solution filling and buffering. (2) Control to switch of state machine state. (3) Feedback the controller status information and interrupt information to the register control module. (4) Provide an overload and error detection mechanism.

The data transferred on the bus is either a data frame or a remote frame. The format of these frames has been described in detail above. According to the characteristics of the CAN bus communication protocol, we designed the state machine of data transmission. The state machine of data transmission is shown in Figure 6.

The CRC check module checks the data sent and received according to the protocol CRC algorithm. The calibration results will serve as the basis for the controller to determine the error.

The Acceptance filter module completes the filter, according to the register control modules in the acceptance code register and the acceptance mask register, and the filter results will be stored as RXFIFO in order to judge the basis of the message.

The RXFIFO module stores messages when the filter passes and the controller is correct.

The Error Management Logic module performs error statistics and generates the corresponding overload frame and error frame according to the protocol when the controller is overloaded or an error occurs. The RTL view of the CAN bus interface controller is shown in Figure 7.

4 SIMULATION AND RESULTS ANALYSIS

In order to verify the correctness of the design, we simulated the CAN bus interface controller. The result of the simulation are shown in Figure 8.

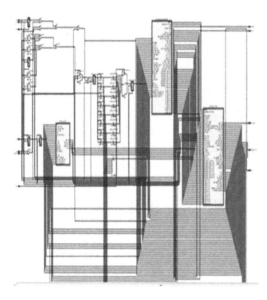

Figure 7. The RTL architecture of the CAN bus interface controller.

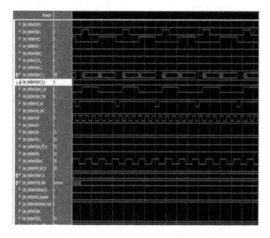

Figure 8. The results of the simulation.

From the simulation results in Figure 7, it can be seen that the data transmission in the CAN bus interface controller is correct.

5 CONCLUSIONS

A CAN bus interface controller is presented in this paper. It consists of a Register logic module, a Bit Timing Logic Module and a Bit stream processor module. By analysing the simulation results, it can be seen that the design implemented in the CAN bus interface controller is functional, so it meets the expectant design requirements.

ACKNOWLEDGEMENT

In this paper, the research was sponsored by the Natural Science Youth Foundation of Heilongjiang Province (Project NO. QC2016089) and the Natural Science Foundation of Heilongjiang Province (Project NO. F2015042).

REFERENCES

Buchanan, W. J. (2004). *CAN BUS: The handbook of data communications and networks*. 705–714.
Guo, S. (2011). The application of CAN-bus technology in the vehicle. *Proceedings of the 2011 International Conference on Mechatronic Science* (pp. 755–758).
Lv, Y., Tian, W. & Yin, S. (2015). Design and confirmation of a CAN-bus controller model with simple user interface. *Proceedings of the 5th International Conference on Instrumentation and Measurement* (pp. 640–644).
Tong, G., Chen, L., Yang, A., Ma, F. & Zhao, F. (2012). Research on CAN BUS-based electronic and electric platform of automobile. *Proceedings of the FISITA 2012 World Automotive Congress* (pp. 297–310).

Electronic Engineering – Wang (ed)
© 2018 Taylor & Francis Group, London, ISBN 978-1-138-60260-1

Research on the technology development model of China Mobile Communications Corporation based on three networks convergence

R. Zhang, Y. An & Y.S. Li
School of Management, Harbin, Heilongjiang, China

ABSTRACT: Due to the rapid development of three networks convergence, China Mobile Communications Corporation actively deploys this in order to promote the development process of national information. This research puts forward the technology development index system of China Mobile Communications Corporation by analysing China's mobile broadband access technology. At the same time, the technology development dynamic model of China Mobile Communications Corporation is built according to the index system, and it discusses the development suggests of China Mobile Communications Corporation.

Keywords: three networks convergence; broadband access technology; system dynamics model; development strategy

1 INTRODUCTION

Based on the background of three networks convergence, the access technology of China Mobile Communications Corporation (CMCC) mainly includes the xDSL access technology, PON + DSL access technology, PON + LAN access technology and FTTH access technology. At present, China Mobile Communications Corporation has largely achieved 'fibre instead of cable'. It has made a great breakthrough in both cable broadband access technology and wireless broadband access technology. At the same time, China Mobile Communications Corporation cooperates can provide the service of radio and television actively to seek the new top point of technology innovation. Some scholars have built and analysed the model according to the technological evolution of three networks convergence, based on user utility. It has been concluded that this may rapidly promote the development of three networks convergence if the company puts extra money into the research and development of the technology (Ke et al., 2013). Through the technology diffusion model, it is seen that the spread of the technology cycle includes four stages: generation period, development period, mature period and decline period (Rui, 2008; Xiuhui, 2011a,b). With regards to the network environment, some scholars point out that information transmission can increase the number of ways of obtaining information between the diffusion source and potential users. It can increase the speed and reduce the cost needed to promote the process of technology diffusion, based on the technology diffusion mechanism of the network environment (Guofang & Juan, 2002). Some scholars used the model to research the effects of technology spillover on technological

progress and development (Qiuzhen & Min, 2016). Some scholars think that the factors that influence an enterprises financing ability for high and new technology mainly include: enterprise size, enterprise holdings, debt paying ability, profitability, and so on (Yilin & Wenbo, 2016). But few scholars have analysed and forecast the development direction from the perspective of technology. This research thoroughly analyses and builds the system dynamics model for China Mobile Communications Corporation from the perspective of technology. Also, this research discusses the future technology development trends of China Mobile Communications Corporation, so as to promote the development of China Mobile Communications Corporation within the background of three networks convergence.

2 THE CURRENT TECHNOLOGY DEVELOPMENT SITUATION

2.1 *The mobile communication technology*

The communications technology of China Mobile Communications Corporation mainly includes 1G, 2G, 3G and 4G technology. 1G is the simulative system for providing telephone services (Wei, 2014). It realises the combination of the microcomputer and mobile communications. It is a large capacity cellular mobile communications system (Yalan, 2011). 2G is a digital cellular mobile communications system. Digital is the main characteristic of 2G, which can transmit business data (Xiaorui, 2013). The characteristics of 3G are that 3G gives priority to multimedia businesses, which can support mobile multimedia data transmission (Dahlman et al., 2010), such as voice, images, video, and so on. 4G integrates 3G and WLAN, which includes TD-LTE and FDD-LTE (Huaan, 2014). 4G can increase the signal transmission distance by the adoption of smart antenna technology. Not only does this ensure the transmission quality, but it also enlarges the coverage area and reduces the operation costs of the system (Zhigang, 2015).

At present, there are three methods used to access the network in China Mobile Communications Corporation, and the optical fibre network deployment is popular.

2.2 Co-operate with radio and television to realise the EPON + EOC access scheme

China Mobile Communications Corporation can realise the EPON + EOC access scheme by co-operating with radio and television. EOC access technology has several problems, such as network structure, authentication billing, application security and management issues (Renxiang et al., 2005; Wenbin, 2011). However, the EPON + EOC access scheme can effectively improve their status. EPON + EOC can be divided into the single line household way of EPON + active EOC and EPON + passive EOC. These two approaches are placed OLT equipment into the machine room and placed the ONU equipment erected in optical node.

3 BUILDING THE TECHNOLOGY INDEX SYSTEM FOR CHINA MOBILE COMMUNICATIONS CORPORATION

3.1 The construction of the index system

The construction of the technology index system plays a vital role in the technical development of China Mobile Communications Corporation, due to the rapid development of technology. Therefore, this research considers three issues, which include technical personnel, technical investment and new technology influences, for the technical development of China Mobile Communications Corporation.

First of all, the salaries of the technical personnel and the number of technical personnel determine the level of technology research and development. China Mobile Communications Corporation needs to invest a lot of money to enhance the development of its infrastructure and strengthen its ability for research and development. Therefore, technology investment is one of the key indexes. Technology convergence is slow. The main reason for this is that every industry has formed its inherent technology between inter-industry and intra-industry. The development of new technology is needed in order to accelerate the process of technology convergence and promote the progress of China Mobile Communications Corporation. Therefore, this research builds a system for new technology influence. The technology index system construction

Table 1. Technology index system construction for CMCC based on three networks convergence.

The influencing factors of three networks convergence (the first grade index)	The influencing factors of technology (the second grade index)	The influencing factors (the third grade index)
Technology	Technical personnel	Technical personnel salaries, the number of technical personnel.
	Technology investment amount	New technology investment, equipment upgrades and maintenance, others.
	New technology influences	Optical fibre access new users, technological applicability, price-affect rate, speed- affect rate.

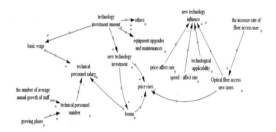

Figure 1. Cause and effect diagram of CMCC technology index.

for CMCC, based on three networks convergence, is shown in Table 1.

3.2 The cause and effect diagram of CMCC technology based on three networks convergence

The cause and effect diagram is helpful for analysing the relationship between each index. Also, it can provide an effective basis for the development of CMCC.

3.2.1 The construction cause and effect diagram of the technology index for CMCC

This study builds a technology index system, and also a cause and effect diagram according to the relationship of the indexes. The technology index cause and effect diagram is shown in Figure 1.

The development of the technology requires the participation of technology personnel. Many indexes include the number of technical personnel, the level of the technical personnel, and so on. The indexes play a decisive role in the development of the project and the technology research and development. The causes and effects are: ① the number of technical personnel increases—>technical personnel salaries will fall; ② the technology investment amount increases—>technical personnel salaries increase; ③ new technology investment increases—>bonuses increase—>technical personnel salaries increase.

CMCC can put a lot of money into technology research in order to enhance the comprehensive competitiveness of the enterprise because they have abundant capital and it does not affect the normal operation of the enterprise. At the same time, CMCC can obtain benefits for enterprise staff by increasing the salary and benefits of their personnel, upgrading their existing equipment and maintenance, and so on. Therefore, investment is related to a company's capital ability, and this determines the development of technology. The causes and effects are: ① technology investment amount increases—>technical personnel salaries increase; ② technology investment amount increases—>new technology investment increases—> the price of the new technology increases; ③ technology investment amount increases—>new technology investment increases—>bonuses increase—> technical personnel salaries increase; ④ technology investment amount increases —>the price of equipment upgrades and maintenance increases; ⑤ technology investment amount increases—>other expenses increase.

When the technology develops to a certain level, it will generate new technology, and this new technology has a certain influence. The price will affect the transmission capacity of the technology. A low price will strengthen the propagation speed of the technology, and this is easily attracted by the large number of users. Optical fibre access technology is the future technology development trend of three networks convergence. It has an absolute advantage over the technologies of LAN, DSL, and so on, with regards to network speed and price. Optical fibre access technology has attracted a customer base. In recent years, the number of optical fibre access users has proportionally increased year by year. Optical fibre access technology is also one of the main research and development technologies for CMCC and is aimed at various indexes, including price-affect rate, speed-affect rate, technological applicability, the number of optical fibre access new users, and so on. This research uses the above indexes to explore the new technology influences for CMCC. The causes and effects are: ① price-affect rate decreases—> the new technology influence increases; ② speed-affect rate increases—>the new technology influence increases; ③ technological applicability increases—> the technology influence increases; ④ the number of optical fibre access users increases—>the new technology influence increases.

3.2.2 The main calculation method of the technology indexes and instructions of CMCC

The operation of the dynamic model needs the index to be supported and to create corresponding equations according to each index. The specific calculation method and instructions of the technical indexes are shown as follows:

$$
\begin{aligned}
&\textit{Technology investment amount} = \textit{technology}\\
&\quad \textit{investment proportion} * \textit{total investment}\\
&= \textit{new technology investment} +\\
&\textit{equipment upgrades and maintenance} + \textit{others}
\end{aligned}
\quad (1)
$$

CMCC will give a certain amount to technology investment. The amount is commonly carried out allocation in accordance with the proportion. It can be divided into new technology investment, equipment upgrades and maintenance and other expenses.

$$
\textit{Technical personnel salary} = \textit{basic wage} + \textit{bonus} \quad (2)
$$

This research assumes that the salary of all of the technical personnel is the same. Technical personnel salaries mainly include a basic wage and a bonus. The basic wage is the foundation wage of every employee. Bonuses depend on the benefit. Because of the progress of the technology, the enterprise gains more profits. The enterprise then gives extra income to the technical personnel.

$$
\begin{aligned}
&\textit{The new technology influence} = \textit{optical fiber}\\
&\textit{access new users} * (\textit{price - affect rate} + \textit{technology}\\
&\textit{practical} + \textit{speed - affect rate})
\end{aligned}
\quad (3)
$$

Optical fibre access technology is the key technology that needs to be developed and invested in by CMCC. So this research uses this index to measure the influence of new technology. The level of technology will affect the price-affect rate. If the technical level is higher, the price will rise and so the technology will have a stronger influence. When the technology practical increases, it can meet the demands of more and more users and will increase the number of optical fibre access users. The speed-affect rate has a certain influence for the new technology influence. It can make the number of optical fibre access users increase when the network speed is accelerated, which satisfies the users' requirements.

$$
\begin{aligned}
&\textit{Optical fiber access new users} = \textit{the increase}\\
&\textit{rate of the optical fiber access user} * \textit{total customers}
\end{aligned}
\quad (4)
$$

In this model, the optical fibre access user acts as a belt to connect technology with the market. This research reflects the current situation with regards to the development of optical fibre access technology by viewing the number of optical fibre access users.

CMCC can enhance the network capacity of their enterprise by increasing the capacity of technology. CMCC should operate more new and integrated businesses and services to improve their network operation capacity, utilisation and connectivity.

4 CONCLUSIONS

With regards to the aspects of technology, China Mobile Communications Corporation should intensify the power of their research and development. CMCC should commit funding to the research of broadband access technology. Optical fibre access technology is not only beneficial to the development of CMCC. It can also improve the network speed, the possessive power and the competitive power of the market. Although optical fibre access technology has entered public life, CMCC has achieved the fibre to households in most areas of our country. It has also made many great breakthroughs in the aspects of tariffs and business to bring more convenience for the users. However, CMCC cannot let the users own 'the feeling of satisfaction' because of the low network speed. Therefore, CMCC should pay more attention to the ill effects of the network speed. The analysis of this research suggests that it should also undertake the following aspects in order to improve the enterprise's technical ability and promote the development of three networks convergence.

Firstly, it should strengthen the innovation consciousness of the technical personnel. CMCC should gain innovative ideas by training the company employees. CMCC can advocate public entrepreneurship and innovation by the establishment of the incentive mechanism. CMCC can provide substantial rewards for innovative staff.

Secondly, it should strengthen its investment in technology innovation. Capital investment is not the more the better, but it is fit for the development of CMCC. First, CMCC should produce a budget in advance to make sure that a specific amount of funds are allocated for the research and development of the technology. Second, the investment should have a certain amount of flexibility. CMCC can use the funds flexibly if the capital can promote the development of technology as long as the funds can show the money use clearly and make the money on practical application of the technology research and development.

Finally, it will strengthen the popularisation and application of the new technology. On the one hand, CMCC should improve their network platform and establish a platform to interact with users. Also, CMCC should adjust and improve their technology, business and services according to the users' requirements. On the other hand, CMCC should pay attention to the ranges and methods of promotion. At the beginning of the technical promotion, technology is not suitable for a wide range of promotion. CMCC should choose a small scale or a small region to be promoting.

ACKNOWLEDGEMENT

This research was financially supported by Harbin Special Fund for Science & Technology Innovation Talents (2013RFLXJ009).

REFERENCES

Cai, W. (2014). Optical access network energy saving optimization technology research. Nanjing University of Posts and Telecommunications.

Dahlman, E., Parkvall, S., Skold, J. & Beming, P. (2010). 3G Evolution: HSPA and LTE for mobile broadband. *Academic Press*, 4(2), 12–15.

Kuang, G., Lin, X. & Wan, L. (2006). Industrial cluster technology diffusion research based on the Hotelling price competition model. *Science and Technology Management Research*, 11(30), 203–206.

Liu, Y. (2005). Based on the Ethernet broadband access technology. *Television Engineering*, 9(30), 50–51.

Luo, R., Liu, W. & Chen, S. (2005). The cable broadband access technology comparative analysis and development. *China Cable Television*, 3(10), 418–422.

Ouyang, Q. & Zhang, M. (2016). Channels of technology spillover efficiency comparison from the perspective of eastern provinces based on panel cointegration analysis of the model. *Modern Business Trade Industry*, 3(2), 10.

Qiao, Y. (2011). *Mobile phones and social interaction*. Wuhan University.

Wang, H. (2014). Hd wireless video monitoring the change of 4G. *China Public Security*, 7(14), 30–35.

Wu, Y. & Sun, W. (2016). High and new technology enterprise financing ability influence factors analysis: Empirical research based on classification and regression tree model. *Modernization Management*, 2(10), 5–7.

Xu, K., Ling, S. & Wu, J. (2013). Based on user utility evolution of model and analysis of the three networks convergence. *Journal of Computer*, 5(15), 903–914.

Xu, X. (2013). *The mobile communication network resource management system research and implementation*. Nanjing University of Posts and Telecommunications.

Yang, G. (2005). Ethernet access technology problems and discussed in this paper. *Gansu Science and Technology*, 1(30), 67–68.

Yao, Z. (2015). 4G mobile communication key technology application and development prospect. *China New Telecommunications*, 4(20), 75–76.

Yi, W. (2011). Comprehensive analyses on the Ethernet access technology. *Science & Technology Information*, 7(13), 26.

Yu, X. (2011). *Based on transmission model of digital content "micro" network diffusion research*. Beijing University of Posts and Telecommunications.

Zhang, G. & Zeng, J. (2002). The research of technology diffusion mechanism based on network environment. *Scientific and Technological Progress and Countermeasures*, 8(30), 69–71.

Zhang, R. (2008). *The policy selection of three networks convergence universal access to technology and market*. Harbin Institute of Technology.

Design and implementation of UART interface based on RS232

H. Guo
School of Software, Harbin University of Science and Technology, Harbin, China

ABSTRACT: The data communication between the processor and the peripheral device is completed by UART, its function is more and more important. In this paper, UART is divided into three parts: data transmitting module, data receiving module and baud rate generator control module. According to the design goal, the circuit is improved and optimized. The functional verification of the design is completed through the joint simulation of Quartus and Modelsim. The simulation results show that the designed UART has good performance. The design is then logically synthesized, formalized for validation, placement and routing, and static timing analysis. It is proved that the function and timing of the design meet the requirements.

Keywords: UART; RS232; Transmitting Module; Receiving Module; Baud Rate Control Block

1 INTRODUCTION

With the rapid development of integrated circuits, the communication between processor and peripheral device is more and more important. However, the processor must be UART-collated for asynchronous transfer.

In this paper, a UART serial interface based on RS232 communication protocol is designed. First of all, the module structure of the UART is divided, the design hierarchy and port signals are given. UART is divided into three parts: data transmitting module, data receiving module and baud rate generator control module. According to the design goal, the circuit is improved and optimized. After completing the design of each module and analyzing the data communication between serial port and external/internal, the design of top-level module is completed. The simulation results show that the designed UART has good performance. Then the logical synthesis of the design is completed to generate gate-level netlist. Through formal verification, the functional consistency between RTL-level code and integrated gate-level netlist is compared. The wiring of the circuit cells is realized by placement and routing. Finally, through static timing analysis, the design after the placement and routing is verified whether the timing requirements are met.

2 MODULE DIVISION OF UART INTERFACE

Modular design concept was adopted in this paper, The design of the top-level module is implemented on the basis that the submodule designs are completed.

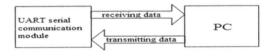

Figure 1. UART structure diagram.

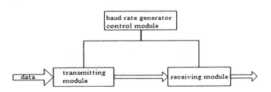

Figure 2. Logic functional block diagram.

2.1 System structure diagram

As shown in figure 1, UART data input is parallel data that is sent by the PC. When the start bit of the PC transmit data frame is detected by the UART, it starts to receive the data stream and then output it in parallel. At the same time, the UART can also transfer serial data to the PC.

2.2 System module division

According to the logic function of the system, UART is divided into three parts: The bit rate divider clock module provides the baud rate 9600 bit clock signal to the transmitting module and the receiving module; data transmitting module; data receiving module. On this basis, the top-level module is designed, its main function is to complete the three sub-module instantiation. As shown in figure 2.

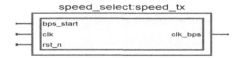

Figure 3. Baud rate generator control module.

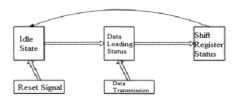

Figure 4. State transition diagram of data transmitting module.

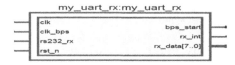

Figure 5. Data transmitting module.

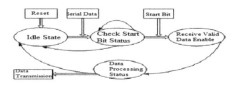

Figure 6. State transition diagram of data receiving module.

3 DESIGN OF UART INTERFACE

The specific implementation of each module is given in this section.

3.1 Baud rate generator control module

Baud rate generator is actually a clock divider in the design, the baud rate of 9600 bits is adopted and the baud rate factor is 1. 1000000 microseconds is required to transfer 1 bit. It is divided by 38400 = 260.40 microseconds. Clock of 200 MHZ clock frequency and 5 nanoseconds clock cycle is used in this paper. Then the number of clock cycles is required to transfer one bit is: 260.40 microsecond is divided by 5 nanosecond = 5208. The RTL level simulation of the RS-232 baud rate controller is shown in figure 3. When the count is 5208/2 clock, the data is stable and suitable for sampling.

3.2 Data transmitting module

Data transmission module's main function is to achieve data conversion from parallel input to serial output, state transition diagram is shown in figure 4.

The RTL level simulation of data transmitting module is shown in figure 5.

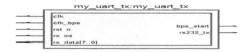

Figure 7. Data receiving module.

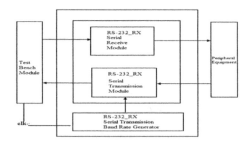

Figure 8. Block diagram of the top-level module.

When the RS232_rx falling edge signal (start bit) is acquired, in the speed_select module, cnt starts counting when bps_start is pulled high, and cnt is cleared when bps_start is low.

3.3 Data receiving module

Data receiving module's main function is to achieve data conversion from serial input to parallel output. The state transition diagram is shown in figure 6.

The RTL level simulation of data receiving module is shown in figure 7.

3.4 Top-level module

The top-level module implements the communication of the data signals between the sub-modules by connecting all ports, which are the baud rate generating module, the data receiving module and the data transmitting module to. At the same time it also carries on the data exchange between test modules and external devices. The block diagram of the module is shown in figure 8.

4 CO-SIMULATION AND RESULT ANALYSIS

From the data of three waveforms in figure 9, it can be shown that the data which is sent over by test file, as it can be sent to the RS-232 transmission module asynchronously and serially. The received signal also can be seen. This indicates that the sending module is working properly. The signal of the receiving module is updated in real time, and the signal is not in error. It denotes that the receiving module is working normally. At the same time, it can be seen that the register Tran_data receives the data 10100011 which is sent by the test module. Through this, the serial data exchange has been achieved in the UART interface.

Simulation results which are sent by the top module and receiving ports are shown in Figure 10. It can be shown from figure 10, the data that received by Port

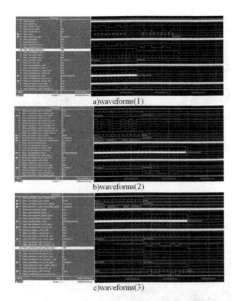

Figure 9. Simulation waveforms of signals in all interface port.

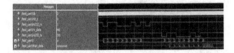

Figure 10. Simulation results of the top module and receiving ports.

rs232_rx from the outside is sent to the Port rs232_tx. And the data information is received and displayed in the register Tran_data. It indicates that the function of UART designed in this paper is correct.

5 BACK-END VERIFICATION

This section focuses on the design of the back-end simulation. It includes logic synthesis, formal verification, place & route and static timing analysis. The simulation results show whether the function of the design, timing and others to meet the needs.

5.1 *Logic synthesis*

The synthesis process can be divided into related library files, the relevant constraints which are established by the function circuit of the design, the integrated simulation of the design code, as well as checking if the timing report and slack meet the requirements. It can be seen from the time report of figure 11 that setting up time and holding time is to meet the design requirements. And the allowance of time is zero, conform to the requirement of time.

It can be shown from the figure 12, there is no violation report after the synthesis process. It shows that the design accord with the constraint under the condition of achieving the logic functions.

```
clock clk (rise edge)                                      0.00      0.00
clock network delay (ideal)                                0.00      0.00
my_uart_tx/tx_data_reg[5]/CK (DFFRHQX2)                    0.00      0.00 r
my_uart_tx/tx_data_reg[5]/Q (DFFRHQX2)                     0.18      0.18 f
my_uart_tx/U16/Y (NOR2BX1)                                 0.15      0.33 f
my_uart_tx/U15/Y (OR2X2)                                   0.18      0.51 f
my_uart_tx/U14/Y (NOR2X4)                                  0.09      0.61 r
my_uart_tx/U25/Y (NAND2X4)                                 0.04      0.64 f
my_uart_tx/U13/Y (AND2X4)                                  0.13      0.78 f
my_uart_tx/U18/Y (NOR2X4)                                  0.06      0.84 r
my_uart_tx/U10/Y (MXI2X2)                                  0.08      0.91 f
my_uart_tx/rs232_tx_r_reg/D (DFFNSXL)                      0.00      0.91 f
data arrival time                                                    0.91

clock clk (fall edge)                                      1.50      1.50
clock network delay (ideal)                                0.00      1.50
clock uncertainty                                         -0.50      1.00
my_uart_tx/rs232_tx_r_reg/CKN (DFFNSXL)                    0.00      1.00 f
library setup time                                        -0.08      0.92
data required time                                                   0.92
-------------------------------------------------------------------------
data required time                                                   0.92
data arrival time                                                   -0.91
-------------------------------------------------------------------------
slack (MET)                                                          0.00
```

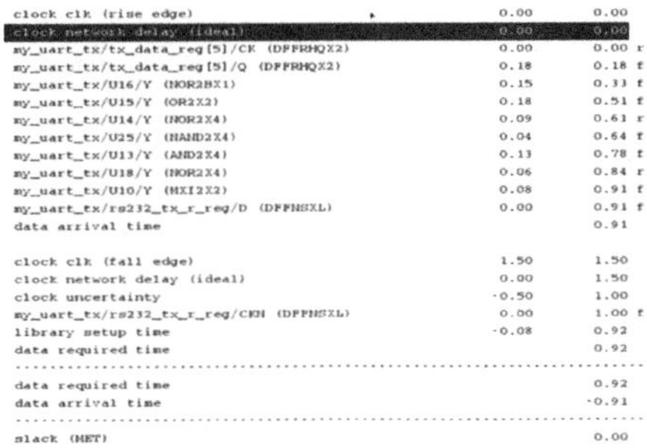

Figure 11. Timing report.

```
****************************************
Report : constraint
Design : uart
Version: D-2010.03-SP1
Date   : Sat May 23 16:25:48 2015
****************************************

                                                            Weighted
    Group (max_delay/setup)      Cost        Weight          Cost
    -----------------------------------------------------------------
    clk                          0.00        1.00            0.00
    default                      0.00        1.00            0.00
    -----------------------------------------------------------------
    max_delay/setup                                          0.00

                               Total Neg     Critical
    Group (critical_range)       Slack       Endpoints       Cost
    -----------------------------------------------------------------
    clk                          0.00           0            0.00
    default                      0.00           0            0.00
    -----------------------------------------------------------------
    critical_range                                           0.00

                                                            Weighted
    Group (min_delay/hold)       Cost        Weight          Cost
    -----------------------------------------------------------------
    clk (no fix_hold)            0.00        1.00            0.00
    default                      0.00        1.00            0.00
    -----------------------------------------------------------------
    min_delay/hold                                           0.00

    Constraint                                               Cost
    -----------------------------------------------------------------
    max_transition                                           0.00 (MET)
    max_capacitance                                          0.00 (MET)
    max_delay/setup                                          0.00 (MET)
    critical range                                           0.00 (MET)
```

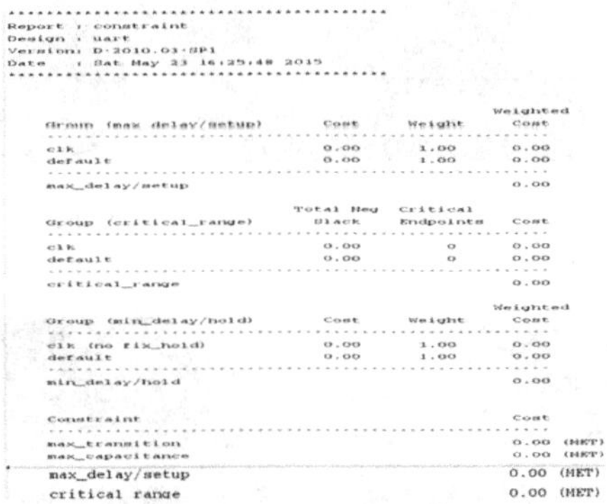

Figure 12. Violation report of the synthesis.

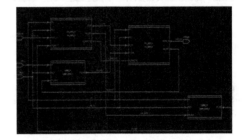

Figure 13. Design module after logic synthesis.

The conclusion can be drawn from figure 13, the representation of the circuit has been changed from the RTL level to the gate level. This makes clear that the synthesis implement the optimization of the time, area and layout of the design.

5.2 *Formal verification*

This design adopts GUI graphical interface for verification. The verification results are shown in figure 14. The results show that the logic functions of the design are the same before and after synthesis.

5.3 *Placement and routing*

The placement and routing satisfies the requirements of the timing of the design through the optimization of logical units and logic gates. It was placed

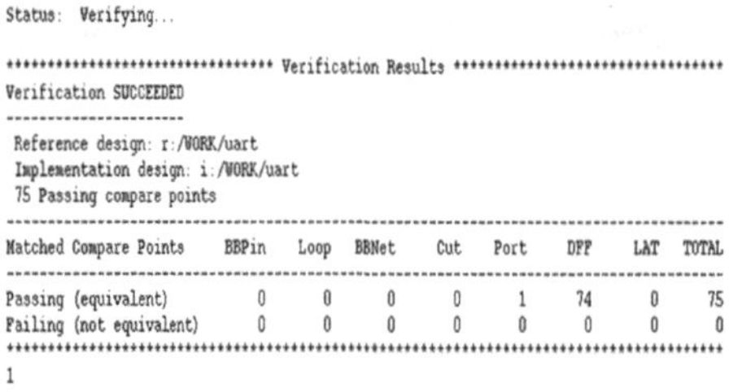

Figure 14. The verification results.

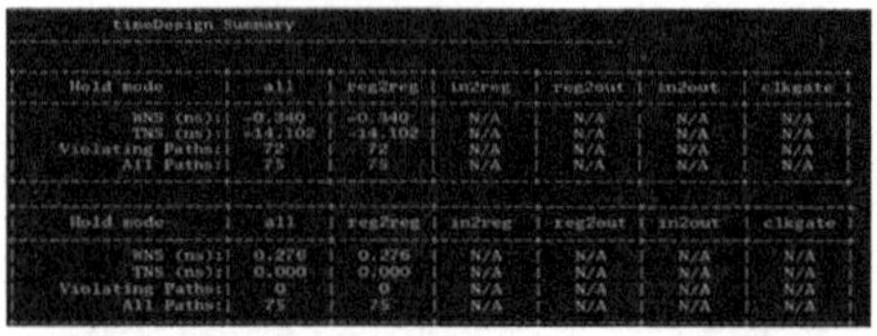

Figure 15. The comparison of time of setting before and after optimization.

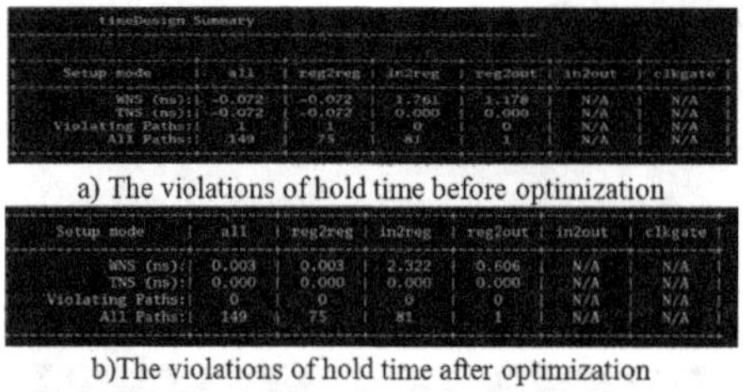

a) The violations of hold time before optimization

b)The violations of hold time after optimization

Figure 16. Contrast of holding time before and after optimization.

detailedly according to the placement parameters and other requirements. It is routing after the placement. First, the clock signal lines and power lines are routed, and then the routing of other signal is done.

It can be shown from figure 15 and figure 16, after the optimization, the routing has no violation, the purpose of optimization is achieved.

5.4 Static timing analysis

It can be shown from the report of figure 17, there is no violation phenomenon in the clock, setup time and hold time. This indicates that the design after the placement and routing is still to meet the design requirements of the timing.

A detailed report of hold time and setup time is shown in figure 18.

As shown in figure 18, the hold time is identified as the following formula: hold time slack = data required time $-(tco + path$ delay$) = 2.45 - (0.1 + 0.02) = 2.33$ ns. This shows that the holding time meets the timing requirements and according to the design criteria. Among them, the establishment of time as: setup time slack = data required time–data arrival time $= 4.85 - 3.10 = 1.76$ ns. It satisfies the design requirement as well.

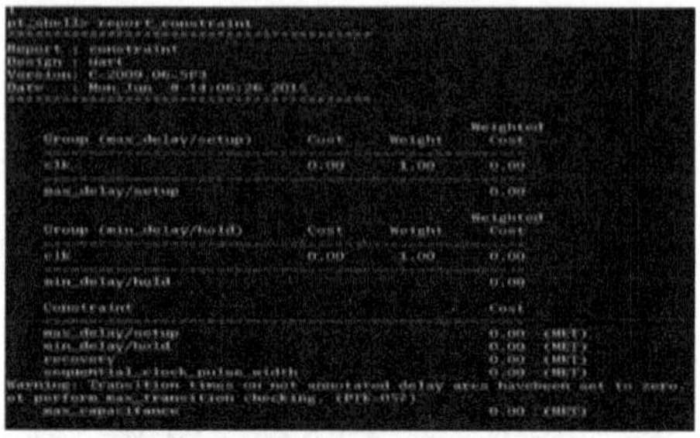

Figure 17. The violation report of timing.

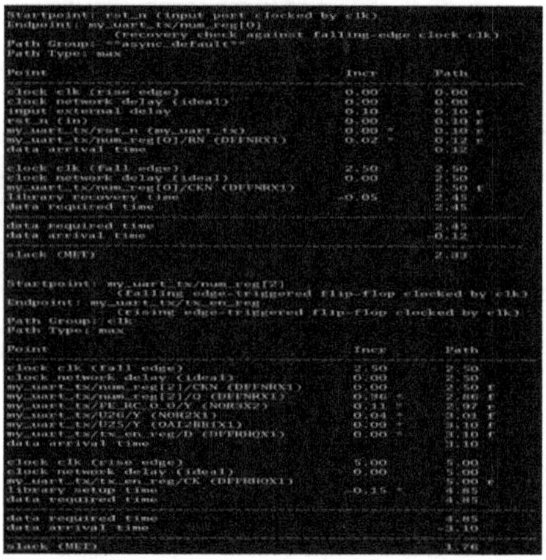

Figure 18. A detailed report of hold time and setup time.

6 CONCLUSIONS

In this paper, the design of UART serial interface can improve the speed of serial communication well. On the basis of the design of each module, the design of the top level module is completed. The correctness of each function are verified by joint simulation. Finally through the logic synthesis, RTL level code is simulated. The integrated gate level network is used into the formal verification. According to the specific function of the design, the overall layout of the chip is placed and routed. By data analyzing of the waveform and the report, it is proved that the design meets the time sequence and so on.

REFERENCES

Ming Lu. The Extension and Implementation of Multi-UART Based on FPGA[J]. Scientific Journal of Control Engineenng, 2013, 3(3).

Xi Ji. Verilog HDL Based Design Methodology for Uart [J]. Control and Automation. 2012(168).

Yu H. H. Programmable digital signal processors: architecture, programming, and applications [M]. Marcel Dekker. 2007: 147–187.

The analysis of the stability of the repetitive controller in an Uninterruptible Power System (UPS) inverter

J.J. Ma, W.W. Kong, J. Xu, X.W. Zang & Y.H. Qiu
Harbin University of Science and Technology, China

ABSTRACT: The repetitive controller can improve the quality of output voltage in an Uninterruptible Power System (UPS) inverter. Because of joining as a feed forward, the repetitive controller can make the system unstable, especially in parallel systems with different output cable lengths. We tend to adjust the volume of the repetitive control, but this method is not adaptable. Firstly, this paper analyses the stability of the repetitive controller, and then designs the compensator based on the small gain theorem and a prototype of the zero-phase filter. Finally, it uses the $1+1$ inverter parallel system for validation. According to this design method, the compensator can offset resonant peaks due to repetition and improve the stability of the system.

Keywords: UPS; repetitive control; stability; parallel system

1 INTRODUCTION

With the development of industrial technology, power electronics, computer technology and automatic control technology have become the three most important technologies. Among these technologies, the Uninterruptible Power System (UPS) lays a solid foundation for industrial development. UPS is a kind of energy storage device and its main component is an inverter, which allows the power to be output by voltage stabilisation and frequency stabilisation. UPS is mainly used in a single computer, computer network systems or other electronic power devices to provide an uninterrupted power supply (Josep et al., 2009).

In the study of UPS, the literature (Chiang & Chang, 2001; Guerrero et al., 2005; De Brabandere et al., 2004) introduced the concept of 'virtual impedance', and added an inverter output impedance adjustment module outside the closed-loop control of the inverter in order to realise the output power split by adjusting the inverter output impedance. The literature (Tuladhar et al., 2000) refers to the injection frequency harmonic in industrial frequency reference voltage, and the adjustment of the output voltage amplitude according to the harmonic power; this method is also called harmonic injection. With regards to repetitive control, the literature (Qiu et al., 2009) used an advanced link to realise phase compensation, and adopted a second order filter to realise high-frequency attenuation, which makes the system phase frequency characteristic complicated. The literature (Escobar et al., 2007a; Escobar et al., 2007b) adopted a comparatively complicated adaptive algorithm in order to ensure stability. This paper focuses on the analysis of the stability of the repetitive controller, gives the adaptive solution under the inverter parallel system and, finally, uses the $1+1$ inverter parallel system for validation.

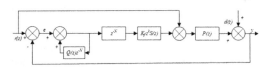

Figure 1. A typical repetitive control system structure.

2 THE STRUCTURE OF THE REPETITIVE CONTROLLER

In order to simplify the analysis, we temporarily do not consider the coupling between the double-loop and the repetitive control. The following is a typical repetitive control system structure, as shown in Figure 1.

This is an analysis model of a general repetitive controller with given voltage feed forward. $P(z)$ is the control object, and it is requested to be stable in itself. $d(z)$ is the repetitive disturbance to the system, such as load current and so on. $r(z)$ is the given output of the system, the shadow part of the figure is the repetitive controller, and $u_r(z)$ is the output of the repetitive controller. We will make a specific analysis of the function of the various parts of the repetitive controller.

2.1 $Q(z)$ filter

Before adding $Q(z)$, the repetitive control is purely an integral process based on cycle for the step. In theory, this pure integration can achieve controlled volume, but it is adverse to the stability and robustness.

Therefore, if we use the internal model, it will bring N pieces of open-loop poles on the unit circle for the system, so that the open-loop system is in critical oscillation. At this time, as long as there is a slight deviation to the object of modelling, or the object parameters vary slightly, the closed-loop system is likely to lose stability.

Therefore, the actual system mostly uses an improved repetitive controller containing $Q(z)$. In order to abate the integration effect, $Q(z)$ can be a low-pass filter, or simply take a constant slightly smaller than 1. In the meantime, with the joining of $Q(z)$, when the error $e(z)$ is down to an output of $(1 - Q(z))$, repetitive control will lose its integration effect. So essentially the improved repetitive controller changes the 'pure integral' of the error to the 'quasi integral', which means that the improvement in stability is at the expense of scarifying floating.

2.2 Compensator $k_r z^k S(z)$

The compensator is set up based on the characteristics of object $P(z)$, and its function is to provide phase compensation and amplitude compensation in order to ensure the stability of the repetitive control system, and to improve the effect of waveform correction on this basis. The compensator has many types, depending on different phase compensation methods, and $k_r z^k S(z)$ is one type of common compensator used as an advanced link to realise phase compensation. Among them, z^k is the advanced link of the phase compensation and the proportional k_r is the gain of repetitive control. The filter $S(z)$ is used to offset the high resonance peak of the object and not spoil the stability, while it also enhances the high-frequency attenuation characteristics of the forward path, and improves the stability and the ability of the anti-high-frequency interference. Because of setting up the filter $S(z)$, the advanced link z^k is required to compensate the total phase lag between the filter $S(z)$ and the object $P(z)$.

3 STABILITY ANALYSIS OF THE REPETITIVE CONTROLLER

From Figure 1 we can get:

$$\frac{u_r(z)}{e(z)} = \frac{z^{-N} K_r z^k S(z)}{1 - Q(z) z^{-N}} \quad (1)$$

$$e(z) = r(z) - (r(z) + u_r(z)) P(z) - d(z) \quad (2)$$

After simplifying:

$$e(z) = \frac{(1 - P(z))(z^N - Q(z))}{z^N - (Q(z) - K_r z^k S(z) P(z))} r(z)$$

$$+ \frac{(Q(z) - z^N)}{(Q(z) - K_r z^k S(z) P(z))} d(z) \quad (3)$$

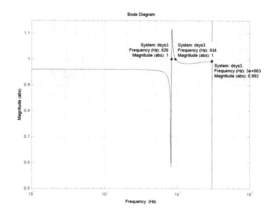

Figure 2. 30KVA machine's T.

System's characteristic equation is:

$$e(z) = z^N - (Q(z) - K_r z^k S(z) P(z)) \quad (4)$$

According to the control system's stability theory, as long as N pieces roots of system's characteristic Equation 4 are located in the unit circle with the origin point as circle centre, the system is stable. But the order N of Equation 4 is usually very high. For example, for 6 kHz sampling and 50 Hz outputting, the order of Equation 4 is 120, so it is almost impossible to derive sufficient conditions of system stability by directly solving the characteristic equation. Also, for a designed system, it is very difficult to judge the stability by adopting conventional methods similar to *Louts Criterion*. However, if we only need sufficient conditions for the stability of the system, this can be obtained by using the small gain theorem in the control theory:

$$\left| Q(e^{j\omega T}) - K_r e^{j\omega k T} S(e^{j\omega T}) P(e^{j\omega T}) \right| < 1$$
$$\omega \in [0, \pi/T] \quad (5)$$

$$T = Q(e^{j\omega T}) - K_r e^{j\omega k T} S(e^{j\omega T}) P(e^{j\omega T}) \quad (6)$$

If within the whole frequency band which the sampling holder can reproduce (from DC or zero frequency to Nyquist frequency, and that is half of sampling frequency), and if the gain of the transfer function $(Q(z) - K_r z^k S(z) P(z))$ is less than 1, we can fully guarantee that the repetitive control system shown in Figure 1 is stable.

Taking the experimental prototype 30KVA machine as an example, $P(s) = \frac{1}{LCs^2 + RCs + 1}$, $Q(z) = 0.9921875$, $k = 2$, $S(z) = 1$. $C = 165\text{e}^{-6} F$, $R = 0.045 \Omega$ (actual measurement), considering that the repetitive control of the rectifier load peak point is the maximum, then $L = 230\text{e}^{-6} H$, $K_r = 0.0315$. The figure of T is shown in Figure 2.

Let $H(z) = Q(z) - K_r z^k S(z) P(z)$. As shown in figure, in the frequency band of $0 \leq f \leq \frac{1}{2\pi \sqrt{LC}}$, $H(z)$ is a constant value of less than 1. In the frequency band of $\frac{1}{2\pi\sqrt{LC}} < f \leq \frac{1}{2}f_s$, H(z) is also less than 1, and it is changing. In the position of $\frac{1}{2}f_s$, $H(z) = Q(Z)$. Nearby the resonant frequency of LC, $H(z) > 1$. If the excitation signal is large enough in this frequency range, it may cause the system to be unstable. To inhibit $H(z)$ in this frequency range, we must carefully design the $S(z)$ filter.

4 DESIGN OF THE S(Z) FILTER

As mentioned in the above section, one of the main functions of the $S(z)$ filter is to offset the higher resonant peaks of the object and improve the stability of the system.

In order to eliminate the resonant peak of the object, we can simply set $S(z)$ to the second order low-pass filter. By adjusting the filter's parameters, the amplitude-frequency gain can be attenuated -20–$30\,\text{dB}$ at the inverter resonant frequency, that is, the harmonic peak of the inverter can be offset. But because the descending slope of the second order filter's amplitude-frequency characteristic is $-40\,\text{dB}/10$ double frequency, to generate -20–$30\,\text{dB}$ gain at the resonant frequency of the inverter, the cut-off frequency of the second order filter must be set at a lower level when the second order filter offsets the inverter resonant peak. Meanwhile, it will significantly reduce gain in a very wide frequency range below the inverter cut-off frequency, which greatly reduces the harmonic suppression effect in the corresponding frequency range. The $S(z)$ filter can be realised by a zero-phase shift notch filter, the expression of which is:

$$S(z) = \frac{a_m z^m + a_{m-1} z^{m-1} + \cdots + a_0 + \cdots + a_{m-1} z^{-(m-1)} + a_m z^{-n}}{2a_m + 2a_{m-1} + \cdots + a_0} \quad (7)$$

Many filters with excellent performance can be derived by allocating the parameters reasonably, for example $S(z) = \frac{z + 2 + z^{-1}}{4}$. The notch point of this notch filter is allocated at the position of the Nyquist frequency, which is a low-pass filter in nature. Another example is $S(z) = \frac{z^3 + 2 + z^{-3}}{4}$. The notch frequency of this notch filter is about 1 kHz, and all of the phase lags are 0. Such characteristics are very suitable for offsetting the resonance peak of the inverter, but they have little effect on gain in the other frequency bands, which will obviously not affect the harmonic suppression ability of the other frequency bands.

As the notch frequency of $S(z) = \frac{z^3 + 2 + z^{-3}}{4}$ is very close to the resonant frequency of the experimental prototype inverter, so $S(z) = \frac{z^3 + 2 + z^{-3}}{4}$ is selected.

In addition, the adopted compensate or $k_r z^k S(z)$ realises the phase compensation based on advanced links, which uses the characteristics of that the phase lag of inverter increases gradually with the increase of

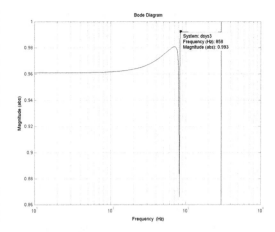

Figure 3. T of 30KVA machine adding the complete $S(z)$.

frequency, such compensator shows a good phase compensation effect on low and medium frequency range, but for the high-frequency range, the compensation effect is not ideal, which may even cause the system to be unstable, so we need the second major function of $S(z)$ filter: enhance the high-frequency actuation characteristic of forward path, and improve the stability and anti-high-frequency interference ability.

As the notch filter shown in Equation 7 does not have a high-frequency actuation characteristic, except $\frac{z + 2 + z^{-1}}{4}$, and to enhance the high-frequency actuation characteristic of the forward path, an additional second order filter can be set to be used with it, which can constitute a complete $S(z)$ filter. For a 30KVA prototype, the complete $S(z)$ filter is:

$$S(z) = \frac{z^3 + 2 + z^{-3}}{4} \cdot \frac{z + 2 + z^{-1}}{4} \quad (8)$$

After adding the $S(z)$ filter, re-check the stability of the repetitive controller of the 30KVA machine, which is shown in Figure 3.

It can be seen that $S(z)$ has offset the resonance peak of the inverter well, which assures the stability of the repetitive control in all of the frequency ranges.

5 EXPERIMENTAL VERIFICATION

A test is carried out on the 30KVA $1+1$ parallel machine. In order to solve the instability problem of the long parallel machine line system, we specially add a 50 metre long line at the output of a machine.

(1) The paronomastic repetition of the long parallel machine line system's instability: we increase the repetitive control volume in the control system; the output current of the No. 1 single machine has a high-frequency ripple, as shown in Figure 4.

(2) We add $S(z) = \frac{z^3 + 2 + z^{-3}}{4} \cdot \frac{z + 2 + z^{-1}}{4}$ under the same conditions, and the output current of the No 1. single machine is shown in Figure 5. It can be seen

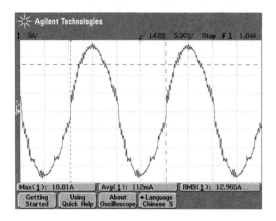

Figure 4. The output current adding a 50 metre long line.

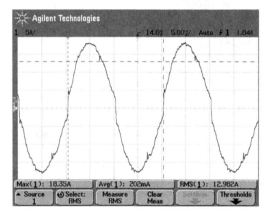

Figure 5. The output current after adding the $S(z)$ filter.

that, after adding $S(z)$, the stability of the system is significantly improved.

Finally, we increase the volume of repeated control in the codes in duplication, and the parallel machine system is still able to run stably.

6 CONCLUSIONS

Under the inverter control structure of the 'double-loop + repetitive control', the fastness of the double-loop can effectively inhibit the resonant peak of the inverter, so that it does not spoil the stability of the system. When designing the $S(z)$ compensator for the repetitive controller, we may not consider eliminating the resonant peak of the inverter, but focus on the second main function of $S(z)$. We enhance the high-frequency attenuation characteristics of the forward path, and improve the stability and the anti-high-frequency interference capability.

In view of the zero-phase shift characteristics and easy implementation of $S(z)$, we can consider using it as a low-pass filter. In the experiments, the performance change of the system is compared before and after adding the $S(z)$ link. The experimental results show that $S(z)$ has almost no effects on voltage regulation accuracy, phase contrast and other performances of the system.

REFERENCES

Chiang, S. J. & Chang, J. M. (2001). Parallel control of the UPS inverters with frequency-dependent droop scheme. IEEE *Annual Power Electronics Specialists Conference, Vancouver, Canada.*

De Brabandere, K., Bolsens, B., Van den Keybus, J., et al. (2004). A voltage and frequency droop control method for parallel inverters. *IEEE Annual Power Electronics Specialists Conference*, Aachen, Germany.

Escobar, G., Martinez, P. R. & Ramos, J. L. (2007a). Analog circuits to implement repetitive controllers with feed forward for harmonic compensation. *IEEE Trans. on Industrial Electronics*, 54(1), 567–573.

Escobar, G., Valdez, A. A., Olguin, R. E. T., et al. (2007b). Repetitive based controller for a UPS inverter to compensate unbalanced and harmonic distortion. *IEEE Trans. on Industrial Electronics*, 54(1), 504–510.

Guerrero, J. M., De Vicuna, L. G., Matas, J., et al. (2005). Output impedance design of parallel-connected UPS inverters with wireless load-sharing control. *IEEE Trans. on Industry Applications*, 52(4), 1126–1135.

Josep, M. G., Juan, C. V., Jose, M., et al. (2009). Control strategy for flexible microgrid based on parallel line-interactive UPS systems. *IEEE Trans. on Industrial Electronics*.

Qiu, Z., Yang, E., Kong, J., et al. (2009). Current loop control approach for LCL-based shunt active power filter. *Proceedings of the CSEE*, 29(18), 15–20 (in Chinese).

Tuladhar, A., Jin, H., Unger, T., et al. (2000). Control of parallel inverters in distributed AC power systems with consideration of line impedance effect. *IEEE Trans. on Industrial Applications*, 36(1), 131–138.

Electronic Engineering – Wang (ed)
© *2018 Taylor & Francis Group, London, ISBN 978-1-138-60260-1*

Realisation of a remote video monitoring system based on embedded technology

X.Y. Fan, M.X. Song & S.C. Hu
Harbin University of Science and Technology, Weihai, Shandong, China

ABSTRACT: With the development of embedded technology, video monitoring systems play an important role in our daily lives. Surveillance video also has high requirements in terms of fluency and image quality. This paper proposes a design method for remote video monitoring to achieve good surveillance video results. We use an OV7740 module camera in this monitoring system for image acquisition and transferring the Motion JPEG (MJPEG) streamer to ARM for the server. Socket programming method is adopted to build the client on a PC. The server communicates with the client over the network. This kind of design method can build a remote video monitoring system of C/S structure to display surveillance videos efficiently through a polling mechanism, which shows a good development trend.

Keywords: embedded; monitor; ARM; socket

1 INTRODUCTION

Video monitoring has entered the digital network with the speedy development of technology (Zhang, 2013). Video monitoring systems have the advantages of small volume, easy installation, flexible configuration and lower costs resulting from the embedded technology (Zhang, 2007). When embedded technology is combined with network technology, the corresponding remote video monitoring can be realised (Chu, 2013).

Currently TCP/IP protocol has the function of the leap over network in the network communication protocols. It also has strong flexibility in various network environments. The TCP/IP protocol is the first choice of embedded remote video monitoring systems with the advantages of both high speed and transport security (Sergio, 2007). However, monitoring data encounters problems during transmission when depending only on the TCP/IP protocol. When the socket connection is set up, the server and client will not stop transmission until the connection is dropped. Nevertheless, in practical applications, the network communication has many internodes including router, gateway, firewall and so on (Jiao, 2012). The polling mechanism needs to be used for checking the connection status in case the connection of the state of torpor is dropped.

In this design, the application framework for software development can significantly reduce the cost of software development and improve software quality in the Linux system. A special structural design is adopted in the camera to reduce the cost of ARM processor resources. The video data collection can be realised by using the V4L2 API functions provided by Linux. Socket programming method is adopted to build the client in a PC. A communication network is established between the server and client, which ensures the achievement of accurate, real-time, ordered data transmission.

2 DESIGN ARCHITECTURE OF THE REMOTE VIDEO MONITORING SYSTEM

The overall design of the remote video monitoring system has three parts: the construction of the server; the configuration of the network environment; and the design of the client program. The Arm9-based S3C2440 microprocessor is adopted in this system design. The open source operation and application design need to be done under Linux. The main benefits are summarised as follows: the server receives a video stream from the camera; the video transmission module transmits the surveillance video to the client by TCP/IP; and the client program completes the tasks of data decoding and display. Ultimately, a remote video monitoring system based on embedded technology can be achieved. The total functional architecture is shown in Figure 1.

3 SYSTEM HARDWARE PLATFORM

The hardware part of this system mainly includes the ARM and some peripherals. The motherboard contains common facilities, such as CPU, FLASH, SDRAM, camera interface and so on. The S3C2440 is manufactured by Samsung and serves as the core component of the hardware. The structure of the S3C2440 provides many capabilities as shown in Figure 2.

57

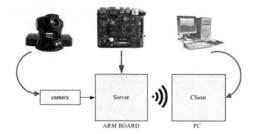

Figure 1. Functional architecture.

Table 1. V4L2 callback functions.

Function	Features
VIDIOC_QUERYCAP	Check the video device
VIDIOC_ENUM_FMT	Query the supported format
VIDIOC_S_FMT	Set the compressed format
VIDIOC_REQBUFS	Apply for buffers
VIDIOC_QBUF	Put the data into the queue
VIDIOC_STREAMON	Start the transmit
VIDIOC_DQBUF	Take out the data from the queue
VIDIOC_STREAMOFF	Stop the transmit

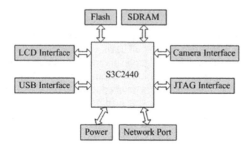

Figure 2. S3C2440 chip structure.

By using the OV7740 module camera, this system can monitor target scenes. The video data can be compressed by an embedded DSP whose main chip is Ip2970. In this system, this camera has the capability of stable image capture in a temperature range of 0–50°C, supporting output resolutions including VGA, QVGA, CIF or smaller resolutions and the output format is RAW RGB or YUV. The data transmission rate is 60 fps which meets the requirements of real-time and accurate image capture.

4 SYSTEM SOFTWARE PLATFORM

4.1 Construction of server

In accordance with the requirements of the microarray handbook, the software of the drive is tailored and transplanted under the Linux system. The camera can be used on the development board without installing the drivers. In order to realise the remote video monitoring system based on C/S, the Motion JPEG (MJPEG) streamer needs to be transplanted to the development board to build the server.

1) Download and decompress the source:

#tar xvf mjpg-streamer-r63.tar.gz

2) Turn compilation mode to cross-compile by modifying Makefile:

#vi Makefile
CC = arm-linux-gcc

3) Compile and debug the project, copy the executable programs and library files to file system:

#make
#cp mjpg_streamer /work/nfs_root/
#cp *.so /work/nfs_root/

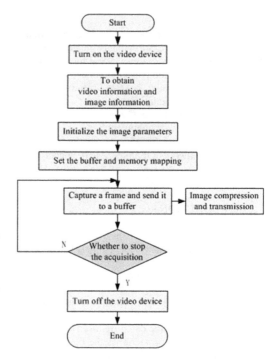

Figure 3. Process of video acquisition.

V4L is a video capture application programming interface for Linux. With higher flexibility and expansibility, V4L2 is adopted in the MJPEG streamer. In this interface, a series of callback functions are used as the receiver of the video. The callback functions are listed in Table 1.

The server uses V4L2 to collect surveillance video and the process is illustrated in Figure 3.

4.2 Configuration of the network environment

This system proposes a solution to realising video control and video network transmission. It is necessary to configure the network environment. The configuration will work for a virtual machine as follows: network connection should be set as bridging by orderly clicking VM, Settings and Network Adapter; in the Network Setting screen, selecting wireless device as the default gateway device. The network development environment is shown in Figure 4.

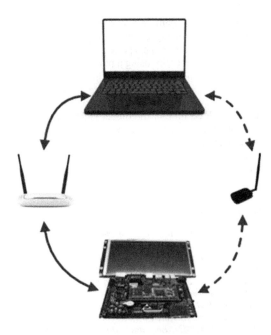

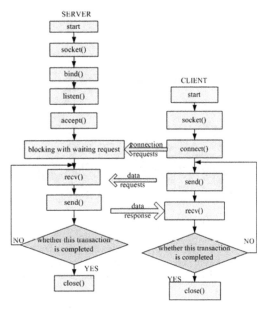

Figure 4. Network development environment.

Figure 6. Interaction style.

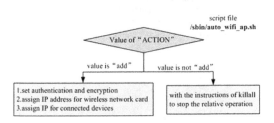

Figure 5. Execution process.

The MJPEG streamer sets a value for the server port number, which by default is 8080. The server and client work on the same network segment. The IP of development board is set to 192.168.7.16, the IP of PC is set to 192.168.7.100, the IP of the virtual machine is set to 192.168.7.124 and the subnet mask is set to 255.255.255.0. In the script file, authentication and encryption are set up with the MDEV mechanism and DHCP assigns IPs automatically for any device with internet access. Mdev.conf is set as wlan0 0:0 777 * /sbin/auto_wifi_ap.sh. The execution process is shown in Figure 5.

4.3 Design of the client program

Through analysis of the server-side source code, it is conducive to design a client-side code for mastering the polling mechanism. The socket programming method is adopted in the application design flow. After the successful establishment of communication, the client receives a frame each time and the data size are 1024 bytes. Thus, the full video goes through a process including video capturing, video receiving, video merging and video looping. In that interaction, the C/S mode is shown in Figure 6.

Figure 7. Instructions and phenomena.

5 EXPERIMENTAL RESULTS AND DISCUSSION

The remote video monitoring system aims at setting up a channel between server and client. The client needs to connect with a wireless network. The USB interface managing the data collection connects with the camera. The server and client contain enough intelligence to assume responsibility for the polling mechanism. Finally, video data are displayed on the PC. The initiator is started on a serial port tool, the relevant instructions and experimental phenomena are shown in Figure 7.

In the text-mode of Linux, the specific instructions are inputted to make the connection between the server

Figure 8. Monitor image.

and client with the IP. On entering the user name and password, the user sees a video on screen and embedded in devices. The video image is shown in Figure 8.

6 CONCLUSIONS

In this paper, the S3C2440 was chosen as the main chip and a design method for remote video monitoring system is stated. The connection between server and client is established by using a wireless network. The monitoring system exchanges data with the server in the data processing centre through WLAN for a long-distance monitor. This project reduces the cost of processor resources, by using a polling mechanism to ensure the connection is active. Through the experimental test, it is shown that the image collection and display system realised by this project have high stability and high definition. Intelligent video surveillance technology will improve the degree of automation of traditional video surveillance systems. This has significance for economy national defence and public security.

REFERENCES

Chu, C.A. (2013). *Detection of embedded remote device via wireless and development of monitoring system.* Hunan: Central South University of Forestry and Technology.
Jiao, S.W. (2012). *Research and implementation of intelligent family-house control system based on TCP/IP protocol.* Nanchang: Nanchang Hangkong University.
Sergio, S. (2007). TCP/IP Basics, implementation and applications. *Trans-Atlantic Pubns 14*(11) 15–16.
Zhang, W.W. (2007). *Embedded video monitoring system design based on ARM.* Nanjing: Nanjing University of Science and Technology.
Zhang, K. (2013). *Linux embedded video monitoring system design and implementation based on S3C2440.* Xian: Xi'an University of Science and Technology.

Electronic Engineering – Wang (ed)
© 2018 Taylor & Francis Group, London, ISBN 978-1-138-60260-1

Design and implementation of a remote monitoring mobile client in a litchi orchard, based on an Android platform

G.X. Yu,[1,2] J.X. Xie,[1,2,4,5] W.X. Wang,[1,2,4,5] H.Z. Lu,[3,5] X. Xin[6] & Y.H. Wang[1]

[1]*College of Electronic Engineering, South China Agricultural University, Guangzhou, China*
[2]*Key Laboratory of Information Acquisition and Application in Agriculture, Guangzhou Science Technology and Innovation Commission, Guangzhou, China*
[3]*College of Engineering, South China Agricultural University, Guangzhou, China*
[4]*Guangdong Engineering Research Centre for Monitoring Agricultural Information, Guangzhou, China*
[5]*Key Laboratory of Key Technology on Agricultural Machine and Equipment, Ministry of Education, South China Agricultural University, Guangzhou, China*
[6]*Wuzhou Vocational College, Wuzhou, China*

ABSTRACT: In order to realise remote monitoring and intelligent management in a litchi orchard, a remote monitoring mobile phone client for a litchi orchard, based on the Android platform was designed. With the development of technologies in socket programming, multiple tab mode, the Java language and the SQL database, this client realises the functions of real-time monitoring of environmental parameters of litchi orchards, querying historical data and operating hardware. Thus, users can do many things in real time, such as obtaining soil environmental information for the litchi orchard, determining the residual energy of each node, controlling the irrigation conditions and acquiring knowledge of litchi cultivation. Experiments show that the mobile client is stable, user-friendly and easy to operate. What is more, experiments also indicate that this client is able to achieve real-time remote monitoring of a litchi orchard environment and make the irrigation decisions in time.

Keywords: remote monitoring; Android; Java; SQLite; litchi orchard; irrigation

1 INTRODUCTION

Since the late 1980s, the litchi planting area and output in our country have increased considerably. Litchi cultivation covered an area of 127.5 thousand hm^2 in 1987, which soared to 553.3 thousand hm^2 in 2012. Besides that, the production of litchi was 11.68 thousand tonnes in 1987, which was much less than the 1.9066 million tonnes in 2012 (Liu, Zhou & Wan, 2008; Pang, Zhang & Zhang, 2014). However, the vast majority of litchi orchards still currently adopt the traditional manual management mode, which uses the primitive methods of flooding or canal irrigation. Thus, low efficiency of water usage and serious waste ensue (Ding, Yang & Wu, 2015). Modern precision agriculture requires higher and higher standards of crop information, which indicates that managers need to control the crop growth environment in real time (Qi, 2014). With the gradual advances and improvements in remote monitoring technology, orchard managers can remotely monitor and control the environment for growing fruit (Wang, Yang & Shi, 2005). The remote monitoring mobile client in a litchi orchard, based on the Android platform, connects the litchi orchard environmental monitoring and intelligent irrigation with the Internet, which ensures that orchardists can remotely control the real-time environmental information for the litchi orchards and make decisions regarding irrigation in time.

2 APP DEVELOPMENT NEEDS AND THE MAIN TECHNICAL ANALYSIS

2.1 Development needs analysis of the app

At present, to obtain information about the environment of litchi orchards, growers must go to the litchi orchard fields in the flesh and judge the growth of the litchi mainly by experience, which makes their decisions somewhat late and inaccurate. Developing a remote monitoring mobile client for a litchi orchard, based on the Android platform, not only helps growers to obtain real-time information for the orchard but also scientifically guides the fruit production, which has a conspicuous significance in modern precision agriculture.

2.2 The mainly technical analysis of the app

Based on the Android operating system and written in the Java language, the remote monitoring mobile client development platform uses socket communication, multiple tab mode, SQLite database technology and ListView.

2.2.1 *Android operating system*

Developed and led by Google and the Open Handset Alliance, Android is a kind of free and open-source operating system based on Linux (He & Gong, 2015), which is mainly used in mobile devices, such as smartphones and tablets.

2.2.2 *Java high-level programming language*

Java is a kind of user-oriented programming language which can be used to write cross-platform applications (Lou, 2015). With remarkable versatility, efficiency, security and platform portability, Java technology is widely used in many fields including PC, data canter, game console, science supercomputer, mobile phone and Internet. What is more, Java is now supported by the largest professional developer society in the world.

2.2.3 *Socket communication*

Based on the TCP protocol, socket communication is a connection-oriented stream (He, Shao & Yuan, 2015). Before socket flow sends data, the client socket and a server socket connection is established, which ensures that data can be received safely and orderly. When the transmission is finished, the socket should be turned off to disable connection. The socket is used to establish two objects that connect the client and the server, and each object is packed with an output flow and an input flow. Once the connection between the object of the client and the object of the server is established, the transmission can be finished when the data are written to the output flow of the combined socket object and the reception is done when the data are read from the input flow of the combined socket object.

2.2.4 *SQLite*

SQLite, a lightweight database that includes a relatively small C library, is a relational database management system that complies with ACID (Sun, Xiong & Zhang, 2015). It is designed to be embedded and has so far been adopted by many embedded systems. SQLite supports Windows/Linux/Unix and other main operating systems and can be combined with many programming languages including Tcl, C#, PHP, Java and so on. Compared with MySQL and PostgreSQL, two open-source database management systems are enjoy worldwide fame. IteOpenHelper class to create a database, which is a very helpful and powerful SQLite auxiliary class (Shabtai Fledel & Elovici, 2010; Enck, Ongtang & McDaniel, 2009). With the help of this class, databases of applications can be built easily and quickly.

2.2.5 *Multiple tab mode*

The tab label window is a common User Interface (UI) element. Due to the limitation of the size, full utilisation of the cell phone screen is vital. What is more, cell phone users usually use the in-app interfaces alternately. The existence of a multiple tab not

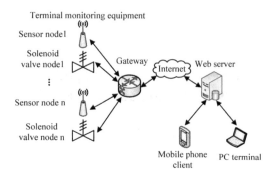

Figure 1. Overall system architecture.

only solves this problem perfectly but also optimises the user experience.

2.2.6 *ListView*

As a senior Android UI component, ListView can display the data in the form of a list. The coordination between an adapter and its subclass decides what and how the data are shown (Jia, Wu & Zhang, 2015). As a data display control with obvious advantages, ListView can display more than one item on an interface simultaneously, which is a great help for the user in the operation of data.

3 SYSTEM DESIGN

3.1 *The overall architecture*

Base on the Android OS, the terminal monitoring device, gateway, web server and mobile phone client. The structure of the whole monitoring system is shown in Figure 1. PostgreSQL, two open-source database management systems are enjoying worldwide fame, SQLite runs even faster than both of them. Android provides users with the SQL.

3.1.1 *Terminal monitoring device*

Terminal monitoring equipment includes multiple sensor nodes and electromagnetic valve nodes. Sensor nodes form a wireless sensor network through an ad hoc network (Xie, Wang & Lu, 2014) to automatically obtain the environmental information from litchi orchards, such as temperature, soil moisture content and so on. Then, data packets are sent to the gateway via a wireless sensor network. Electromagnetic valve nodes receive commands from the gateway and control the switching of electromagnetic valves.

3.1.2 *Gateway*

As the protocol conversion equipment (Xie, Yu & Wang, 2015), the gateway receives data packets from sensor nodes or sends commands to control electromagnetic valve nodes via the ZigBee network (Yu, Xie & Wang, 2016; Yu, Wang, Xie, Lu & Lin, 2016), which sends sensor nodes to the network service system via the Internet or receives control commands

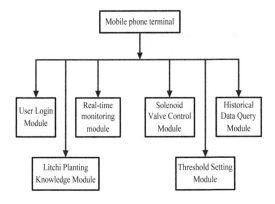

Figure 2. Mobile client module design.

from the network service system via the Internet. Thus, the ZigBee–Internet protocol conversion is completed.

3.1.3 *The web server*

The web server receives the gateway connection request and saves the environmental information from the litchi orchard to the local database. In addition, the web server receives orders from the mobile client and transmits commands to the gateway via the Internet.

3.1.4 *Mobile client*

The socket communication between the mobile client and the web server enables users to obtain much useful information in real time, such as the soil environment and the residual energy of each node. What is more, the mobile client realises the functions of controlling the irrigation conditions and helps users to acquire knowledge of planting litchi.

This passage will focus on the design and implementation of the mobile client.

3.2 *The modular design of the mobile client*

Based on the Android platform, the design of the remote monitoring mobile client in litchi orchards includes six modules, namely: user login module; real-time monitoring module; electromagnetic valve management module; historical data query module; litchi planting knowledge module; and system setting module.

The block diagram of the design is shown in Figure 2.

3.3 *Database design and implementation*

3.3.1 *The establishment of a database table*

Firstly, a database of the orchard named My Data is built up in the SQLite. After that, tables named Information, Warning and User are respectively built up in the database. The table Information contains the main information for the litchi orchard environment. With the node number and time as the main index, the received data will serve as a new record and can be inserted into the data table. This process is shown in Table 1.

Table 1. Litchi orchard environment information.

Column name	Data type	Allow null	Instructions
node	integer	no	node number (primary key)
type	varchar(50)	yes	node type
time	varchar(50)	yes	time
package	varchar(50)	yes	packets number
tem	varchar(50)	yes	temperature
hum	varchar(50)	yes	humidity
lx	varchar(50)	yes	light
water	varchar(50)	yes	soil moisture content
battery	varchar(50)	yes	battery voltage
sam	integer	yes	monitoring period
sol	varchar(50)	yes	electromagnetic valve state
auto	varchar(50)	yes	whether control solenoid valve automatically

Table 2. Upper and lower limits of the environmental information of a litchi orchard.

Column name	Data type	Null allowance	Instructions
node	integer	no	node number (primary key)
type	varchar(50)	yes	node type
tu	varchar(50)	yes	upper limit of temperature
td	varchar(50)	yes	lower limit of temperature
hu	varchar(50)	yes	upper limit of humidity
hd	varchar(50)	yes	lower limit of humidity
lu	varchar(50)	yes	upper limit of illumination
ld	varchar(50)	yes	lower limit of illumination
wu	varchar(50)	yes	upper limit of soil moisture content
wd	varchar(50)	yes	lower limit of soil moisture content
bat	varchar(50)	yes	lower limit of voltage

The table User contains the table Warning contains the upper limit and the lower limit of the main environmental information of the litchi orchard. Using the node number as the main index, this table compares the received data with the data in the table to determine whether to turn on the alarm. The concrete design is shown in Table 2.

The table User contains the user information of the network monitoring system. Using the user name as the main index, it compares the inputted user login information with the user login information in the table to determine whether to allow login. The specific design is shown in Table 3.

3.3.2 *Connecting to the database*

In Android, by inheriting the SQLiteOpenHelper, rewriting the internal construction using on Create

Table 3. User information.

Column name	Data type	Allow null	Instructions
user	varchar(50)	no	user name (primary key)
pass	varchar(50)	no	user password
type	varchar(50)	no	user types

and on Upgrade and adopting the method of getReadableDatabase and getWritableDatabase of the SQLiteOpenHelper entity class, database commands like insertion, delete and upgrade can be executed. The process of connecting to the database is as follows:

1) Create a new class to inherit the SQLiteOpenHelper;
2) Rewrite the construction method and use on Create and on Upgrade;
3) Materialise class objects that inherit the SQLiteOpenHelper;
4) Use the SQLiteOpenHelper. GetReadableDatabase () or SQLiteOpenHelper. GetWritableDatabase () to obtain an SQLite Database object;
5) Use of SQLite Database insert (), update (), the query (), raw Query () to complete the operation of the database;
6) Execute commands;
7) Return the operation to the result;
8) Disable the database connection.

4 MOBILE CLIENT MODULES

4.1 *Communicate with a web server*

The web server constantly monitors whether there are connection requests from the client (gateway or cell phone), if there are, the server will create a socket object to store the gateway information such as the IP address and port number, and use this socket to transmit data to the client and save the data of the litchi orchards to the related database table.

Based on the TCP/IP protocol, mobile communications between the client and the web server use socket and multithreading technology. The realisation of the socket program of the communication between server and client is as follows:

1) Instantiate a socket object via the IP address and port number, this object will send requests to remote hosts with a designated IP and port:
 The socket = new Sock (String dstName, int dstPort);
2) Call the getOutputStream () method of the socket object to obtain the output stream, send information to the server socket object, then getInputStream () method to get the input stream and receive information sent from the server socket:
 DataInputStream in = new DataInputStream (Socket. GetInputStream);
 DataOutputStream out = new DataOutputStream (Socket. GetOutputStream);

Figure 3. Server settings interface.

3) Turn off flow and sockets after the completion of communication:
 Socket. The close ();

In the mobile client server setup screen, once users set up the server's IP address and port number, the mobile client and the server socket will connect and communicate with each other. The server settings interface is shown in Figure 3.

4.2 *User login module*

In order to protect system security, only the administrator can register an account. If ordinary users obtain the administrator account, they are required to input the correct user name and password to get authenticated and gain access to the network monitoring system. The login interface is shown in Figure 4.

4.3 *Real-time monitoring module*

The real-time monitoring interface mainly displays some real-time parameters from the litchi orchard, such as temperature, humidity, illumination, soil moisture content, voltage of each node and electromagnetic valve state. The software obtains data from the server via socket communication and then uses Android control tools to display data on the interface. The real-time display interface is shown in Figure 5.

4.4 *Electromagnetic valve control module*

For the convenience of users to manage their orchards, the mobile client includes an interface for the operation

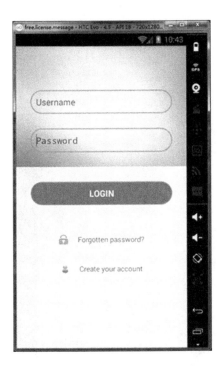

Figure 4. The user login interface.

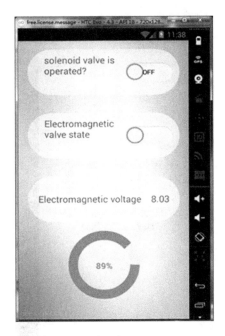

Figure 6. Electromagnetic valve.

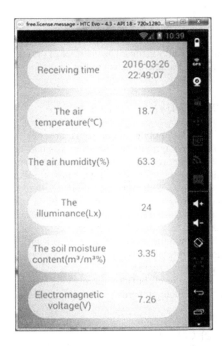

Figure 5. The current information display interface.

Node	Receiving time	temperature	The air humidity	The illuminance	The soil moisture content	Electromagnetic voltage
1	2016-03-26 22:49:07	18.7	63.3	24	3.35	7.26
1	2016-03-26 22:49:07	18.7	63.3	24	3.35	7.26
1	2016-03-26 22:49:07	18.7	63.3	24	3.35	7.26
5	2016-04-22 22:40:34	26.4	92.2	10	31.75	8.2
6	2016-04-23 10:32:4	26.2	93.1	66	32.04	8.2

Figure 7. Historical data query interface.

separately. The interface of the electromagnetic valve control module is shown in Figure 6.

4.5 Historical data query module

Users can view historical data by choosing the sensor node number and time of the litchi orchard. Firstly, the SQLite obtains local data information. If there is no information for the periods chosen by the users, data will be obtained remotely via socket communication.

All data are displayed in a list if the data obtained via SQLite are loaded on ListView, which is convenient for users to browse. The interface for the historical data query module is shown in Figure 7.

of the electromagnetic valves. If related buttons are pressed, the mobile phone will send the corresponding data to the server. After that, the server will send data to the gateway through the server to realise the function of controlling the on or off of the electromagnetic valves

Figure 8. Threshold settings interface.

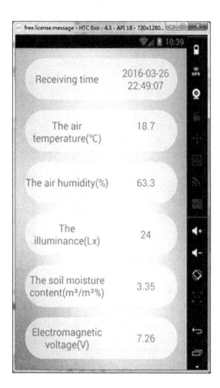

Figure 9. Real-time monitoring interface.

4.6 Litchi planting knowledge module

In this module, the user can view some knowledge of litchi cultivation. For example, what should be noticed regarding irrigation, fertilisation and medication of different litchi varieties in different periods. Besides that, knowledge of litchi diseases and insect pests is included, such as the different diseases and insect pests that different litchi varieties may encounter in different periods, the effects on litchi exerted by diseases and pests, the drugs to be applied and the effects of these drugs on litchi.

4.7 System setting module

The threshold setting module controls the water flow mainly by setting electromagnetic valves and various parameters of the litchi, which include varieties, growth stage, upper and lower threshold and sampling time. The interface of the threshold setting module is shown in Figure 8.

5 CLIENT TESTING

From 7 April to 24 April, the remote monitoring client for litchi orchards, which is based on Android, was tested in the Longxiang Litchi Orchard in Xinyu Town, Yangxi County, Yangjiang. Through the client, users can learn the environmental information of the litchi orchard in real time. When the environmental information exceeds the scope, the client will send a warning signal, which enables users to make corresponding countermeasures in time. The interface of the real-time monitoring module is shown in Figure 9.

6 INNOVATIVE FEATURES

1) Currently, mobile phone apps are not widely applied in agriculture. However, as smart devices become more and more popular, agriculture will also face enormous reforms. Thus, this kind of mobile phone client is strongly innovative;
2) The software interface utilises multiple tabs and ListView, which makes it friendly in displaying data and in terms of user experience;
3) Combining the traditional wireless sensor network and the Internet, this app not only realises the functions of environmental monitoring and smart irrigation remotely but also improves the management level of litchi orchards.

7 CONCLUSION

Based on the Android platform, the design of a remote monitoring mobile client for litchi orchards realises intelligent monitoring and smart irrigation, saves manpower, material resources and time, provides a large amount of data support for the growth of litchi trees and improvement of litchi production. The system has remarkable real-time properties, stable performance, flexible and simple operation, a friendly interface and outstanding extensiveness. Therefore, the intelligent production management in litchi orchards is a good application prospect, with certain referential significance to the design and implementation of congener clients.

ACKNOWLEDGEMENTS

The research content of this article comes from special funds for the construction of a modern agricultural technology system (No. CARS-33-13), Guangxi Science and Technology Project (No. AB16380286) and Science and Technology Planning Project of Guangdong Province (No. 2015A020209161; No. 2016A020210088). The authors would like to gratefully thank these support funds. Thanks also goes to my instructor.

REFERENCES

Ding, X.L., Yang, C.C. & Wu, Y.H. (2015). Based on the wireless network environment monitoring and intelligence control irrigation system design study. *Water Saving Irrigation*, 7, 86–89.

Enck, W., Ongtang, M. & McDaniel, P.D. (2009). Understanding Android security. *IEEE Security & Privacy*, 7(1), 50–57.

He, H. & Gong, C.Y. (2015). Based on the Android intelligent terminal iot gateway design. *Industrial Instrumentation and Automation Device*, 41–43.

He, C., Shao, Q.F. & Yuan, H. (2015). Based on the Socket implementation Android (Java) with c # synchronous communication. *Journal of Wireless Technology*, 2, 15–16.

Jiang, X.F. (2015). Smart home control system based on Android mobile terminal design. *Science and Technology and Innovation*, 20, 91–92.

Jia, G.X., Wu, B.M. & Zhang, W.L. (2015). Based on the design and development of plant diseases and insect pests of Android mobile query system. *Industrial Instrumentation and Automation Device*, 5, 44–48.

Liu, Y., Zhou, C.F. & Wan, Z. (2008). Annual Guangdong litchi Longan industry development present situation analysis. *Journal of Guangdong Agricultural Science*, 2, 110–112.

Lou, D. (2015). Based on computer software development of the JAVA programming language to explore. *Journal of Wireless Technology*, 15, 50–51.

Pang, X.H., Zhang, J. & Zhang, Y. (2014). The research progress of litchi industry in our country and the countermeasures. *Journal of Agricultural Research and Applications*, 4, 58–61.

Qi, W. (2014). Litchi situation characteristics and influencing factors of market analysis in 2013 in China. *China Tropical Agriculture*, 2, 33–36.

Sun, X.Q., Xiong, Y.L. & Zhang, Y.N. (2015). Android SQLite database in the system efficiency study. *Journal of Electronic Design Engineering*, 8, 22–24.

Shabtai Fledel, Y. & Elovici, Y. (2010). Securing Android mobile devices – powered using SELinux. *IEEE Security & Privacy*, 8(3), 36–44.

Wang, J.X., Yang, S.F. & Shi, Y.J. (2005). The development status and trend of the technology of remote monitoring. *Journal of Foreign Electronic Measurement Technology*, 4, 9–12.

Xie, J.X., Wang, W.X. & Lu, H.Z. (2014). Based on CC2530 litchi orchard intelligent irrigation system design. *Journal of Irrigation and Drainage*, 4-5, 189–194.

Xie, J.X., Yu, G.X. & Wang, W.X. (2015). Based on wireless sensor network intelligent water-saving irrigation, two-way communication and control system in Litchi orchard. *Journal of Agricultural Engineering*, 31(2), 186.

Yu, G.X., Xie, J.X. & Wang, W.X. (2016). ASP.NET-based design and implementation of remote monitoring and control system for litchi orchard irrigation. *Fujian Journal of Agricultural Sciences*, 31(7), 770–776.

Yu, G.X., Wang, W.X., Xie, J.X., Lu, H.Z. & Lin, J.B. (2016). Mo Haofan. Information acquisition and expert decision system in litchi orchard based on internet of things. *Transactions of the Chinese Society of Agricultural Engineering (Transactions of the CSAE)*, 32(20), 144–152. (in Chinese with English abstract).

Electronic Engineering – Wang (ed)
© 2018 Taylor & Francis Group, London, ISBN 978-1-138-60260-1

Calculation and simulation of the impact torsional vibration response of a stator current for a local fault in a rolling bearing

X.J. Shi, Q.K. Gao, W.T. Li & H. Guo
School of Mechanical and Power Engineering, Harbin University of Science and Technology, Harbin, China

ABSTRACT: This paper researches the induction mechanism of rolling bearing fault by stator current analysis method in depth. The impact torque calculation model of a rolling bearing local fault is established according to the elastic collision theory of Hertz. The fluctuation component of DFIG generator system speed is derived by the dynamic equation of motor. In the off-grid or grid-connected environment, the response process of the bearing local fault in the generator stator current is analysed in various conditions. SIMULINK simulation and analysis shows that the fault characteristics of the rolling bearing can be distinguished clearly in the stator current, but it is related to some parameters such as impact torque, load and rotational speed. Therefore, the influence of these factors must be considered in the actual application process.

Keywords: DFIG generator; rolling bearing; stator current; impact torque; fault diagnosis

1 INTRODUCTION

The method of stator current analysis has the characteristics of low cost and simple operation, and it is a non-contact, non-destructive testing method which replaces the traditional vibration diagnosis method of rolling bearings. At present, there is a large volume of research and literature regarding motor or generator diagnostics (mainly wind turbine generators) (Zhou et al., 2007; Obaid et al., 2003; Irahis & Roberto, 2006). However, the current signal contains a variety of mechanical and electrical information components, so one of the difficult problems that hinders the practical application of this method is how to identify effective information correctly, especially when mechanical fault information is weaker than power grid information. Additionally, it is also a research hotspot to determine how to effectively separate and extract the effective information (Ibrahim et al., 2006; Gong et al., 2010). The key to solving these problems is to study the theory of fault induction deeply, but the research on these aspects is scarce (Zhou et al., 2007). The research object of this paper is a common doubly fed generator and its rolling bearing. The calculation model of the bearing impact torque and the simulation model of the generator are established. Then, we use the theoretical simulation method to research the theory of mechanical fault induction in a motor current signal. Meanwhile, this paper provides some theoretical supports and references for application and research using the method of electrical signal analysis.

2 FUNDAMENTALS

At present, there are two main motor current detection models for rolling bearing fault detection. One is the radial vibration transmission model which was first proposed by Schoen et al. (1995), who believed that a rotating eccentricity will be generated in the stator when the bearing is under a fault frequency. This eccentricity will lead to periodic variation of the motor induction current. Another is the torque vibration transmission model. This considers that a transient fluctuation of torque will be generated when the ball passes through a pit in the inner and outer rings. Most scholars believe that there is an amplitude modulation relationship between this fluctuation component and power grid frequency. Blodt et al. (2008) further deduced that the periodic variation of torque can also cause the phase modulation of the stator current, which is named frequency modulation. No matter what kind of model, an additional frequency f_{bf} is generated in the stator current:

$$f_{bf} = |f_s \pm k f_c| \qquad (1)$$

where f_s is the motor stator power supply frequency, f_c is the rolling bearing fault frequency, $k = 1, 2, \ldots$.

This paper adopts the torque vibration transmission model. Outer ring faults are simplified to a small pit. The rolling body will be hindered by the force when it goes through the small hole. As a result, the load torque is increased. The fault model is shown in Figure 1.

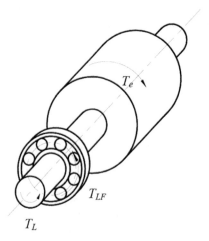

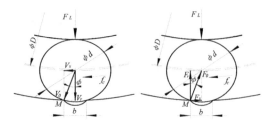

Figure 2. Torque dynamic model of a rolling bearing.

It can be seen that the mechanical angular velocity of the rotating shaft consists of a constant ω_{r0} and a changing part $\sin(\omega_c t)$, and the amplitude of the changing part is related to T_c, J and rotational speed fluctuation frequency ω_c.

Figure 1. Torque model of a faulty bearing.

3 ANALYSIS OF THE IMPACT TORQUE FLUCTUATION OF A ROLLING BEARING FAULT

When the bearing is in the normal working state, the load torque T_L is equal to the electromagnetic torque T_e generated by the motor. If the bearing is faulty, an additional torque T_{LF} will be generated when the rolling body goes through the faulty parts. At the same time, the corresponding induction current is generated in the stator coil, leading to the appearance of an additional electromagnetic moment T_{eF} to balance the additional torque T_{LF}.

When a bearing fault occurs, the load torque function can be regarded as the sum form of a fixed value T_{L0} and the additional torque T_{LF}. For convenience, the load torque can be considered as the cosine law, so the expression is as follows:

$$T_L(t) = T_{L0} + T_{LF} = T_{L0} + T_c \cos(\omega_c t) \quad (2)$$

where T_c represents the torque amplitude change related to the fault of the rolling bearing inner and outer rings, $\omega_c = 2\pi f_c$.

The kinetic equation of the motor drive system is as follows:

$$T_e - T_L = J \frac{d\omega_r}{dt} \quad (3)$$

where T_e is the electromagnetic torque, T_L is the load torque and J is the moment of inertia.

Torque change will lead to the change of motor speed ω_r according to the dynamic Equation 4:

$$\omega_r(t) = \frac{1}{J} \int_t T_e(\tau) - T_L(\tau) d\tau \quad (4)$$

In the steady state, assuming that the electromagnetic torque of the motor is equal to the constant part of the load torque: $T_e(t) = T_{L0}$, the expression is as follows:

$$\omega_r(t) = -\frac{1}{J} \int_{t_0}^{t} T_c \cos(\omega_c \tau) d\tau + C_0$$
$$= -\frac{T_c}{J\omega_c} \sin(\omega_c t) + \omega_{r0} \quad (5)$$

The torque kinetic model of a bearing is shown in Figure 2. When a fault occurs in the outer ring, we assume it happens at the bottom of the bearing outer ring. Due to the existence of the outer ring depression, the rolling body is in a suspended state. Under the action of the radial force, the inner ring has a small deformation, making the rolling body drop into the defect site. However, the contact point between the rolling element and outer ring has deviated from the normal working position. The supporting force, F_r, has a certain angle α with a vertical direction. F_r can be decomposed into a force F_v pointing to the centre of a circle along the radial direction and a force F_x which is perpendicular to the force F_v. In the next movement of the rolling body, it needs to overcome the resistance torque T_{Lx} generated by F_x. Additionally, the corresponding additional torque T_{ex} is generated in the motor.

When a fault occurs in the outer ring of the rolling bearing, the rolling body will have a great impact force with the outer ring each time it goes through the fault site. The size of the impact force can be obtained by using the theory of Hertz regarding elastic body collision (Fan & Yin, 1985). The maximum impact force, $P_{\max}$, between the colliding spheres can be obtained by the elastic body impact equation. The expression is as follows:

$$P_{\max} = n^{\frac{2}{5}} \left(\frac{5}{4} \frac{V_0^2}{m_r} \right)^{\frac{3}{5}} \quad (6)$$

$$m_r = \frac{m_1 + m_2}{m_1 m_2}, \quad n = \frac{4}{3\pi(\theta_1 + \theta_2)} \sqrt{\frac{r_1 + r_2}{r_1 r_2}},$$

$$\theta_i = \frac{\lambda_i + 2v_i}{4\pi v_i(\lambda_i + v_i)}, \quad \lambda_i = \frac{2\mu_i v_i}{1 - 2\mu_i}.$$

where m_1 and m_2 are the quality of two collision spheres, V_0 is the relative velocity of the collision moment, r_1 and r_2 are the radii of two collision spheres, μ_i and ν_i are the Poisson's ratio and the shear modulus.

The rotation speed of the rolling body is ignored, assuming that the rotation speed of the rolling body is V_c and the diameter is d, the width of the pit is b, so the instantaneous collision velocity V_0 when the rolling body goes through the fault site is as follows:

$$V_0 = V_c/\sin\phi = V_c \frac{d}{b} \qquad (7)$$

The rotation speed V_c of the cage can be calculated according to Equation 8:

$$V_c = \frac{\pi}{2} Df_r \left(1 - \frac{d}{D}\cos\alpha\right) \qquad (8)$$

where f_r is the rotation frequency of the shaft, D is the pitch circle diameter of the bearing, α is the nominal contact angle of the bearing.

The material of the bearing is steel, so we get $\mu_1 = \mu_2 = 1/3$, $\nu_1, \nu_2 = 8.1 \times 10^5$ kg/cm^2. The following result is obtained:

$$\theta_1 = \theta_2 = 1/3\pi \times 8.1 \times 10^5 \text{ cm}^2/\text{kg}$$

Taking the 307 bearing as an example, the diameter of the rolling element is $d = 13.494$ mm, the pitch circle diameter is $D = 58.56$ mm, the contact circle diameter of the outer ring is $D' = D + d$. Thus, the rolling body radius is $r_1 = d/2 = 6.745$ mm and the outer ring radius is $r_2 = D'/2 = 36.03$ mm.

The quality of the rolling body is:

$$m_1 = \frac{4}{3}\pi\left(\frac{d_1}{2}\right)^2_1 \rho/g = 1.023 \times 10^{-5} \text{ kg s}^2/\text{cm}$$

The quality of the outer ring is:

$$m_2 = 1.97 \times 10^{-4} \text{ kg} \cdot \text{s}^2/\text{cm}$$

$$n = \frac{4}{3\pi(\theta_1 + \theta_2)}\sqrt{\frac{r_1 + r_2}{r_1 r_2}} = 1.221 \times 10^7$$

$$m_r = \frac{m_1 + m_2}{m_1 m_2} = 1.028 \times 10^5$$

Assuming that the width of the outer ring fault is $b = 8$ mm, the nominal contact angle of the bearing is $\alpha = 0$, the rotating frequency of the bearing is $f_r = 15$ Hz, then we can work out:

$$V_c = \frac{\pi}{2}Df_r\left(1 - \frac{d}{D}\cos\alpha\right) = 1.014 \text{ m/s}$$

$$V_0 = V_c d/b = 1.71 \text{ m/s}$$

Putting the parameters into Equation 6, then we can obtain the maximum support force:

$$F_0 = n^{\frac{2}{5}}\left(\frac{5}{4}\frac{V_0^2}{m_r}\right)^{\frac{3}{5}} = 146.3 \text{ kg}$$

The resistance moment is obtained as follows:

$$T_{Lx} = F_x \frac{D}{2}g = F_0\frac{b}{d}\frac{D}{2}g = 24.9 \text{ N} \cdot \text{m} \qquad (9)$$

Once the ball passes through the fault point, the resistance moment will appear once. The frequency of its occurrence is the outer ring fault frequency of the rolling bearing (Shi et al., 2013).

$$f_{op} = \frac{n}{2}f_r\left(1 - \frac{d}{D}\cos\alpha\right) = 30.744 \text{ Hz} \qquad (10)$$

For convenience, we consider that the fluctuation of torque is a harmonic function as shown in Equation 2. The expression of motor speed fluctuation generated by it is shown in Equation 5. Its amplitude is related to the frequency of T_c ($T_c = T_{Lx}$), J and rotational speed fluctuation ω_c ($\omega_c = 2\pi f_{op}$).

4 SIMULATION MODEL OF A DOUBLY FED INDUCTION GENERATOR

The study object of this paper is a doubly fed induction generator (DFIG), the simulation model is established as shown in Figure 3. The model includes a DFIG generator, AC-DC-AC bidirectional pulse width modulation (PWM) module, speed analogue input module, three-phase supply, three-phase circuit breakers, a load and measurement module. A wound-rotor induction motor is used instead of the doubly fed induction generator model. The bidirectional PWM power frequency conversion module changes the AC power to DC power, and then turns the DC into AC power to supply power to the rotor. The power supply phase can be regulated by the PWM generator, in order to realise the synchronous grid-connected function. Due to this paper only needs to simulate the running state when wind speed is stable, in order to make the model simple, there is no automatic adjustment part in the simulation model. The mechanical input type of the generator is selected as the speed adjusting type. The calculation results of Equations 9–12 are substituted in Equation 5 to calculate the input speed. The reference coordinate system is selected as the synchronous reference coordinate system and the values of all the initial states are 0.

The rated power of the generator in the simulation model is 3 kW, the rated voltage of the AC power supply is 220 V, the number of the magnetic pole is $p = 2$, the resistance value of the stator side is $R_s = 1.9188 \Omega$, the value of inductance is $L_s = 0.24122$ H, the resistance value of converting to the rotor side is $R_r = 2.5712 \Omega$, the value of inductance is $L_r = 0.24122$ H and the mutual inductance between stator and rotor is $L_m = 0.234$ H. Assuming the rotor excitation frequency is $f_{er} = 20$ Hz and angular velocity is $\omega_r = 94.24778$ rad/s, the generator output frequency f_e is worked out as follows:

$$f_e = p\frac{\omega_r}{2\pi} + f_{er} = 2 \times \frac{94.25}{2\pi} + 20 \approx 50\text{Hz} \qquad (11)$$

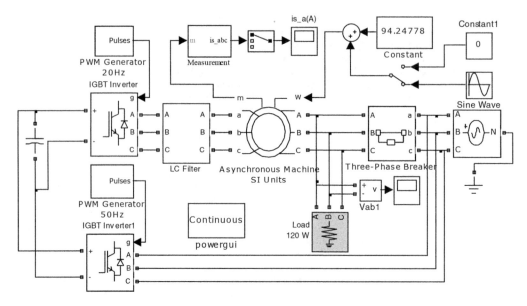

Figure 3. Simulation model of the doubly fed induction generator and control system.

5 SIMULATION RESULTS OF A ROLLING BEARING FAULT SIMULATION IN A DFIG

5.1 Simulation results in the normal operation state

When the rotor is supplied with a 30 Hz frequency field current by the PWM frequency converter and the speed of the doubly fed induction generator is stable at $\omega_r = 94.24778$ rad/s, a steady state signal of the stator current is collected. Then, using the FFT transform and Hilbert transform demodulation algorithms (Shi et al., 2013) to calculate the amplitude spectrum and the amplitude envelope spectrum, the results are requirements of the power generation frequency. Although there is no characteristic frequency of the rolling bearing in the envelope spectrum, there are 89.56 Hz and 109.4 Hz frequency components with some equal interval frequency components with low amplitude and low frequency. It can be seen from Figure 5 that most of these components are harmonic components of the rotor excitation frequency which is induced in the stator current shown in Figure 4. It can be seen that the amplitude spectrum of the stator current signal is mainly based.

5.2 Simulation results when the simulated rolling bearing outer ring is faulty

The wave component of Equation 5 can be superimposed on the stable speed to simulate the bearing fault. According to the impact torque calculated by Equation 9 and considering that the moment of inertia of the motor is $J = 0.03/\text{N·m}^2$. When the motor rotating frequency is $f_r = 15$ Hz, the bearing outer ring fault frequency is $f_c = 45.69$ Hz. At this moment the amplitude of the wave component is calculated to be

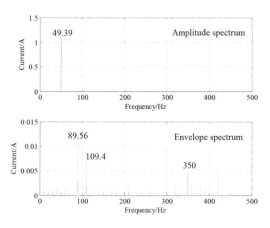

Figure 4. Amplitude spectrum and envelope spectrum of generator stator current in normal time (power supply frequency of rotor $f_{er} = 20$ Hz, rotor speed $\omega_r = 94.24778$ rad/s).

2.28. The cosine wave component of that frequency is 45.69 Hz and the amplitude is 2.28, which is superimposed to the rotating circular frequency of the fixed rotor in Figure 3. At this time the simulation results are shown in Figure 6. The fault characteristic frequency of 45.13 Hz can be clearly observed in the envelope spectrum.

5.3 Simulation results at different rotational frequencies

A waterfall plot of the stator current envelope spectrum is shown in Figure 7 when the generator is in a different frequency. The broken line is the fundamental frequency amplitude of the fault frequency in different conditions. We can observe that the fault

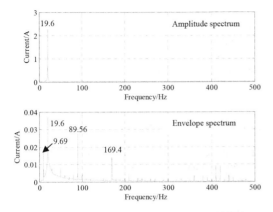

Figure 5. Amplitude spectrum and envelope spectrum of generator rotor current in normal time (power supply frequency of rotor $f_{er} = 20$ Hz, rotor speed $\omega_r = 94.24778$ rad/s).

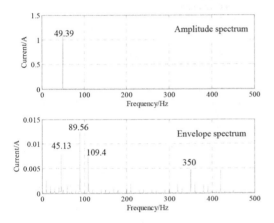

Figure 6. The spectrum of generator stator current when the outer ring of the rolling bearing is faulty (power supply frequency of rotor $f_{er} = 20$ Hz, rotor speed $\omega_r = 94.24778$ rad/s).

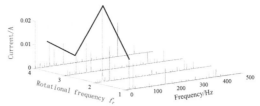

Figure 7. Stator current envelope spectrum waterfall of outer ring fault simulation at different rotational frequencies (1: $f_r = 5$ Hz; 2: $f_r = 10$ Hz; 3: $f_r = 15$ Hz; 4: $f_r = 20$ Hz).

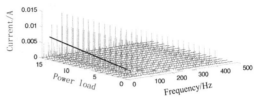

Figure 8. The spectrum of the generator stator current at different loads (power supply frequency of rotor $f_{er} = 20$ Hz, rotor speed $\omega_r = 94.24778$ rad/s).

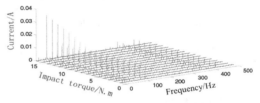

Figure 9. The spectrum of the generator stator current at different impact torques (power supply frequency of rotor $f_{er} = 20$ Hz, rotor speed $\omega_r = 94.24778$ rad/s).

frequency ($f_{op} = 30.46$ Hz) of the second spectral line ($f_r = 10$ Hz) is the largest. At this time the rotor power supply frequency is also 30 Hz. In addition to participating in the synthesis of the power generation frequency, a 30 Hz harmonic component will be left in the envelope spectrum. The harmonic component coincides with the fault frequency of the rolling bearing and thus, a resonance response is generated. This will cause a mismatch in practical applications, so we need to pay more attention to it.

5.4 Simulation results at different loads

A waterfall plot of the stator current envelope spectrum is shown in Figure 8 when the generator is under different loads. The size of the three-phase load is from 60 W to 3 kW. We can observe that the amplitude of the fault characteristic frequency decreases with the increase of load. There is no change in the amplitude of the other ingredients related to the rotor winding frequency harmonic. Therefore, we can identify the source of some frequency components according to this characteristic.

5.5 Simulation results at different impact torques

A waterfall plot of the motor stator current envelope spectrum is shown in Figure 9 when the impact torque is different and the fluctuation amplitude changes from 0.1 to 15.0 Nm. It is different from Figure 8. At this time, the amplitude of all the frequency components are related to the fluctuation range. Therefore, it can be shown that the impact wave is the main factor to cause the harmonic component of the stator current.

5.6 Simulation results at grid connection

In grid-connected and off-grid cases, the simulation results are similar. The main conclusions are presented in previous papers. Due to the limitation of space, this paper only introduces the simulation results of the grid-connected case under the conditions of Figure 6. In order to satisfy the grid-connected conditions, before the grid-connected simulation, the three-phase circuit breaker is closed and the initial phase of the

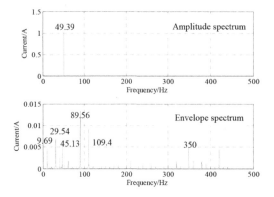

Figure 10. The spectrum of the generator stator current in the simulation of the rolling bearing outer ring fault (grid connection, power supply frequency of rotor $f_{er} = 20$ Hz, rotor speed $\omega_r = 9\,4.24778$ rad/s).

rotor side converter is adjusted so that the power generation signal and the power grid have the same phase. The simulation results are shown in Figure 10. We can observe that it has the same frequency component as the off-grid case, but the amplitude distribution has changed. For example, the amplitude of the 9.69 Hz and 29.54 Hz components is increased obviously, and the amplitude of the 45.13 Hz fault characteristic is reduced.

6 CONCLUSION

Through the theoretical calculation of the local fault impact torque of the rolling bearing, and the computer simulation, it was shown that the generator stator current signal can accurately reflect the fault characteristic information of the rolling bearing. However, in the stator current signal, it also has some independent harmonic frequency components, especially the DFIG wind power generator. In order to adapt to the influence of wind power, the rotor excitation frequency should be adjusted at any time in the course of operation. The excitation harmonic of the rotor has a great influence on the stator current harmonics. So, the error will generate in some particular cases (second spectral line of Figure 6). Therefore, we need to pay attention to avoid these areas which are prone to cause interference when test conditions are selected. In addition, in the off-grid environment, the amplitude of the bearing fault characteristic decreases with the increase of the load and decreases with the decrease of the amplitude of impact torque. These laws will help us to recognise and identify the fault characteristic frequency and the system interference frequency correctly.

ACKNOWLEDGEMENT

This article was funded by the National Natural Science Foundation of China (51275136).

REFERENCES

Blodt, M., Granjon, P. & Raison B. et al. (2008). Models for bearing damage detection in induction motors using stator current monitoring. *IEEE Transaction on Industrial Electronics*, 55(4), 1813–1822.

Fan, J.S. & Yin, A.M (1985). The calculate of the impacting force on railway joint by locomotive wheel and the stress concentration. *Journal of Yunnan University*, 12(2), 51–55.

Gong, X., Qiao, W. & Zhou, W. (2010). Incipient bearing fault detection via wind generator stator current and wavelet filter. *Proceedings-IECON 2010, 36th Annual Conference of the IEEE Industrial Electronics Society*, 2615–2620.

Irahis, R. & Roberto, A. (2006). Bearing damage detection of the induction motors using current analysis. *2006 IEEE Transmission and Distribution Conference and Exposition Latin America*, 1–5.

Ibrahim, A., Badaoui, M.E. & Guillet, F. et al. (2006). Electrical signals analysis of an asynchronous motor for bearing fault detection. *IECON 2006 – 32nd Annual Conference on IEEE Industrial Electronics*, 4975–4980.

Obaid, R.R., Habetler, T.G. & Stack, J.R. (2003). Stator current analysis for bearing damage detection in induction motors. *IEEE International Symposium on Diagnostics for Electric Machines*, 182–187.

Schoen, R.R., Habetler, T.G. & Kamran, F. et al. (1995). Motor bearing damage detection using stator current monitoring. *IEEE Trans. on Industrial Applications*, 31(6), 1274–1279.

Shi, X.J., Wang, G.R. & Si. J.S. (2013). *Mechanical fault diagnosis and typical case analysis*. Chemical Industry Press.

Zhou, W., Habetler, T.G. & Harley, R.G. et al. (2007). Stator current based bearing fault detection techniques: A general review. *2007 IEEE International Symposium on Diagnostics for Electric Machines, Power Electronics and Drives*, SDEMPED, 7–10.

Electronic Engineering – Wang (ed)
© 2018 Taylor & Francis Group, London, ISBN 978-1-138-60260-1

An analysis of the electromagnetic field of the winding inter-turn short-circuit of an asynchronous motor

S.Y. Ding & Y. Wang
Harbin University of Science and Technology, China

ABSTRACT: The inter-turn short-circuit faults of stator winding are the most common electrical faults of an asynchronous motor. When an inter-turn short-circuit occurs, the short-circuit winding generates a short-circuit current, which causes a lot of motor problems, resulting in motor performance deterioration or damage. Therefore, it is of great theoretical and practical significance to analyse the short-circuit fault of an asynchronous motor by means of a numerical method. This paper was based on the basic theory of electrical engineering, to carry out a numerical analysis of the electromagnetic field in the different short-circuit conditions of asynchronous motor stator winding, which are under a PWM (Pulse-Width Modulation) inverter power supply. Through a detailed analysis of the numerical and experimental results, the rule of the asynchronous motor performance with the turn-to-turn short-circuit of the stator winding changing is defined, providing a theoretical basis for the detection of asynchronous motor running status and the prediction of similar faults.

Keywords: induction motor; inter-turn short-circuit; electromagnetic field analysis

1 INTRODUCTION

An asynchronous motor has the characteristics of simple manufacture, reliable operation and high efficiency, which provides a wide range of applications in industry and agriculture (Fanlin & Trutt, 1988; Gao & Li, 1994). The most common motor faults during the running of a motor are a short-circuit of the motor stator winding, broken rotor bars and so on, with the stator winding inter-turn short-circuit fault accounting for about 40% of all motor faults. Motor stator winding failure will inevitably cause an air-gap magnetic field distortion, so that the motor electromagnetic torque and magnetic flux density will change, affecting the running performance and service life (Hu & Huang, 2003; Hu, 2011). With the motor capacity and the motor reliability requirements increasing, the study of stator winding faults has attracted wide attention from scholars in China and abroad (Liang & Tang, 2003; Qiao & Zhang, 2004).

The turn-to-turn short-circuit is mainly due to a short-circuit caused by insulation breakdown in the windings between the in-phase windings (Rangarajan, 1996). When an inter-turn short-circuit occurs, the three-phase currents of the windings will not be symmetrical and three-phase reactance will also no longer be equal, which generates a great short-circuit current, causing the motor to shock and the reactive power of the motor to decrease. A short-circuit current will

make the winding temperature rise too high, leading to a ground short-circuit and partial discharge, so the motor cannot operate normally. There are many methods of motor fault analysis domestically and abroad, the most important are: (1) The classical method of coordinate transformation which is widely applied to the analysis of the asymmetry of the motor (Sun et al., 2004; Tamer et al., 1986), but this method is only for linear analysis. If you do not select an operating point, it will affect the analysis accuracy; (2) The multiloop method (Wang et al., 2009), where the key to application is the accurate calculation of the motor parameters, especially the reactance parameters of the motor; otherwise, it is easy to produce a large calculation error; (3) The finite element method, whose analysis can effectively deal with the characteristics of a nonlinear medium, and has a higher calculation accuracy which has been widely used to solve the problem of the electromagnetic field of motors (Wang, 1990; Trutt et al., 2001; Wang, 1999).

In this paper, through the comparative analysis of the finite element of a 55 kW asynchronous motor in normal operation and the occurrence of an inter-turn short-circuit, the waveform characteristics of the magnetic flux density, stator three-phase current and torque of the motor when the fault occurs are determined, which can provide a reference and theoretical basis for quick judgement of short-circuit faults in a stator winding.

Table 1. Basic data of the motor.

Parameter	Value
Rated power	55 (kW)
Rated voltage	380 (V)
Rated current	103 (A)
Rated speed	1475 (r/min)
Number of poles	4
Phase	3
Power factor	0.83
Effectiveness	93%

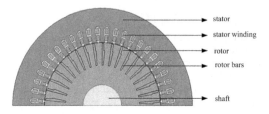

Figure 1. Two-dimensional physical model of the motor.

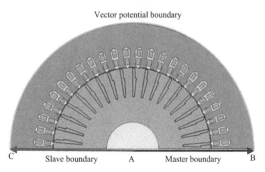

Figure 2. Boundary conditions imposed.

2 PHYSICAL MODEL

A 55 kW asynchronous motor was used as an example to analyse the characteristics in this paper. The basic parameters of the motor are shown in Table 1.

Due to the symmetry of the motor structure and magnetic field, in order to facilitate the calculation and analysis, we selected 1/2 model of the motor for simulation and make the following fundamental hypotheses.

Because there is no axial change of magnetic field, this paper makes the 2D model as the study object. And the inductance of the stator windings end is calculated into its end effect. The inductance and resistance of the rotor end ring is constant. The core of the stator and rotor is seeing as isotropic. Electromagnetic field within the motor changes cyclically, and the effect of temperature on magnetic field is ignored. The model is shown in Figure 1.

When the motor is calculated by a finite element, in order to gain the only solution of the equation, we regulate the specified area as the simulation range, and provide the following boundary conditions:

(1) Ignore the parallel plane field without displacement current in the electromagnetic field, there are only the Z-axis components in the electromagnetic potential;
(2) Imposing boundary condition on the boundary of the solution domain, to divide the solution range, to the magnetic field lines which parallel to a given boundary, the Dirichlet boundary condition is imposed, and the value is 0;
(3) In order to reduce the computation and improve efficiency, the motor is reduced to a pair of pole models, with one side of the motor solution model defined as the main boundary condition, the other side as the sub-boundary condition, and think that $A_M = A_S$.

On the basis of the above boundary conditions, we set a boundary condition for the two-dimensional model, as shown in Figure 2.

The equation expression of the mathematical model which is based on the boundary condition is shown in Equation 1:

$$\begin{cases} \dfrac{\partial}{\partial x}\left(\dfrac{1}{\mu}\dfrac{\partial A_z}{\partial x}\right) + \dfrac{\partial}{\partial y}\left(\dfrac{1}{\mu}\dfrac{\partial A_z}{\partial y}\right) = -J_z \\ A_z\big|_{\overline{BC}} = 0 \\ A_z\big|_{\overline{BA}} = A_z\big|_{\overline{CA}} \end{cases} \quad (1)$$

In the formula: A_Z is the axial electric vector position; μ is the permeability of material.

3 ESTABLISHMENT OF A TURN-TO-TURN SHORT-CIRCUIT MODEL

To get the change curve of the motor magnetic parameters in the case of short circuit, the model of inter turn short circuit is needed, and the finite element simulation is carried out.

When the asynchronous motor is under normal operating conditions, in order to simulate the short-circuit of the stator winding, first, we divide the stator winding of the A-phase into two parts, the normal winding and short-circuit winding. At both ends of the short-circuit winding we parallel a voltage control switch. After the motor starts running for 0.4 seconds, we close the switch. The short-circuit winding is short-circuited by the switch. The short-circuit position of the stator winding is shown in Figures 3 and 4.

It can be seen from Figure 4, that before the short-circuit occurs, the short-circuit winding and normal winding are in series. When the short-circuit occurs, both ends of the short-circuit winding are not directly shorted, but shorted by the contact resistance. Where the short-circuit occurs, there is a contact resistance. This contact resistance is 1/10 of the short-circuit resistance and even smaller.

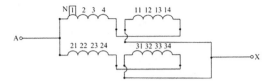

Figure 3. Diagram of the short-circuit position.

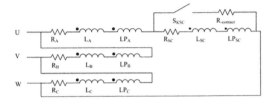

Figure 4. Schematic diagram of the external circuit.

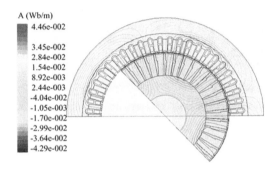

Figure 5. Diagram of the magnetic field lines.

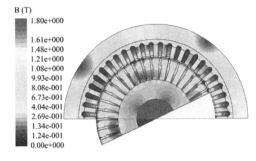

Figure 6. Diagram of the magnetic flux density.

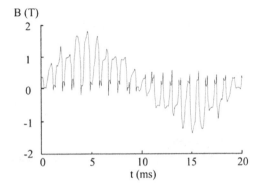

Figure 7. The magnetic flux density of the air-gap.

4 RESULTS AND ANALYSIS OF THE SIMULATION

According to the basic parameters of the motor and the basic assumptions, in accordance with the boundary condition, the normal and short-circuit operation of the motor can be solved and we obtained diagrams of flux density, magnetic field lines, rotational speed and torque curve of the motor under two running states. By comparing and analysing the data under the two states we can get the characteristics of the magnetic field under the short-circuit condition.

4.1 Electromagnetic field characteristics of the motor running in normal state

Figures 5 and 6 show the magnetic field lines and magnetic flux density under the normal operation of the motor. It can be seen that the magnetic field lines are with four poles closed in the circumferential direction from the figure. Most of the magnetic field lines appear in the core area of the stator and rotor, almost not passing through the stator winding and the rotor bars. Part of the stator and rotor teeth have a maximum flux density at the pole centre line, which decreases toward both poles and the lowest is at the geometric centre line.

Figure 7 shows the waveform curve of the air-gap magnetic flux density in one period under the normal operating condition. As shown in Figure 7, the average air-gap magnetic flux density is approximately sinusoidal. Because the stator and rotor core are slotted, the air-gap flux density waveform appears harmonic. Excluding the effect, the waveform of the average air-gap magnetic flux should be consistent with the sine wave and the crest and valleys of the average air-gap magnetic flux density are one to one corresponding with a pair poles of the motor, which is in line with the theory of AC excitation magnetic field distribution.

Figure 8 is the waveform of the stator three-phase current with the motor running in the normal state. It can be seen from the figure that the phase current is in 120° different with each other phase and symmetrically distribution. The amplitude of each phase is substantially equal.

Figures 9 and 10 show the curve of torque and speed for the motor in normal running. When the motor reaches a steady state in normal running, the torque of the motor is 354.99 N·m, the speed of the motor is 1479 rmp, and the fluctuation of the torque and the rotational speed waveform are small. The motor achieves the target of normal and stable operation.

4.2 Characteristics of the electromagnetic field in short-circuit operation of the motor

In the case of a rated load, a seven turns short-circuit occurred when the motor was running for 0.4 s. The

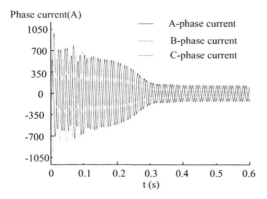

Figure 8. Three-phase current waveform of the stator winding.

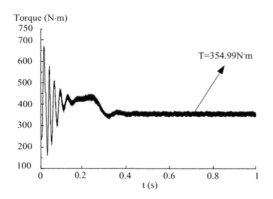

Figure 9. Torque curve in the normal condition.

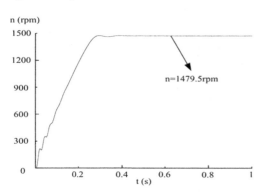

Figure 10. Speed curve in the normal condition.

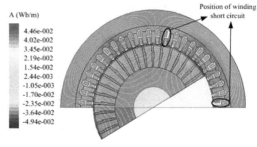

Figure 11. Distribution of magnetic field lines in the short-circuit.

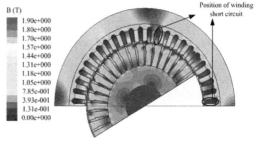

Figure 12. Diagram of magnetic flux density in the short-circuit.

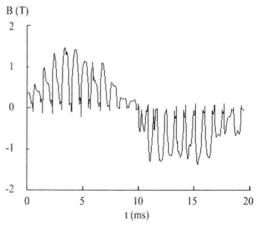

Figure 13. Magnetic flux density in the short-circuit case.

magnetic field lines and magnetic flux densities of the motor after the short-circuit are shown in Figures 11 and 12. The magnetic flux density and the magnetic field line distribution of the induction motor have obviously changed and the magnetic flux density near the position of the short-circuit is significantly larger than that in the normal case. Besides, it increases in varying degrees at other position, especially near the air-gap.

Figure 13 shows the waveform of the magnetic flux density of the air-gap in the short-circuit case. The waveform of the magnetic flux density and magnetic pole are one to one correspondingly in the normal case and the corresponding waveform for each pole are also substantially similar and the waveform is symmetrical. In the short-circuit condition, magnetic flux density of air-gap lose the original symmetry and the amplitude of the air-gap magnetic flux density also increased significantly. This shows that the turn-to-turn short-circuit winding will affect the motor's air-gap rotating magnetic field.

Figure 14 shows the three-phase current waveform of the winding in a short-circuit condition. In the normal condition, three-phase current has symmetry and the amplitude is also equal with a difference of 120°

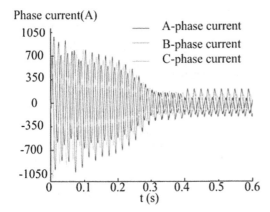

Figure 14. Three-phase current waveform of the stator winding in the short-circuit.

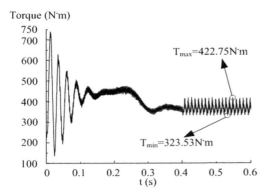

Figure 15. Curve of the torque in the short-circuit state.

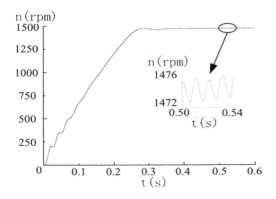

Figure 16. Curve of speed in the short-circuit state.

between each of the two phases. Under the short-circuit state of the stator winding, the short-circuit winding cuts the air-gap of the rotating magnetic field and an electromotive force is produced, then the induced current is also produced. The current of A phase increased highly, the current amplitude of the other two phases is basically same with the current amplitude under the normal situation. The waveform of three-phase current is obvious asymmetry.

Figures 15 and 16 show the curves of torque and speed in the short-circuit state of the motor. When the motor reaches a steady state in the short-circuit running status, compared with the normal status, the average values of torque and speed are basically similar, but at the time of stabilisation, fluctuations of torque and speed are significantly enhanced and the stability of the motor decreases.

5 CONCLUSION

In this paper, an asynchronous motor is simulated by the finite element method. The simulation results are in line with AC motor theory. The following conclusions can be drawn by comparing the magnetic field curves of the two states:

(1) When the induction motor occurs an inter-turn short-circuit the magnetic flux density rises significantly at the short-circuit position, especially in the position of the stator and rotor core tooth. The uniformity of the magnetic field of the motor is affected, with the air-gap flux density waveform losing the original space symmetry.
(2) The A-phase current amplitude of the stator increases and the amplitude of the A-phase current is significantly higher than that of the B and C-phases. The three-phase current waveform loses its original symmetry.
(3) Under the short-circuit fault, compared with the normal state, the motor's torque and speed are basically similar. However, the volatility of the torque and speed of the motor was increased. At the same time, motor stability was reduced.
(4) In the case of the PWM (Pulse-Width Modulation) inverter power supply, the results of the magnetic field analysis under turn-to-turn short-circuit faults are in line with AC motor theory and the reasonableness of the analysis method is tested and verified, as well as the correctness of the magnetic field calculation results.

ACKNOWLEDGEMENT

This work is supported by Harbin Science and Technology Innovation Talent Project under No. 2016RAX XJ026.

REFERENCES

Fanlin, D. & Trutt, F.C. (1988). Calculation of frequency spectra of electromagnetic vibration for wound-rotor induction machines with winding faults. *Electric Power Components and System, 14*(3), 137–150.

Gao, Jingde. & Li, Fahai. (1994). *Analysis of AC motor and system.*

Hu, Minqiang. & Huang, Xueliang. (2003). *Numerical calculation method and application of motor running performance.* Nanjing: Press of Southeast University.

Hu, Nade. (2011). Brief analysis of generator stator winding short circuit fault. *Science & Technology and Management of Lianyuan Steel, 05*, 40–42.

Liang, Zhenguang. & Tang, Renyuan. (2003). Fault simulation of power transformers using 3D finite element model coupled to electric circuit equations. *Proceedings of the Chinese Society for Electrical Engineering, 03*, 137–140.

Qiao, Mingzhong. & Zhang, Xiaofeng. (2004). Dynamic analysis of short circuit fault in stator winding of multi phase permanent magnet motor. *Transactions of China Electrotechnical Society, 04*, 17–22

Rangarajan, F., Franceschini G, Tassoni C. (1996). A simplified model of induction machine with stator shorted turns oriented to diagnostics. *International Conference on Electrical Machines, Vigo, Spain*, 410–413.

Sun, Yuguang., Wang. Xiangheng, Wang. Weijian, (2004). Transient calculation of stator' internal faults in synchronous generator using fem coupled with multi-loop method. *Proceedings of the Chinese Society for Electrical Engineering*, 136–141.

Tamer, P.J. Graydon B G, Ward D M. (1986). Monitoring generators and large motors. IEE *Proceedings B of Electric Power Applications, 133*(3), 169–180.

Wang, yanwu. Yang li, and Sun. fengrui. (2009). Simulation and analysis of 3D temperature field for stator winding short-winding short-circuit in asynchronous. *Proceedings of the Chinese Society for Electrical Engineering, 24*, 84–90.

Wang, Yudong. (1990). *Electrical machinery*. Hangzhou: Press of Zhejiang University.

Trutt, F.C. Sottile J, Kohler J L. 2001.Detection of AC machine winding deterioration using electrically excited vibrations.IEE *Transactions on Industry Application, 37*(1), 10–14.

Wang, Weijian (1999). Some theoretic and operating problems for electric main equipment protection. *Automation of Electric Power Systems, 23*(11), 1–5.

Electronic Engineering – Wang (ed)
© 2018 Taylor & Francis Group, London, ISBN 978-1-138-60260-1

Continuous max-flow medical image segmentation based on CUDA

B. Wang, Y.H. Wu & X. Liu
School of Automation, Harbin University of Science and Technology, Harbin, China

ABSTRACT: Medical image segmentation is an important part of medical image analysis and processing, which also continues to be a large challenge in this field. While continuous max-flow algorithm has been demonstrated a great potential for medical image segmentation, it has a large computational burden thus is not suitable for real-time processing. In this paper, a parallel accelerated method based on CUDA was proposed to implement multiplier-based continuous max-flow algorithm. Designing the scheme of thread allocation and kernel function design, the reduction algorithm was also used to optimize the data transfer between the host and device-side memory. The experimental results show that the parallel algorithm achieves a significant improvement of segmentation speed while preserving the qualitative results, when compared with the CPU serial algorithm. This method is viable and makes the fast medical image segmentation come true.

Keywords: Medical image segmentation; continuous max-flow; CUDA; parallel image processing

1 INTRODUCTION

Medical image segmentation is one of the most important parts of medical image analysis and processing, which also continues to be a large challenge in the field of medical image processing. While many algorithms have been proposed and applied to implement image segmentation, their computational cost is too large to satisfy the real-time processing requirement, which also limits their application. Therefore, it is a meaningful problem to solve the algorithm's computing speed to meet the requirements of real-time processing.

In recent years, the continuous max-flow method has been widely concerned and applied to image segmentation. The continuous max-flow algorithm based on graph theory was firstly proposed to minimize the energy function in computer vision (Greig et al. 1989). Meanwhile, the interactive image segmentation method based on the max-flow was proposed (Boykov & Jolly 2000). To solve the shortcoming of measure errors implied discrete max-flow methods which cannot achieve parallel processing, continuous max-flow methods have been widely concerned and presented many studies (Appleton & Talbot 2006; Bae et al. 2014; Chan et al. 2006; Kaba et al. 2015; Liu 2012; Strang 2008; Yuan et al. 2010a; Yuan et al. 2010b; Yuan et al. 2011). Parallel processing of the massive data in the CUDA environment can be amazingly accelerated. There are a lot of studies made a detailed comparison between this platform and the CPU environment (Chen & Hang 2008; Pieters et al. 2011; Si & Zheng 2010; Zhang et al. 2014).

In this paper, we use the NVIDIA's GPGPU model (Sanders & Kandrot 2010) and propose a parallel accelerated method based on CUDA to implement multiplier-based continuous max-flow algorithm. In taking this approach, there are two mainly purposes in our study: one is to ensure the segmentation effect of continuous max-flow model, another is to improve image processing speed in order to meet the real-time requirement.

2 CONTINUOUS MAX-FLOW MODEL

Firstly, an S-T network is defined in 2D or 3D space Ω, including source point s and sink point t. Suppose that point $x \in \Omega$, let $p(x)$ be the spatial flow through x, $p_s(x)$ be the source flow that from s to x, $p_s(x)$ be the sink flow from t to x, then the continuous max-flow model can be defined as follows:

$$\sup_{p, p_s, p_t} \left\{ P(p, p_s, p_t) = \int_\Omega p_s(x)dx \right\} \tag{1}$$

s.t.

$$\left| p(x) \right| \leq C(x) \tag{2}$$

$$p_s(x) \leq C_s(x) \tag{3}$$

$$p_t(x) \leq C_t(x) \tag{4}$$

$$div \quad p(x) - p_s(x) + p_t(x) = 0 \tag{5}$$

where $C(x)$, $C_s(x)$ and $C_t(x)$ are the limits of spatial flow, source flow and sink flow, respectively.

To simplify the continuous max-flow model, we introduced the Lagrange multiplier $\lambda(x)$ for Equation 5. Then the continuous max-flow model can be described as follows:

$$\sup_{p,p_s,p_t} \inf_{\lambda} \left\{ E(p, p_s, p_t; \lambda) = \int_\Omega p_s(x)dx + \int_\Omega \lambda(x)\,(div\ p - p_s - p_t)dx \right\} \quad (6)$$

After organization, Equation 1 can be described as follows:

$$\sup_{p,p_s,p_t} \inf_{\lambda} \left\{ E(p, p_s, p_t; \lambda) = \int_\Omega \{\lambda\ div\ p + (1-\lambda)p_s + \lambda p_t\}dx \right\} \quad (7)$$

The continuous max-flow problem is converted into the saddle point problem. The model could be gained by optimizing the flow variable p, p_s and p_t:

$$\min_{\lambda(x)\in[0,1]} \left\{ D(\lambda) = \int_\Omega \{(1-\lambda(x))C_s(x) + \lambda(x)C_t(x) + C(x)|\nabla\lambda(x)|\}dx \right\} \quad (8)$$

3 CONTINUOUS MAX-FLOW MEDICAL IMAGE SEGMENTATION BASED ON CUDA

3.1 Thread scheme allocation

It needs corresponding between medical image segmentation algorithm of maximum flow based on CUDA with original image and variable matrix by iteration, as well as threads in GPU. Each of the image elements was involved in one calculation from older to get a new variable matrix. For the fast iterative, at least one thread was applied to ensure the full use of physical resources in GPU. By using the corresponding relations between pixels and thread, the image would be divided into several thread block as integrals. Then each thread block was partitioned a mass of threads, while each pixel was corresponded with the only thread. The allocation schemes were changed with the uncertain scale of the image that would be split. It was the better operation effect with the 16 × 16 size of the thread which included 16 × 16 thread in each one.

3.2 Kernel function

The total parameters of equipment were set by host, including thread, thread blocks, the internal memory distribution, transmission in GPU and so on, all of which were prepared for the kernel function. CUDA was mapped vast threads parallel execution through computing task distributed by host when it invoked kernel. The size of thread was decided by the host application and system resources allocation. Figure 1 shows the flow diagram CUDA program.

In this paper, 11 kernels were designed for the side of devices. Per-function served for the different work

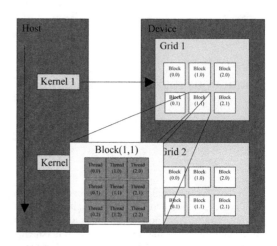

Figure 1. CUDA program flow diagram.

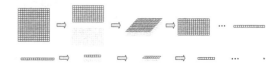

Figure 2. Flowchart diagram of reduction algorithm.

in the process of iteration, and was assigned different threads and memory space. The shared memory and texture memory were employed in the design of kernels to maximize the ascension of parallel computing ability. The common access memory was supplied by shared memory for the threads in the same thread block. The buffer shared memory is within the GPU, which can speed up the calculation. While race condition is the greatest problem in the shared memory, we used synchronization to avoid this issue. Texture memory within the chip as the RAM is similar to constant memory. It can reduce the requests for memory effectively and provide higher memory bandwidth to improve the computing speed in the case of frequently accessed.

3.3 The optimization of reduction algorithm

As mentioned above, the memory devices of host terminal and device terminal are relatively independent. Data required to operate on device terminal need the host terminal pass the data to the device terminals for operation. The data is frequently exchanged between host and device, with a large amount of data, resulting in a longer delay. Therefore, it adopts the reduction algorithm to optimize data transfer, which can effectively solve this problem. Figure 2 shows the flowchart diagram of reduction algorithm.

The calculation error used reduction method and divided into two steps. The first step was the internal thread blocks reduction. The same size threads and shared memory were allocated within each thread block, which made the thread sequentially correspond with shared memory. The shared memory was divided into two parts (i.e. the upper and lower part). The elements of each memory unit in upper part have its corresponding lower memory unit. Half size of threads which use application makes its corresponding position to be accumulated. The size of original shared memory can be reduced to half. From that time, the rest of the shared memory is divided into two parts for reduction, and the shared memory can be reduced by half again. Eventually, the shared memory which passes through the column direction reduction will be reduced to one line. This shared memory is passed through the row direction for reduction, the method is similar to the column direction reduction, and the shared memory in the thread block will eventually be reduced about a memory unit. The second step was the reduction between threads blocks. The reduction results of different thread blocks would be rearranged, which were assigned to the shared memory, then passing through the reduction of the line and column direction to get the final result. After the final result was passed to the host terminal, it went on the follow-up processes after judgment and other operations.

4 EXPERIMENT AND RESULT

4.1 Experimental setup

All of the dataset used in our experiments was from hospital that contains 153 kidney images, resolution is 512 × 512. The segmentation core functions are using the C language and CUDA C to compile. The experiments were implemented in the CPU of Intel Core I5 4200M that internal memory is 8 GB. The type of GPU is NVIDIA GeForce GT 755M, and its memory is 2 GB. Software environment: the operating system is Windows 8.1, MATLAB 2013a, Microsoft Visual Studio 2010, and CUDA V6.0.

4.2 Experimental results and analysis

Figure 3 shows the experimental results of a 5 × 5 image. Figure 3(a) is the original image, Figure 3(b) is the segmentation result which is used to prepare the partition function using C language, and Figure 3(c) is the segmentation result which is improved by using CUDA architecture.

From Figure 3, we can see that the segmentation algorithm based on continuous max-flow can better identify the details of the kidney image, less being susceptible to be interfered by the small target area. Compared to the experimental results of two kinds of core segmentation functions, there is no difference between the two segmentation images, and it won't cause a loss of precision CUDA to accelerate the image segmentation.

In this paper, we used the concept of acceleration rate (α) to describe the acceleration degree after and before optimization between algorithms.

$$\alpha = \frac{t_{after}}{t_{before}} \qquad (9)$$

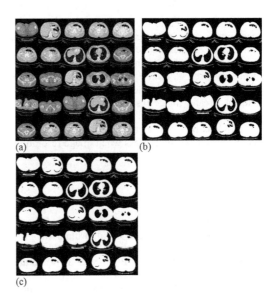

Figure 3. Segmentation effect diagram.

Table 1. Experimental result which uses the acceleration rate to describe the acceleration degree of algorithm.

IR*	AR1	AR2	AR3	SD1	SD2
256 × 256	10.22	15.18	1.45	5.65	5.06
512 × 512	11.53	14.20	1.24	4.95	6.19
1024 × 1024	11.56	15.42	1.34	3.10	4.01
2048 × 2048	8.64	11.22	1.30	1.72	2.23

* Abbreviated words: *IR* image resolution; *AR*1 acceleration rate α_1; *AR*2 acceleration rate α_2; *AR*3 acceleration rate α_3; *SD*1 standard deviation of acceleration rate σ_1; *SD*2 standard deviation of acceleration rate σ_2.

where α represents the acceleration rate, the running time of algorithm after optimization, the running time of algorithm before optimization in which the time unit is seconds (s). Table 1 is the experimental result which uses the acceleration rate to describe the acceleration degree of algorithm in this paper.

In Table 1, α_1 is the acceleration rate between algorithms optimized by the GPU and non-optimized, α_2 is the acceleration rate between the algorithms without optimization and optimized by GPU and reduction method, and α_3 is the acceleration rate between the algorithm optimized by GPU and reduction method. σ_1 and σ_2 are the standard deviations of the acceleration rate α_1 and α_2, respectively.

From Table 1, we can see that it has a good acceleration effect in four different resolutions through the GPU optimization algorithm, the average acceleration rate has achieved 8.6 or higher. The maximum and minimum acceleration rate was 26.02857 and 3.75, respectively. Then the acceleration capability have further been improved through the reduction optimization algorithm, which has more than 23.5% algorithm acceleration compared to the previous average algorithm, and the average value has reached more than 11.2 compared to the non-optimized algorithm acceleration rate. In the experiment, the maximum acceleration rate has reached 39.26, and the minimum is also 5, which fully demonstrates the superiority of the proposed algorithm in time. Compared to the standard deviation of acceleration rate, we can obtain that, the greater image resolution, the smaller standard deviation value, that is to say that our algorithm not only has better acceleration effect, but also has better acceleration robustness when image resolution increases.

5 CONCLUSIONS

In this paper, a paralleled algorithm improvement of the medical image segmentation algorithm was proposed and implemented in the CUDA environment which is the NVIDIA's GPGPU model. To solve the transfer time-consuming problem between the host memory and the device-side memory, the reduction idea was used to improve the efficiency of continuous max-flow medical image segmentation algorithm. Experimental results show that the continuous max-flow algorithm based on CUDA could be well used in medical image segmentation. Meanwhile, this method can be executed on the general PC devices with little of money investment under the premise of no affecting the image processing effect, which provides a theoretical reference for quick medical image processing technology.

ACKNOWLEDGEMENTS

This work is supported by the National Natural Foundation of China, under Grant No. 61672197; the University Nursing Program for Young Scholars with Creative Talents in Heilongjiang Province, under Grant No.UNPYSCT-2015045; the Natural Science Foundation of Heilongjiang Province of China under Grant No. F201311; the Foundation of Heilongjiang Educational Committee under Grant No. 12531119.

REFERENCES

Appleton, B. & Talbot, H. (2006). Globally minimal surfaces by continuous maximal flows. *IEEE Transactions on Pattern Analysis and Machine Intelligence* 28(1): 106–118.

Bae, E., Yuan, J. & Tai, X.C., et al. (2014). A fast continuous max-flow approach to non-convex multi-labeling problems (eds). *Efficient Algorithms for Global Optimization Methods in Computer Vision*: 134–154.

Boykov, Y. & Jolly, M. (2000). Interactive organ segmentation using graph cuts. *In: Proc. of 3rd International Conference on Medical Image Computing and Computer-Assisted Intervention, Pittsburgh, PA, USA: Springer*: 276–286.

Chan, T., Esedoglu, S. & Nikolova, M. (2006). Algorithms for finding global minimizers of image segmentation and denoising models. *SIAM Journal on Applied Mathematics* 66(5): 1632–1648.

Chen, W.N. & Hang, H.M. (2008). H.264/AVC motion estimation implementation on Compute Unified Device Architecture (CUDA). *In: Proc. of 2008 IEEE International Conference on Multimedia and Expo*: 697–700.

Greig, D., Porteous, B. & Sheult, A. (1989). Exact maximum a posteriori estimation for binary images. *Journal of the Royal Statistical Society, Series B*, 51(2): 271–279.

Kaba, D., Wang, Y. & Wang, C., et al. (2015). Retina layer segmentation using kernel graph cuts and continuous max-flow. *Optics Express*, 23(6): 7366–7384.

Liu, S.T. (2012). The basic principle and its new advances of image segmentation methods based on graph cuts. *Acta Automatica Sinica*, 38(6): 911–922.

Pieters, B., Hollemeersch, C.F.J. & De Cock J., et al. (2011). Parallel DE blocking filtering in MPEG-4 AVC/H.264 on massively parallel architectures. *IEEE Transactions on Circuits & Systems for Video Technology*, 21(1): 96–100.

Sanders, J. & Kandrot, E. (2010). CUDA by example: an introduction to general-purpose GPU programming. *Addison-Wesley Professional*: 387–415.

Si, X. & Zheng, H. (2010). High performance remote sensing image processing using CUDA, *In: Proc. of 2010 Third International Symposium on Electronic Commerce and Security*: 121–125.

Strang, G. (2008). Maximum flows and minimum cuts in the plane. *Advances in Mechanics and Mathematics* 3(1): 1–11.

Yuan, J., Bae, E. & Tai, X.C., et al. (2010a). A continuous max-flow approach to potts model (eds). *Computer Vision – ECCV* 2010: 379–392.

Yuan, J., Bae, E. & Tai, X.C. (2010b). A study on continuous max-flow and min-cut approaches. *IEEE Conference on Computer Vision & Pattern Recognition, IEEE Computer Society*: 2217–2224.

Yuan, J., Bae, E. & Boykov, Y., et al. (2011). A continuous max-flow approach to minimal partitions with label cost prior. *Scale Space & Variational Methods in Computer Vision*: 279–290.

Zheng, J.W., An, X.H.& Huang M.S. (2014). CUDA-based PCG algorithm optimization for a large sparse matrix. *Journal of Tsinghua University (Science and Technology)* 54(08): 1006–1012.

Electronic Engineering – Wang (ed)
© 2018 Taylor & Francis Group, London, ISBN 978-1-138-60260-1

Vulnerabilities detection in open-source software based on model checking

Y. Li, S.B. Huang, X.X. Wang, Y.M. Li & R.H. Chi
College of Computer Science and Technology, Harbin Engineering University, Harbin, China

ABSTRACT: Open-source software is known as being unsafe due to some vulnerabilities. The investigation of vulnerabilities detection in open-source software has attracted interest in the software detection area. An automatic detection method for vulnerabilities in open-source software has been developed, which expands the detectable vulnerabilities set and improves detection efficiency. The method is based on conditional model checking. The formal description and algorithm flow of configurable program analysis on a CPA checker are proposed. The experimentation indicates that the algorithm is correct and efficient, using an open-source web server, Tomcat.

Keywords: open-source software; model checking; Tomcat; vulnerabilities detection

1 INTRODUCTION

Recently, with the development of open-source software (DiBona et al., 1999), open-source software begins to use wildly. However, the security of open-source software could not appear grimmer (Xia, 2013). Researchers have proposed many effective methods to improve security, such as using model checking tools for source code vulnerability detection (Goldwasser et al., 1989).

A new model checking method – conditional model checking – has been proposed (Beyer, 2011). Research has shown that conditional model checking is very suitable for the detection of security vulnerabilities in open-source software. In this paper, a method of vulnerabilities detection in open-source software based on conditional model checking has been developed. The modules of Catalina and Coyote in open-source software Tomcat were chosen for further checking. The experimental results demonstrate that the vulnerabilities detection method in this paper is correct and efficient.

2 PRELIMINARIES

2.1 Conditional model checking

The principle of model checking (Baier & Katoen, 2008) has been widely researched and applied both in the academic circle and industry. It has major significance in hardware detection, protocol verification and so on. With the development of the principle of model checking, a variety of model checkers were produced (Cavada et al., 2014; Holzmann, 2015; Legay & Viswanathan, 2015), such as the conditional model checker.

Similarly, the process of conditional model checking consists of three stages: the modelling phase; running phase; and analysis phase. However, unlike traditional model checking, the conditional model checker returns a special condition (usually a state predicate). In this condition, the tested program satisfies the property specification. That is, if $\Phi = true$, it means that the model checker completed the testing procedures in the absence of additional conditions; if the detected program does not meet the specifications, the model checker does not return the conditions $\Phi = false$, but points out which parts of the program do not contain errors and which parts contain errors. The counterexamples and all the states that have passed validation are returned. If the model checker fails, the condition Φ summarises the testing work before a storage space runs out of space, the time exhausted and the detection tools cannot be detected. For example, there is an effective program to be detected containing two branches, one of which can be verified by the model checker, and another branch will lead to an infinite loop. Making use of the traditional model checking method, any feedback cannot be got from the detection process. The conditional model checker certainly cannot prove the correctness of the program; however, it will give the conclusion that the first branch has been successfully detected. If you have to complete verification of the program, you can use a different model checking method to continue the remaining validation of the program.

2.2 Tomcat

Tomcat is the core component of the Apache project and has a large number of Java users with its leading technology and stable performance. With the increase in the number of users, the case of Tomcat vulnerability invasion gradually increased. Hackers have broken into many websites using the Tomcat server. In the Seebug vulnerability database, there are more than 120 security vulnerabilities regarding Tomcat. Many of these vulnerabilities are very dangerous. Researchers believe that the number of vulnerabilities in the Tomcat project is by no means all. Thus, detecting the security flaws and vulnerabilities in Tomcat is important and challenging.

Testers, however, usually use the penetration method to detect vulnerabilities and defects in Tomcat. Using this method, the testers not only need to rely on testing tools, but also need to understand the module function to be detected. Different from the penetration method, the method using model checking in this paper only needs to rely on automated testing tools, the testing personnel do not need to have a deep understanding of Tomcat and can conduct a comprehensive test on the Tomcat system.

3 VULNERABILITIES DETECTION IN TOMCAT BASED ON THE CPACHECKER

3.1 Modelling

Firstly, the program should be modelled and represented by control flow automaton (CFA). A control flow automaton (CFA) $A = (L, G)$ consists of a finite set of locations $L = \{0, \ldots, n\}$, modelling the program counter of a corresponding sequential code, and edges in $G \subseteq L \times Ops \times L$ labelled with quantifier-free first-order formulas over the set Var of program variables and their next-state primed forms, Var' (Lange et al., 2015). Ops represents all operations in the program. The assembly of the program variables that appear in the operation is represented by X. A specific state c is represented as a variable declaration: $c : X \cup \{pc\} \rightarrow Z$, which assigns an integer value to each variable. All concrete states in the program are represented by C, and $r \subseteq C$ is called a region. $g \in G$ is a transition formula $\rightarrow \subseteq C \times \{g\} \times C$. $c_n \in \text{Reach}(r)$ represents c_n is reachable from the region r.

The module Catalina was chosen for further testing. Catalina is responsible for assembling various containers and components, and does some initialisation work for the server which is used to start Tomcat. In Figure 1, the code snippet is excerpted from Catalina, and Figure 2 is its CFA model.

3.2 Setting the specification

After modelling, the next step is modelling the specification. In the CPAchecker (Beyer et al., 2011), there are 14 standardised documents defined, including error location, error marking and so on. These specifications were described using an automated language description similar to BLAST query language, and were saved in the files with the .Spc format.

In this paper, we use *default.spc* the standard default model of CPAchecker. In this standard model, the

```
791    LoginConfig config = context.getLoginConfig();
792    if ((config != null) &&
793        (Constants.FORM_METHOD.equals(config.getAuthMethod())))) {
794        String requestURI = request.getRequestPathMB().toString();
795        String loginPage = config.getLoginPage();
796        if (loginPage.equals(requestURI)) {
797            if (log.isDebugEnabled())
798                log.debug(" Allow access to login page " + loginPage);
799            return (true);
800        }
801        String errorPage = config.getErrorPage();
802        if (errorPage.equals(requestURI)) {
803            if (log.isDebugEnabled())
804                log.debug(" Allow access to error page " + errorPage);
805            return (true);
806        }
807        if (requestURI.endsWith(Constants.FORM_ACTION)) {
808            if (log.isDebugEnabled())
809                log.debug(" Allow access to username/password submission");
810            return (true);
811        }
812    }
```

Figure 1.   The code snippet excerpted from Catalina.

```
UsedConfiguration.properties ✖    *3546-1.c ✖    3546.c ✖
int m_config = 0;int m_requestURI;int m_loginPage;int m_errorPage;
void transmit1(void)
{       m_config = 1;}
void transmit2(void)
{       m_config = 2;}
void error(void)
{
        ERROR: ;
        goto ERROR;
        return;
}
int start_simulation(void)
{
        if(m_loginPage == 0)
                transmit1();

        if(m_errorPage == 0)
                transmit2();

        if(m_requestURI == 1)
                error();
}
void init_model(void)
{
        {
        m_loginPage = 0;
        m_errorPage = 0;

        m_requestURI = 1;

        return;
        }
}
int main(void)
{ int _result ;
        {
        {
        init_model();
        start_simulation();
        }
        _result = 0;
        return (_result);
}}
```

Figure 2.   The modelling result in CFA.

CPAchecker model checker will search the program model statements with the 'ERROR' label to check whether they are reachable. If any of them are reachable, it means there are loopholes in the program, and 'UNSAFE' will be returned.

4 RUNNING AND EXPERIMENTAL RESULT

In this section, we experimentally demonstrate the effectiveness of the method described in Section 3. The experiments were performed on a dual-core of Intel Core i5-4210 machine with 4GB RAM and

Figure 3. The running result in the Eclipse console.

Ubuntu12.04 32-bit operating system. The version of the conditional model checker is CPAchecker 1.2 executed on Eclipse.

4.1 Running the model checker

In CPAchecker, there are of four methods of implementation, CPAline, ExplicitAnalysis, PredicateAnalysis and PredicateAnalysisWithUFs. CPAline detects the pointers and dynamic data structure of the C language program. ExplicitAnalysis is used to perform an explicit value analysis of model checking. PredicateAnalysis is used to perform model checking of predicate analysis. The PredicateAnalysisWithUFs is used to perform predicate analysis with UF. In this paper, ExplicitAnalysis is chosen to verify the specification using running the manifest-state model checker.

There is a dialogue box to select the validation program and find out the pre-compiled generated CIL files. Considering that the program to be verified in this experiment is very small, the default configuration of the CPAchecker is adopted. The default configuration includes: using the call stack, using the reverse postorder traversal, adding bound and refined in the algorithm, a combination of CPA and LocationCPA by ExplicitCPA CPU and the time limit is 900 s. The CPAchecker running feedback is clearly shown in the Eclipse console, as in Figure 3.

In Figure 3, various information of conditional model checking in the implementation process, including system resource limits, Code detection states and whether there is error path. The last two lines of the output of the CPAchecker test results show that the checking has finished and vulnerability has been found. A detailed directory of the test results output file is given to help the user yield more useful information. In the directory, it can be found that 27 files were generated which cover a variety of data types. For example, there were eight files relevant to CFA, including error program CFA, the whole CFA of programs to be detected, JSON CFA and so on.

Figure 4 is the error path graph generated by this conditional model. The error path diagram gives the part of the program's error, describes how the program enters the error and shows the error conditions. Figure 4 shows that it will encounter an error when the program performs at the sixty-second step. We can conclude that if the value of requestURI is 1, it will enter the error state.

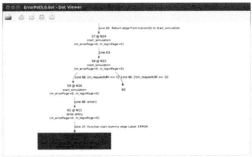

Figure 4. The error path graph.

Table 1. Experimental result of the UNSAFE program.

Program number	LOC Source program	Modelled program	Nums of states	Results
1	25	40	26	known
2	47	64	53	unknown
3	19	51	62	known
4	22	41	12	known
5	22	39	14	false
6	37	62	36	false

Table 2. The serial number of the program files.

Program number	Program name
1	FormAuthenticator.java
2	ChunkedInputFilter.java
3	RealmBase.java
4	InternalNioInputBuffer.java
5	LifecycleMBeanBase.java
6	AsyncStateMachine.java

4.2 Experimental result

In this paper, the core code of Tomcat in two modules was detected, and there were six programs with the detection results UNSAFE, of which two were false alarms, three proved to be known vulnerabilities, and one is an unpublished vulnerability. Table 1 lists some statistics on the test results for the UNSAFE program.

In Table 1, the first column is the file name of the program with errors. The second and third columns give the number of lines of code in the original program and the modelled program, respectively. The next column is the number of states in ARG generated after model checking. The last column gives the type of vulnerability discovered in the experimental process. Table 2 gives the full names of the UNSAFE programs.

5 CONCLUSIONS

In this paper, the vulnerability detection method in open-source software based on model checking theory has been proposed. The method is based on conditional model checking combining two or more model checking algorithms together in serial form. The results of conditional model checking on Tomcat show that, besides the vulnerabilities that other detection methods can detect, the conditional model checking can discover vulnerabilities that other methods cannot find.

REFERENCES

Baier, C. & Katoen, J.P. (2008). *Principles of model checking*. London: MIT Press.

Beyer, D., Henzinger, T.A., Keremoglu, M.E. & Wendler, P. (2011). Conditional model checking. *Computer Science*.

Beyer, D. & Keremoglu, M.E. (2011). CPACHECKER: A tool for configurable software verification. *International Conference on Computer Aided Verification*, 184–190.

Cavada, R., Cimatti, A. & Dorigatti, M. et al. (2014). The nuXmv Symbolic Model Checker. *International Conference on Computer Aided Verification*, 334–342.

DiBona, C., Ockman, S. & Stone, M. (1999). *Open sources: Voices from the open source revolution*. Sebastopol: O'Reilly Media, Inc.

Goldwasser, S., Micali, S. & Rackoff, C. (1989). The knowledge complexity of interactive proof systems. *Journal on Computing, 18*(1), 186–208.

Holzmann, G. (2015). The spin model checker. *IEEE Transactions on Software Engineering*, 124–128.

Lange, T., Neuhauber, M.R. & Noll, T. (2015). IC3 software model checking on control flow automata.

Legay, A. & Viswanathan, M. (2015). Statistical model checking: challenges and perspectives. *International Journal on Software Tools for Technology Transfer, 17*(4), 369–376.

Xia, Q. (2013). Initial analysis of open source software in network and information security system. *Computer Applications and Software, 30*(1), 325–327.

Design and realisation of an intelligent agricultural greenhouse control system

J.F. Liu & Y.X. Wu
Department of Mechanical Engineering, Harbin University of Science and Technology, Rongcheng, Shandong Province, China

H.L. Zhu & C. Liu
Agronomy College of Heilongjiang Bayi Agricultural University, Daqing, Heilongjiang Province, China

ABSTRACT: This paper introduces the overall structure of a greenhouse and the characteristics of a greenhouse control object. It determines the design goals for an intelligent agricultural greenhouse control system, which gathers data through a greenhouse information collection system, and then controls actuators to adjust the environmental factors inside the greenhouse to provide the most suitable living environment for crops, in order to achieve the goals of high quality and high yield. The overall structure of an intelligent greenhouse control system is determined, and the technical skills involved in the process, the overall design scheme of the hardware and software systems, and the system's control scheme are described.

Keywords: agricultural greenhouse; intelligent control system; system hardware; software design

1 GREENHOUSE STRUCTURE

Greenhouses are a form of protection to create crop growth and a development environment under artificial control. Modern agricultural greenhouses can generally be divided into two categories: glass greenhouses and plastic greenhouses. Glass greenhouses have a metal or wooden frame, covered with glass. Plastic greenhouses are covered with plastic film or plastic plates (including reinforced plastic board, glass steel, etc.) (Yang et al., 2006). Glass greenhouses represent the universal application of the modern greenhouse, so the intelligent agricultural greenhouse control system is mainly designed for the well-known Venlo glass greenhouse.

The specific Venlo greenhouse structure is shown in Figure 1: single-building greenhouse; 9.6 m north–south width for three-ridge greenhouse glass roof; east–west length 40.5 m; area 388.8 m^2; shoulder height 5 m; top 5.8 m; 6.3 m high shading; greenhouse frame is a light steel structure; in addition to the windows, the whole room is covered by the glass. At the top are ventilation windows, which are in the same area on both sides of the ridges; the size of the skylight is 2.4 m × 1.1 m × 9 m; the roof ventilation windows are alternately arranged (Li & Shi, 2011). Greenhouse light is uniform and bright.

The actuator is equipped with a variety of devices to regulate the environment in the greenhouse, including visor curtain, skylight, sprinkler irrigation system, circulation fan, heater, lights and CO_2 supply

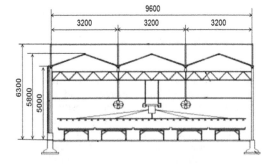

Figure 1. The structure of the greenhouse.

device. Thus, the control components the system needs are: external shading screen motor, sunroof motor, hydraulic pump motor, circulating fan, heater, lights and CO_2 supply device. The visor curtain and sunroof motors are equipped with a limit switch, which will control the forward, reverse and stopping movement to complete the corresponding mechanism of opening and closing. The hydraulic pump motor, circulating fan, heater, lights and CO_2 supply device are switch devices, directly starting and stopping the associated mechanisms. The controls of these devices are concentrated in the central console; an embedded system.

We design a reasonable placement for the sensors, as shown in Figure 2.

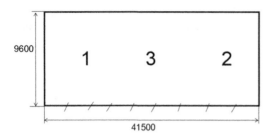

Figure 2. Sensor placements.

2 OVERALL PLAN

2.1 *Design goals and functional requirements*

The environmental parameters of the intelligent greenhouse control system are gathered through the greenhouse information collection system. According to the control scheme, the visor curtain, greenhouse skylight, hydraulic pump, circulation fan, heater, lights and CO_2 supply device are controlled as the executing agency, the environmental factors inside the greenhouse are adjusted, the most suitable survival environment for crops is provided, and the goal of high yield and high-quality crops is achieved.

According to the system design goals, the functions of the greenhouse intelligent control system are as follows:

(1) collecting environmental information in greenhouse, it can control the environmental parameters manually/automatically;
(2) controlling the actuator devices of the visor curtain, sunroof, hydraulic pump, circulation fan, heater, lights and CO_2 supply;
(3) through a touchscreen, setting the upper and lower limits of air temperature and humidity, CO_2 concentration and light intensity; through the touchscreen, monitoring of greenhouse environmental factors, showing the data statistics and trends of environmental parameters, and providing a real-time alarm; manually controlling the starting and stopping of the executive mechanism in the greenhouse through a manual switch on the touchscreen.

2.2 *System structure and technical means*

The hardware core of the intelligent agricultural greenhouse control system is the greenhouse information collection system, with an embedded STM32F103ZET6 microcontroller as the core of the actuator system. Other hardware includes a touchscreen, computer monitor, DL-20 wireless serial module and power supply.

It is necessary to design the software and hardware of the system according to its entire structure. The overall hardware system design ideas are as follows. According to the system signal flow diagram, system hardware will be divided into greenhouse information acquisition system, actuator system, and embedded system using a STM32F103ZET6 microcontroller as the core. The hardware design of the embedded system and wiring design have to be designed and form the core of the research. For the embedded system, 16 digital inputs for the limit switches and the manual switch will be used; ten digital outputs control the corresponding relay system, then the corresponding actuator control interface; eight input-simulating interfaces for greenhouse sensors; two RS232 serial communication interfaces for communication with the touchscreen and the computer monitor.

The overall software system design idea is as follows: the system software module is divided into an embedded system software module and a touchscreen configuration software module. Finally, the whole intelligent greenhouse control system operation will be tested.

By analysing the design of the overall structure of the system, the technology expected to be used can be defined in terms of: hardware design of greenhouse information acquisition system; hardware design of embedded system; hardware design of the executive system; hardware wiring design of the system; software development of embedded system; touchscreen configuration software.

3 HARDWARE DESIGN AND IMPLEMENTATION OF THE SYSTEM

3.1 *Embedded microcontroller*

3.1.1 *STM32 microcontroller*

In terms of the hardware design of the intelligent greenhouse controller, as demand increases for the function, control and program algorithm to be larger and more complex, the storage and data processing capability of common 8-bit microcontrollers cannot meet the requirements of the new greenhouse control systems. Currently, 16-bit microcontrollers are commonly used, and the price is almost the same as 32-bit microcontrollers. Thus 32-bit microcontrollers with better performance offer more advantages. In addition, Inter-Process Communication (IPC) control system wiring is complex and difficult to maintain, while Programmable Logic Controllers (PLCs) are expensive. In our research, we selected a STM32F103ZET6 32-bit embedded microprocessor based on Cortex-M3, licensed from Arm Holdings, as the core. STM32-series microcontrollers have more advantages than the traditional ARM7 microcontroller, mainly in the following areas (Yu et al., 2014):

(1) advanced Cortex-M3 kernel – using the Harvard structure, the Thumb-2 instruction set is built with a fast-interrupt controller;
(2) superior power consumption efficiency – the system optimises three energy consumption requirements;
(3) internal height integration – embedded power supply monitor in the system can reduce the demand for external devices; a master oscillator can drive

the whole system; system also has USB, USART, SPI, GPIO, timer, Analogue-to-Digital Conversion (ADC) channels, and other innovative and outstanding peripherals;

(4) flexible static-storage controller – the operating speed of the external memory is fast; the system supports various memories; it has programmable timing; it can execute code from external memory.

3.1.2 *STM32F103ZET6 chip*

The STM32-series chip internal resources are as follows:

(1) kernel: a high-performance 32-bit ARM Cortex-M3 core processor, CPU frequency of 72 MHz, operating voltage of 2.0–3.6 V.
(2) I/O port: 144 pins, 112 GPIO ports.
(3) timer: four general-purpose timers, two advanced control timers, and two basic timers.
(4) communication interface: three SPI, two I2C, five USART/UART, one USB, one CAN bus.
(5) ADC: three 12-bit ADC modules; DAC: two 12-bit converters.

3.1.3 *STM32 minimum system circuit*

The STM32 microprocessor must provide the peripheral circuit, which can work independently, including: 3.3 V power conversion circuit, 8 MHz clock, reset circuit, digital and analogue conversion between decoupling circuit, I/O interface, and serial communication interface circuit (Lu et al., 2013).

3.2 *Greenhouse information collection system*

The greenhouse information acquisition system is the key to the collection of information by the intelligent greenhouse control system. The real-time and accurate detection of environmental parameters in the greenhouse provides data sources and the basis for the effective control of those environmental parameters. The greenhouse information acquisition system includes an air temperature and humidity acquisition system, a soil moisture collection system, a CO_2 acquisition system, and an illumination acquisition system.

The STM32F103ZET6 chip can accept signals in the range 2–3.6 V; therefore, it is necessary to design the circuit of the sensor accordingly. The numerical value of agricultural greenhouse sensors is output in the form of 4–20 mA current or 0–5 V voltage. After a circuit of current to voltage transfer, or a circuit of voltage division, the current or voltage signal in the greenhouse sensor is converted into a voltage signal that is acceptable to the microcontroller pin.

3.2.1 *Air temperature and humidity collection system*

The air temperature and humidity sensor uses a JWSL-2 series wall temperature and humidity transmitter (Beijing Kunlun Coast Sensing Technology Co. Ltd). This type of sensor is an integrated design for sensing and transmitting. The sensor is supplied with a 24 V DC voltage source; the three output types are two-wire current output, voltage output, and network output. Output type selection is two-wire current output in the system; the temperature range is 0–50°C, humidity range is 0–100% RH; at 25°C the accuracy is ±0.5°C, the humidity is ±3% RH; the working environment temperature is −10–60°C, the humidity is 0–100% RH; the output signal is the current output type, 4–20 mA (Du et al., 2004).

3.2.2 *Soil moisture collection system*

The soil moisture sensor used is from Congtai CHUANGMEI Equipment Co. Ltd., Handan city. This type of sensor is an integration of sensing and transmitting design. The output form of the sensor is divided into voltage of 0–5 V, current of 4–20 mA, and RS232 or RS485 network. The current soil moisture sensor is in the system; sensor working voltage is 12–24 V DC voltage; unit measurement is the soil moisture volume, the measuring range is 0–100% (m^3/m^3), the accuracy is 0–50% (m^3/m^3); the range is ±2% (m^3/m^3), resolution is 0.1% (m^3/m^3); measurement area for cylinder diameter is 3 cm and length is 6 cm; standard cable length is 5 m and can also be customised according to actual needs (Li et al., 2016).

Two soil moisture sensors are required in this system. The first soil moisture sensor is in position 1 in the greenhouse (see Figure 2) and the signal terminal of the embedded hardware circuit board is port AI4; the second soil moisture sensor is in position 2 in the greenhouse and the signal terminal of the embedded hardware circuit board is port AI5. Cable length needs customisation; based on the standard of greenhouse for which this system is designed, cable length is about 15 m.

3.2.3 *CO_2 acquisition system*

The CO_2 sensor is a DCO2-T8-series greenhouse CO_2 sensor from Beijing Di Hui Technology Co. Ltd.; the sensor has a variable integration design. The sensor uses a 24 V DC voltage source and output type sensor for current output; the output signal is 4–20 mA; range is 0–2000 ppm; accuracy is ±30 ppm + 3% FS; the temperature of working environment is in the range 0–50°C, working humidity is in the range 0–95% RH.

The CO_2 sensor is located at position 3 of the greenhouse (see Figure 2) where the CO_2 concentration signal is passed to the ADC-IN6 (AI6) port of the embedded chip.

3.2.4 *Illumination acquisition system*

The light intensity sensor is a DZD-T5 light transmitter, also from Beijing Di Hui Technology Co. Ltd., with a variable integration design. The sensor has a 24 V DC voltage source; the measuring range is 0–2000 Lux; the measurement precision is less than or equal to ±5% FS; the measurement resolution is 2 Lux; the working environment temperature range is −20–85°C; the output type is divided into the current output, the range is 4–20 mA.

An illumination sensor is located in the greenhouse at position 3 (see Figure 2), and the light intensity

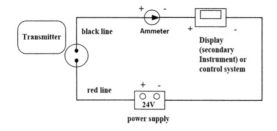

Figure 3. Power supply circuit design.

signal is directed to the ADC-IN7 (AI7) port of the embedded chip. Installation is the wall type.

3.3 Power supply circuit design

The power circuit design of this system is divided into 24 V turns 5 V and 5 V turn 3.3 V. The 3.3 V power is used for the STM32F103ZE microcontroller minimum system, communication circuit and JTAG module; 5 V power is used to improve the stability of the PS2815 interface circuit in the digital input (DI) module and provides a protective effect for the ULN2803 control circuit in the digital output (DO) module. We chose a 24 V external power supply for the greenhouse information acquisition system, embedded system power supply, output control circuit electromagnetic relay power supply and protection. 24 V turns 5 V power circuit and 5 V turns 3.3 V power circuit supply power to the whole embedded system of each module.

4 CONCLUSION

This paper analyses the functional requirements of a temperature control system, and allocates the embedded system I/O ports. The design of the hardware circuit of the embedded system involves: STM32 minimum system circuit design, input and output interface circuit design, control circuit design, power circuit design, the connection of embedded system layout on the executive mechanism, circuit design and wiring.

REFERENCES

Du, S., Li, Y., Ma, C., Chen, Q. & Yang, W. (2004). Current situation on greenhouse environment control system modes in China. *Transactions of the Chinese Society of Agricultural Engineering*, 20(1), 7–12.

Li, L.L. & Shi, W. (2011). Design and realization of intelligent temperature and humidity control system in greenhouse. *Hunan Agricultural Sciences*.

Li, Y.D., Miao, T., Zhu, C. & Ji, J.W. (2016). Design and implementation of intelligent monitoring and control system for solar greenhouse in North China. *Journal of China Agricultural Science and Technology Herald*.

Lu, L.J., Qin, L.L., Shi, C. & Wu, G. (2013). Design and implementation of modern greenhouse remote monitoring system based on web. *Control Conference*, 40, 290–304.

Yang, W.Z., Wang, Y.M. & Hai-Jian, L.I. (2006). Design and implementation of intelligent controller in distributed greenhouse intelligent control system. *Journal of Agricultural Mechanization Research*.

Yu, H., Liu, J., Ma, L., Wu, S. & University, J.A. (2014). Design and implementation of remote-intelligent control system of internet of things for facility agriculture based on web. *Journal of Chinese Agricultural Mechanization*.

Electronic Engineering – Wang (ed)
© 2018 Taylor & Francis Group, London, ISBN 978-1-138-60260-1

Research on intuitionistic fuzzy measure in decision making

S. Zhang, Z.C. Huang & G.F. Kang
School of Informatics and Technology, Heilongjiang University, Harbin, PR China

ABSTRACT: With the growing complication and uncertainty of the socioeconomic situation, the decision-makers often take different degrees of hesitation or show a certain degree of lack of knowledge in dealing with decision-making problems. Intuitionistic fuzzy sets can be increased by adding a non-membership parameter with good description of the decision-making process hesitation information. Compared with the fuzzy sets (FS), the intuitionistic fuzzy sets (IFS) could be described the fuzzy nature of the independent world much better, so it has much flexibility as well as practicability in dealing with fuzziness and uncertainty, and has been widely used in decision-making. Although the research of intuitionistic fuzzy (IF) multiple attribute decision making has made some achievements, there are still many problems that need to be improved and further studied. In this paper, with improving the existing fuzzy entropy measure, this paper proposes similarity measures for intuitionistic FS and interval intuitionistic FS, improves the existing intuitionistic fuzzy sorting functions, and applies it to the intuitionistic fuzzy multiple attribute taking into account the decision maker's behavioral factors decision-making applications. For the multiple attributes decision-making problem, the weights of the attributes were completely unknown and partly unknown. The weights are determined by the entropy weight method and the minimization of the optimization model to solve the optimal weight, and then the ratio of the intuitionistic fuzzy multiple attribute decision. Application examples illustrate the effectiveness and feasibility of the method.

Keywords: Multiple attributes decision-making; intuitionistic fuzzy sets; intuitionistic trapezoidal fuzzy numbers; foreground theory

1 INTRODUCTION

In information theory, entropy represents a measure of average uncertainty. Zadeh first introduced the concept of operator to the fuzzy event in 1986. After the research of DeLuca, Zadeh developed fuzzy operator definition and constructed a class of fuzzy entropy (Dalal et al. 2005). Fuzzy operator is a very important information degree to refer to the fuzzy degree of FS, then has been widely applied to pattern recognition, image segmentation and clustering analysis. IF set contains intuitionistic information, so the fuzzy degree of IF set needs to consider two aspects (Haubl et al. 2002): uncertainty degree and unknown degree, in which the unknown degree is expressed by hesitation degree.

In order to depict the fuzzy degree of the intuitionistic FS, the theory of IF gambit is introduced into the IF set, furthermore a kind of IF is constructed, and the fuzzy set is characterized by fuzzy degree. Intuitionistic FS contain hesitant information, so IF entropy to describe intuitionistic FS of fuzzy degree need to consider two aspects of the factors uncertainty and unknown degree, in which the degree of uncertainty expressed by the degree of hesitation. Since then, the construction of intuition fuzzy operator has been the

concern of many scholars. In the literature, two kinds of IF operator are constructed by using triangular fuzzy function. On the basis of index fuzzy principles, a kind of exponential IF operator is constructed (Yannis et al. 2001). Two kinds of intuitionistic FS are constructed respectively.

However, for any two intuitionistic FS with same absolute deviation of membership and non-membership, the intuition entropy calculated by the above-mentioned literature is the same, which is obviously incompatible with intuition (Zhang et al. 2000). The root cause of this problem is that the fuzzy degree of the IF set contains two aspects of information, uncertainty and unknown degree. While the degree of membership and degree of non-membership reflects the degree of uncertainty, the uncertainty degree reflected the hesitation.

In this paper, we add the hesitancy function to the construction of entropy, construct the IF by using cotangent function and the cosine function respectively to overcome the deficiency. Two types of IF operator are proposed, which include membership degree, non-membership degree and hesitancy degree H are applied to IF multiple attribute decision making. However, the IF entropy of these researches still not distinguish well from the fuzzy degree of some

93

FS. Therefore, in order to overcome for the short-comings of the above research, this paper constructs a new IF operator for intuitionistic FS and interval FS measure. The new IF operator not only considered the deviation of membership degree and non-membership degree (Gao et al. 2015), but also includes the degree of uncertainty in the entropy measure, which can reflect the uncertainty and unknown degree of the IF set. In addition, for the IF multi-attribute decision-making problem which has been paid more attention to now, this paper will give the weight determination method based on the new fuzzy operator, which is based on the IF information more than the unknown and partially known attribute weight information. The using methods of the proposed IFS in multiple attribute decision making is explained, and the efficiency and feasibility of the processes are illustrated by case study.

2 RESEARCH AND APPLICATION OF MULTIPLE ATTRIBUTES DECISION MAKING USING INTUITIONISTIC FUZZY ENTROPY OF H–CLASS

As enterprises pay more and more attention to the production of their own core technology, it is imperative to get the right resources from the outside to participate in the production. As for the general manufacturing industry (Zhao et al. 2015), although the proportion of material procurement cost to product unit cost will vary with different industries, there is a substantial proportion of product quality and delivery delays caused by external suppliers of. Therefore, an efficient supply chain management, in addition to rely on good procurement operations, the supplier's assessment process is to assess the success or failure of enterprises is an important factor. The choice of suppliers needs to consider several difficult to quantify the evaluation attributes, so that the supplier selection problem becomes a complex multiple attributes decision-making problems (Specht et al. 2001). Because of the inaccuracy of information and quantitative and qualitative characteristics of decision criteria, it is often impossible for decision makers to establish an objective and quantitative approach to the evaluation of these criteria in a competitive environment where complex industries are competing with each other. And the interval IF number can be used to characterize the situation.

2.1 New intuitionistic fuzzy entropy method

By study on the problem of IF deterioration measure and its application in multiple attributes Decision Making, this paper face to the shortcomings of the existing IF entropy, and an improved measure for the intuitionistic FS and the interval intuitionistic FS was proposed, not only consider the absolute value of membership and non-membership deviation, but also consider the degree of hesitancy The Influence on the Fuzziness of Intuitionistic FS. As the problem of attribute weight information in multi-attribute

decision-making problem, the entropy weight method for weight determination and the optimal weight are established by minimizing the entropy optimization model, and the trade-off ratio of IF multiple attribute decision making principle. Application examples illustrate the effectiveness and feasibility of the method.

Before proposing new IF entropy, we first analyze the limitations of IF entropy.

1) The measure is not considered the degree of hesitation and the degree of intuition fuzzy uncertainty. Such as Gao's IF entropy measure:

$$w_j \geq 0, \sum_{j=1}^{n} w_j = 1, j = 1, 2, ..., n \tag{1}$$

Zhao's IF entropy measure:

$$E_Z(A) = 1 - \frac{1}{n} \sum_{i=1}^{n} |\mu_A(x_i) - v_A(x_i)| \tag{2}$$

IF Entropy Method

$$E_{ZJ}(A) = -\frac{1}{n} \sum_{i=1}^{n} \left[\frac{\mu_A(x_i)+1-v_A(x_i)}{2} \log\left(\frac{\mu_A(x_i)+1-v_A(x_i)}{2}\right) + \frac{v_A(x_i)+1-\mu_A(x_i)}{2} \log\left(\frac{\mu_A(x_i)+1-v_A(x_i)}{2}\right) \right] \tag{3}$$

Using of equal membership and non-membership deviation, the entropy standards calculated in the above cases are equal, which is not consistent with the intuitionistic fact.

2) The other is that although the effect of hesitancy on the IF entropy is considered, there are some cases that are not well distinguished.

Considering two intuitionistic FS $A_1 \leq \langle 0.2, 0.6 \rangle$ and $A_2 \leq \langle 0.2, 0.6 \rangle$, it obvious that the ambiguity of the fuzzy ratio is large, but by using the fuzzy intuitionistic measure

$$E_W(A) = \frac{1}{n} \sum_{i=1}^{n} \cot\left(\frac{\pi}{4} + \frac{|\mu_A(x_i) - v_A(x_i)|}{4(1+\pi_A(x_i))} \pi\right) \tag{4}$$

And the IF measure

$$E_{WEI}(A) = \frac{1}{n} \sum_{i=1}^{n} \cos\left(\frac{\mu_A(x_i) - v_A(x_i)}{2(1+\pi_A(x_i))} \pi\right) \tag{5}$$

The calculated results are $E_W(A_1) = E_W(A_2) = 0.8458$ and $E_W(A_1) = E_W(A_2) = 0.8660$, and the entropy values are equal, which is not consistent with our intuition.

In order to make up for the deficiency of the existing IF entropy, a new class of IF was constructed.

For any IF set, the definition of it as below:

$$A = \{< x_i, \mu_A(x_i), v_A(x_i) > | x_i \in X, i = 1, 2, ..., n\} \tag{6}$$

$$E(A) = \frac{1}{n} \sum_{i=1}^{n} \cos \frac{(\mu_A^2(x_i) - v_A^2(x_i))(1 - \pi_A(x_i))}{2} \pi$$

It can be seen that $E(A)$ is the deviation of membership degree and non-membership degree is $\mu_A^2(x_i) - v_A^2(x_i)$, the hesitancy degree is $\pi_A(x_i)$ included as well.

The measure defined by Eq. (3) is an intuitionistic obfuscation which satisfies the four conditions of Eq. (6);

i. $E(A) = 0$ only if A is a classical set;
ii. $E(A) = 1$ only if $\mu_A(x_i) = v_A(x_i), \forall x_i \in X$;
iii. $E(A) = E(A^C)$;
iv. if $A \leq B$, then $E(A) \leq E(B)$

The entropy proposed in this paper not only considers the membership degree and non-membership degree deviation, it takes into account the influence of the degree of hesitancy on the entropy value as well, which can more reasonably describe the fuzzy degree of the intuitionistic model. For the two FS in Eq. (6), the new fuzzy entropy (3) is used to get the result $E(A) = 0.8763 < E(A) = 0.8893$, this is in line with our intuition.

Gao et al. proposed a new information retrieval algebraic model – latent semantic index (Latent semantic Index) is mainly used for feature dimensionality reduction. Thought it is widely used in information retrieval, which solves the problem of synonymy and polysemy based on keyword retrieval, and can improve the efficiency of information retrieval by 10%–30%. With the development of application field, LSI has been widely used in many fields such as information filtering, information classification, clustering, cross-language retrieval, information understanding, judgment and prediction.

The traditional text-based vector-space model (VSM), which has the advantage of unstructured text expressed as a vector form, making a variety of mathematical processing possible. However, the vector space model is based on the assumption that the relationship between words is independent of each other (orthogonal hypothesis), in reality it is difficult to be met, and words in the text often appears a certain relevance to each other. The relevance should be calculated. The latent semantic index is based on the assumption that there is some connection between the word and the word in the text, that is, there is some potential semantic structure. This latent semantic structure is implicit in the context of the words in the text. Synonyms have basically the same semantic structure, and polygenes must have different semantic structures.

2.2 Multiple attributes decision-making method using new intuitionistic operator

The existing shortcomings of IF similarity measure are analyzed. For the intuitionistic FS and the interval intuitionistic FS, an improved similarity measure the membership degree, non-membership degree and interval valued of intuitionistic FS are proposed. For the multiple attributes decision-making problem which the attribute weight information is totally unknown and partially known, by establishing the maximum similarity optimization model, this paper proposed the maximum deviation method for weight determination and the optimal weight are established. Application

examples illustrate the efficiency and practicality of the method.

Assuming that the evaluator gives the attribute (attribute value) an IF number $\overline{a_{ij}} = \langle \mu_{ij}, v_{ij} \rangle$ to evaluate the scheme, the result is the satisfaction degree and dissatisfaction $W = (w_1, w_2, \ldots, w_n)$ respectively.

3 CONCLUSION

In order to extent the fuzzy degree of intuitionistic FS more accurately, this paper proposes new fuzzy operators for intuitionistic FS and interval intuitionistic FS respectively. It contains the deviation of membership and non-membership, and contains the information of decision-making degree of hesitation as well, which can better describe the degree of uncertainty and unknown degree of IF set. To overcome the shortcomings of most IF measure which only consider membership degree and non-membership degree without considering the unknown degree. The preprocessing and clustering of Sogou user logs are carried out to obtain query expansion words corresponding to the input search terms. The algorithm extracts the query extension candidate phrases from the title and keyword of the web page, which is different from the traditional text. Because the query words are the knowledge description from the user's point of view, this paper puts forward the method of querying the log After weighting the user query web pages, the weight matrix is established by the vector space model (VSM), and the features are reduced by the LSI algorithm, and then the text is processed by the K-means algorithm Clustering, and get the text clustering center as the category markup words. On this basis, the multiple attributes decision-making method proposed has the following characteristics for the multiple attributes decision-making problem in which the attribute weight is completely unknown and partly unknown.

1) Because the Technique for Order of Preference by Similarity to Ideal Solution (TOPSIS) method has some shortcomings, such as reverse order, the comprehensive evaluation value reflects the respectively degree of the attributes and cannot reflect the true degree of closeness. Based on fuzzy entropy, this paper proposes a compromise value method for the IF multiple-attribute decision-making problem. This method not only reflects the degree of the alternatives approaching the positive ideal point but also discards the negative ideal point, but also introduces the relative distance between the alternatives and the positive ideal solution by introducing the attitudinal factor.

2) For attribute completely unknown, the weight of attribute is determined by entropy weight method. For the known information of attribute weight information, the attribute weight is solved by establishing minimized optimization model.

3) From the application example, it can be seen that the proposed method of IF multiple attribute decision-making is feasible in addition to effective.

The suggested manner could be used in many fields such as image processing, pattern recognition and medical diagnosis. The proposed multi-attribute decision-making method can be applied to multi-attribute decision-making problems for example venture investment project collection, site selection, etc., and has good theoretical value And practical application value.

ACKNOWLEDGEMENTS

This work is sponsored by Harbin Science and Technology Innovation Talent Research Special Funds (2016RAQXJ039).

REFERENCES

Bakos, Y. 2001. The emerging landscape for retail e-commerce. Journal of Economic Perspectives, 15(1): 69–80.

Dalal, N. & Triggs, B., 2005. Histograms of oriented gradients for human detection. IEEE Conference on Computer Vision and Pattern Recognition (CVPR), 33(2):189–191.

Fu, Q. & Xin, D. 2015. A Context-Aware Mobile User Behavior Based Preference Neighbor Finding Approach for Personalized Information Retrieval. Procedia Computer Science, 134(2):239–241.

Gao, Z. 2000. Neural networks for classification: a survey.IEEE Trans on Systems, Man, and Cybernetics-Part C: Applications and Reviews, 30(4):451–462.

Hansen, K. & Salamon, P. 1990. Neural Network Ensembles. IEEE Transactions on Pattern Analysis and Machine Intelligent. 36(4):219–232.

Kaplan, L., 2001. M. Analysis of multiplicative speckle models for template-based SAR ATR. IEEE Transactions on Aerospace and Electronic Systems. 213(5):21–23.

Liu, W.Y., Ju, Y. & Yang, S. 2011. User interest modeling and its application for question recommendation in user-interactive question answering systems. Information Processing and Management. 12(3):213–222.

Minton, S. & Johnston, D. 1992. Minimizing conflicts: a heuristic repair method for constraint satisfaction and scheduling problems. Artificial Intelligence, 58:161–205.

Mueller, A., 1995. "Fast Sequential and Parallel Algorithms for Association Rule Mining," Technical Report CS-TR-3515, University of Maryland, College Park. 36(4):219–232.

Rode, H. & Trifts, V. 2002. Consumer Decision Making in Online Shopping Environments: The effects of Interactive Decision Aids. Marketing Science. 145(4):222–232.

Specht. D., A general regression neural network. 2001. IEEE Transactions on Neural Networks. 136(4):232–234.

Zhao, Q. & Principe, J.C. 2013 Support vector machines for SAR automatic target recognition. IEEE Transactions on Aerospace and Electronic Systems. 34(1):222–223.

Electronic Engineering – Wang (ed)
© 2018 Taylor & Francis Group, London, ISBN 978-1-138-60260-1

Design and research of positioning function of automobile guidance system based on WinCE

N. Wang, B.F. Ao, H.P. Tian & F. Liu
College of Electrical Engineering, Heilongjiang Polytechnic, Harbin, China

ABSTRACT: Our system uses Global Positioning System (GPS) positioning data as the primary source of data for vehicle location and navigation. First, this paper describes the GPS communication protocol and data format. Then the reception and processing of the navigation information is discussed. Under a WinCE environment, vehicle navigation and positioning are achieved using an ARM processor platform to receive and extract GPS navigation information.

Keywords: embedded systems; WinCE; GPS; navigation; location

1 INTRODUCTION

The concept of vehicle positioning systems was originally started in the late 1970s, using dead reckoning and map-matching technology to achieve positioning and navigation of the vehicle. However, this system was prone to error accumulation, low positioning accuracy and limited practicability (Chen & Qin, 2012). In the 1990s, with the rapid development of computer and communication technology, vehicle positioning and navigation systems began to enter the practical stage. The Global Positioning System (GPS) was a new generation of satellite navigation and positioning system with real-time three-dimensional navigation and positioning capabilities for the full spectrum of air, sea and land. Taking 20 years and costing 20 billion dollars to develop, GPS was fully completed in 1994 (Zhu et al., 2014).

GPS provides a cheap and convenient method of location for users worldwide because of all-weather, high-precision, high-efficiency, multifunctional and simple operation, wide use and other significant features. GPS, which has been greatly improved in terms of positioning accuracy and practicality, has taken the development of vehicle positioning and navigation systems to a new level (Chen & Li, 2014).

2 GPS COMMUNICATION PROTOCOL AND DATA FORMAT

There are many GPS communication protocols but most manufacturers follow the NMEA-0183 protocol. This widely used serial communication data protocol was drafted for marine electronic devices by the National Marine Electronics Association: all input and output information of the protocol is in the form of ASCII characters (Baire et al., 2016).

The NMEA-0183 message statements collected from the GPS receiver consist of a number of parameter fields, one of which is called a statement. Each statement begins with $ (0×24) and ends with a line feed (<CR> <LF>, or $0 \times 0D0A$), with a number of fields separated by commas (",," or $0 \times 2C$). The data in the field includes time, latitude and longitude, the number of received stars used for positioning, geometric accuracy, velocity, track direction, magnetic heading, magnetic error, check code and other information. The NMEA-0183 protocol has a dozen statement types, and different types of statements output different data information; the most commonly used of them are GPGGA, GPRMC, GPGSV and GPVTG. Users can select one or more statements as needed. This paper selects the GPS receiver output of some of the following statements:

$GPGGA – GPS positioning data (latitude, longitude, time, location, etc.);

$GPGLL – latitude, longitude, positioning Coordinated Universal Time (UTC);

$GPGSA – GPS receiver operating mode, the satellite used in positioning, Dilution Of Precision (DOP) value, etc.;

$GPGSV – GPS satellite data;

$GPRMC – navigation satellite-specific data.

We take $GPRMC as an example to explain the meaning of a statement. Thus, we have: $GPRMC, <1>, <2>, <3>, <4>, <5>, <6>, <7>, <8>, <9>, <10>, <11>, *hh, <CR>, <LF>.

GP is the receiving data source, RMC is the sentence mark, and GPRMC locates the output statement. The data in each sentence is separated by commas, and the data elements of the output can be chosen according to the needs of users:

<1> Greenwich Time in current position – format is hhmmss;

<2> Positioning status – A is the valid position, V is the non-effective receiving warning that the number of satellites is less than four;

<3> Latitude – format is ddmm. mmmm;

<4> Latitude direction – N represents the northern hemisphere, S represents the southern hemisphere;

<5> Longitude – format is dddmm. mmmm;

<6> Longitude direction – E represents the eastern hemisphere, W represents the western hemisphere;

<7> Ground speed – the unit is knots per hour, in the range 0.0–999.9;

<8> Azimuth angle – in the range 0–359.9°;

<9> UTC date – format is ddmmyy;

<10> Magnetic declination – in the range 000.0–359.9°;

<11> Magnetic declination direction – E or W;

* hh – Checksum;

<CR><LF> Carriage Return/Line Feed–marks end of statement.

3 PLATFORM DESIGN

This research uses an ARM920T processor as the hardware platform, on which the Windows CE .NET operating system is installed and is responsible for the overall scheduling and control system. The ARM920T is connected through a Universal Asynchronous Receiver/Transmitter (UART) interface to the GPS receiver, which is used for determining the positioning information that is required for display. Then, by matching with the loaded map, a map of the area in which the vehicle is currently situated is displayed by LCD.

The WinCE operating system is a multitasking, fully pre-emptive, 32-bit embedded operating system, and supports the WinCE Microsoft Foundation Class (MFC), Advanced Template Library (ATL) and Application Programming Interface (API), as well as some additional programming interfaces and a variety of communication technologies (Cong et al., 2015). Using Embedded Visual C++ (EVC) as the development tool, which is the mainstream Windows CE development tool, provides encapsulation of the underlying network communications, COM interoperability, Remote Application Programming Interface (RAPI), and so on. EVC also supports a subset of the MFC class library, so that a Visual C program on the Win32 platform can be easily migrated to the WinCE platform.

4 RECEPTION OF GPS NAVIGATION INFORMATION

The operating system and hardware platform use the serial port for communication, so the serial port must first be configured. There are two limitations in the communication of serial data: first, EVC does not support the serial communication control MSComm; second, WinCE does not support overlapping I/O operations. Therefore, it is necessary to use the WinCE API function and multithreading technology for the data communication basis of the development. Because the Windows API function treats the serial port as a file, the operations on the serial port are similar to those for ordinary files. Within the EVC environment, the read-write mode is set by calling the CreateFile function to open the serial port (Hess et al., 2015):

```
HANDLE m hCom=CreateFile(strCom,
GENERIC_READ|GENERIC_WRITE, 0, NULL,
OPEN_EXISTING, FILE_ATTRIBUTE_NORMAL,
NULL)
```

Because WinCE does not support overlapping I/O, the sixth parameter of the CreateFile function cannot be set to FILE_FLAG_OVERLAPPED, otherwise the serial communication processing will be blocked by the system information.

The serial port is then opened; the serial port initialisation function SetCommState() can be used to obtain the current serial port configuration, such as baud rate, parity, data bits and stop bits and other detailed information. Finally, the PurgeComm() function is used to complete the operation of the buffer function.

The GPS receiver sends the positioning information to the processor via the RS232 serial port, so in the application, once the serial port settings are complete, the GPS receiver will continuously send navigation and positioning information to the processor through the serial port. However, without further processing, the information is a long stream of bytes in the cache; it cannot be used until it has been classified. Therefore, the information of each field must be extracted from the cache through the program to convert it into meaningful positioning information data (Zhang & Zhang, 2011). The actual data received by the GPS receiver is shown below:

```
$GPGGA, 130050.000, 4037.0811, N,
10949.3204, E, 1, 03, 4.0, 1063.1, M, -
28.0, M, 0000*46
$GPGLL, 4037.0811, N, 10949.3204, E,
130050.000, A*3B
$GPGSA, A, 2, 01, 20, 03,,,,,,,,,,
4.2, 4.0, 1.0*30
$GPGSV, 3, 1, 11, 16, 85, 352, 19, 01,
47, 157, 44, 23, 46, 306, 13, 31, 43,
083, 13*74
$GPGSV, 3, 2, 11, 25, 25, 296, 14, 07,
24, 050, 12, 03, 23, 180, 43, 20, 20,
243, 47*7D
$GPGSV, 3, 3, 11, 13, 17, 318, 15, 06,
14, 042, 22, 14, 03, 147, 35*44
$GPRMC, 130050.000, A, 4037.0811, N,
10949.3204, E, 0.43, 7.10, 100507, *0E
```

5 THE EXTRACTION OF GPS NAVIGATION INFORMATION

The data is processed according to the needs of the vehicle navigation system; information such as the

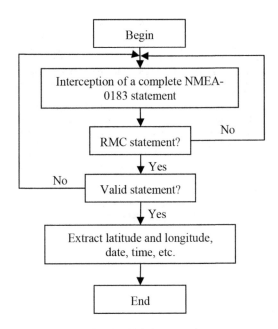

Figure 1. Flow chart for GPS data extraction.

required date, time, latitude, longitude, and height is extracted. Date, time, latitude and longitude information is extracted from the GPRMC statement, and height information is extracted from the GPGGA statement. The extraction flow chart is shown in Figure 1.

The following is an example of a simple procedure for a GPS data decoding process implementation:

```
DWORD WINAPI readthread ( LPVOID lpara )
{
int i=0,length=0,j=0;
DWORD dwstate, dwread, dwerror;
COMSTAT;
HANDLE hprocessthread =INVALID_HANDLE_VALUE;
SetCommMask (hcom,EV_RXCHAR );
WCHAR*point =NULL;
WCHAR tcharbuf [512];
WCHAR latitude [15];
WCHAR north_south,west_east;
WCHAR longtitude[15];
WCHAR flag_gprmc []=_T("GPRMC");
WCHAR totaltude [100];
char recvbuf [512];
memset (recvbuf,0,512*size of (char));
memset (tcharbuf,0,size of (TCHAR)*512);
HWND hwnd=(HWND) lpara;
HWND hstatus=::GetDlgItem (hwnd,
IDC_STATUS);
HWND htext=:: GetDlgItem (hwnd,
IDC_TEXT);
while (m_connect)
{
i=0;j=0;
WaitCommEvent (hcom, & dwstate, NULL);
SetCommMask (hcom, EV_RXCHAR);
if (dwstate&EV_RXCHAR)
{
memset (recvbuf,0,512*size of (char));
memset (latitude,0,sizeof(WCHAR)*10);
memset (longtitude,0,sizeof(WCHAR)*10);
ClearCommError (hcom, &dwerror,
&comstat);
length=comstat.cbInQue;
if (length<512)
continue;
ReadFile
(
hcom,
recvbuf,
512,
&dwread,
NULL
);
Mbstowcs (tcharbuf, recvbuf, 512);
point = wcsstr (tcharbuf, flag_gprmc);
if (*point==NULL|*(point+80)==NULL)
{
continue;
}
else
{
memset (latitude,0,sizeof(WCHAR)*15);
memset (longtitude,0,sizeof(WCHAR)*15);
while (*point++!=',');
while (*point++!=',');
while (*point++!=',');
while (*point!=',')
{
latitude[i++]=*point++;
}
point++;
north_south=*point++;
point++;
while (*point!=',')
{
longtitude[j++]=*point++;
}
point ++;
west_east=*point++;
if (*point!=',')
{
continue;
}
else
{
memset (totaltude,0,sizeof(WCHAR)*100);
wcscpy (totaltude,latitude);
wcscat (totaltude,&north_south);
wcscat (totaltude,T("\r\n"));
wcscat (totaltude,longtitude);
wcscat (totaltude,&west_east);
wcscat (totaltude,T("\r\n"));
::SetWindowText (htext,totaltude);
}
::SetWindowText (hstatus,tcharbuf);
}
Sleep (100);
}
}
return 1;
}
```

6 CONCLUSION

In this paper, a system for vehicle navigation through GPS is established, based on the WinCE environment. The system, which is stable, is expected to be reliable and to satisfy real-time requirements.

REFERENCES

Baire, Q., Bruyninx, C., Legrand, J., Pottiaux, E. & Aerts, W. (2016). Influence of different GPS receiver antenna calibration models on geodetic positioning. *GPS Solutions*, 18(1), 529–539.

Chen, W. & Li, X. (2014) Success rate improvement of single epoch integer least-squares estimator for the GNSS attitude/short baseline applications with common clock scheme. *Acta Geodaetica et Geophysica*, 49(3), 295–312.

Chen, W. & Qin, H. (2012). New method for single epoch, single frequency land vehicle attitude determination using low-end GPS receiver. *GPS Solutions*, 2012(3), 329–338.

Cong, L., Li, E., Qin, H., Ling, K.V. & Xue, R. (2015). A performance improvement method for low-cost land vehicle GPS/MEMS-INS attitude determination. *Sensors*, 15(3), 5722–5746.

Hess, S., Quddus, M., Rieser-Schüssler, N. & Daly, A. (2015). Developing advanced route choice models for heavy goods vehicles using GPS data. *Transportation Research Part E Logistics & Transportation Review*, 77, 29–44.

Zhang, J. & Zhang, M. (2011). A design of embedded multimedia player based on WINCE. *Procedia Engineering*, 16(16), 252–258.

Zhu, J., Hu, X., Zhang, J., Li, T. & Wang, J. (2014). The inertial attitude augmentation for ambiguity resolution in SF/SE-GPS attitude determination. *Sensors*, 14(7), 11395–11415.

Electronic Engineering – Wang (ed)
© 2018 Taylor & Francis Group, London, ISBN 978-1-138-60260-1

Thermal shock resistance of $(ZrB_2 + 3Y\text{-}ZrO_2)/BN$ composites

L. Chen, Y.J. Wang, L.X. Zhou & Y. Zhou
Institute for Advanced Ceramics, School of Materials Science and Engineering, Harbin Institute of Technology, Harbin, China

ABSTRACT: $(ZrB_2 + 3Y\text{-}ZrO_2)/BN$ composites were fabricated by hot-pressing with different ZrB_2 contents ranging from 0 to 30 vol%. The thermal shock resistance of $(ZrB_2 + 3Y\text{-}ZrO_2)/BN$ composites was investigated by single and multiple-cycle thermal shock. The residual strength ratios of 30% $3Y\text{-}ZrO_2/BN$ and $(10\% ZrB_2 + 20\% 3Y\text{-}ZrO_2)/BN$ composites were higher than 100% after single cycle thermal shock at temperature difference more than $1100°C$. After quenching forty cycles, the residual strength ratio of $(10\% ZrB_2 + 20\% 3Y\text{-}ZrO_2)/BN$ composite was 77.6%, whereas that of 30% $3Y\text{-}ZrO_2/BN$ composite decreased remarkably, down to 42.5%. The influence of $3Y\text{-}ZrO_2$ on improving thermal shock resistance of the composites was higher than that of ZrB_2. The oxidation layer was beneficial to the residual strength by the healing-crack when the oxidation layer was density and composed of continuous glass phase.

Keywords: $(ZrB_2 + 3Y\text{-}ZrO_2)/BN$ Composites; Thermal shock resistance; Refractories; Residual strength

1 INTRODUCTION

BN has been regarded as a promising material for structural applications due to its unique properties, including extremely high melting temperature, good solid-state phase stability, no wetting with metal, solid lubricant, excellent thermal shock resistance, chemical inertness, non-toxicity and environmental safety (Lipp et al. 1989). BN is also a machinable material (Li et al. 2005). It can be widely used in the metal industry, high-temperature furnaces, insulator and continuous strip casting. However, the low mechanical properties and high-thermal conductivity of BN materials have severely limited their applications. Adding second phase to the BN matrix is a very useful method for improving the mechanical properties of BN materials. The second phase can be oxide (Li et al. 2002), nitride (Wang et al. 2002, Jung & Baik 2007), boride, or carbide (Kusunose et al. 2007; Jiang et al. 2008). ZrO_2 and ZrB_2 have high melting point, good hardness good solid-state phase stability, no wetting with metal and chemical stability, and also used as refractory materials for steel industry. BN composites incorporated with monoclinic zirconia have been used in steel production as a side-dam material for thin-strip casting and a break ring for horizontal continuous casting (Eichler & Lesniak 2008; Vikulin et al. 2004).

The previous results have shown that adding of appropriate amounts of ZrO_2 or ZrB_2 particles into BN matrix enhances the mechanical properties, and decreases the thermal conductivities (Zhang et al. 2008; Wu & Zhang 2010). BN composites with addition of ZrO_2 or ZrB_2 particles would expand the application in metallurgy industry. Thermal shock resistance is another major issue for refractories using in the high-temperature environment. The purpose of this study was to investigate the thermal shock resistance of $(ZrB_2 + 3Y\text{-}ZrO_2)/BN$ composites with the temperature difference ranging from 700 to $1400°C$. The effect of ZrB_2 content on the thermal shock resistance was valuated by analyzing the microstructure and residual strength after single and multiply-cycle thermal shock.

2 EXPERIMENTAL

2.1 *Preparation*

Commercially available raw materials were used in this study. The starting powders were $3 \, mol\%Y_2O_3$ partially stabilized zirconia ($3Y\text{-}ZrO_2$ particles, $2.32 \, \mu m$, Nanbo Powder Co., Ltd., Shenzhen, China), ZrB_2 particles ($2.0 \, \mu m$, Chemical Research, Xi'an, China), h-BN particles ($1.7 \, \mu m$, Bangde Ceramic Material Co., Ltd.,), SiO_2 particles ($2.0 \, \mu m$, Youli Chemical Co., Xi'an, China). The powder mixtures for the preparation of $(ZrB_2 + 3Y\text{-}ZrO_2)/BN$ composites contained 60% BN, 10% SiO_2, and 30% (ZrO_2 and/or ZrB_2) as shown in Table 1. These powder mixtures were ball-milled for 24 h in a polyethylene bottle using zirconia balls as grinding media. After mixing, the slurry was dried in a rotary evaporator and sieved through a 100-mesh screen. The $(ZrB_2 + 3Y\text{-}ZrO_2)/BN$ composites were fabricated by hot-pressing at $1800°C$ for 60 min under the uniaxial load of 20 MPa in Ar atmosphere.

Table 1. Composition parameters of $(ZrB_2 + 3Y-ZrO_2)/BN$ composites.

Composites	\multicolumn{4}{c}{The designed content of powders (vol%)}
	BN	SiO_2	ZrB_2	ZrO_2
BO03	60	10	0	30
BO12	60	10	10	20
BO21	60	10	20	10
BO13	60	10	30	0

2.2 Characterization

The bulk density of the sintered specimens was measured by the Archimedes method. Fracture strength was determined by a three-point bending test using a rectangular bar (3 mm × 4 mm × 35 mm). The surface of the test bar was grounded with 1200-grit SiC sandpapers.

Thermal shock test was carried out in a tube furnace at temperatures from 700°C to 1400°C in the laboratory air. The furnace was heated to a preset temperature at a heating rate of 10°C/min. The simples were inserted into the furnace when the furnace was at the preset temperature. After the samples were heated for 10 min in the furnace, they were dropped into a water bath. The temperature difference ΔT was chosen as 700, 900, 1100 and 1400°C. All flexural strengths of the samples after thermal shock at each temperature were measured by three points bending.

The phase compositions of simples after thermal shock were detected by X-ray diffractometry (D/max-γB) using Cu kα radiation. The microstructure features of the samples after thermal shock were analyzed by scanning electron microscopy (SEM, S4700, Hitachi) along with energy dispersive spectroscopy (EDS, EDAX Inc.) for chemical composition analysis.

2.3 Parameter of thermal shock resistances

Several parameters of thermal shock resistance for the evaluation of ceramic materials had been proposed and reviewed in several literatures (Aksel & Warren 2003; Cheng et al. 2000; Li et al. 2011). These parameters were divided into two categories, one that was based on criteria that relate to the initiation of cracks and others that described the conditions necessary for propagation of cracks (Buchheit et al. 2009). The thermal shock parameters were calculated as following equations.

$$R = \sigma_f (1-\upsilon)/E\alpha \quad (1)$$

$$R'''' = \gamma E / \sigma_f^2 (1-\upsilon) \quad (2)$$

$$R_{st} = (\gamma / \alpha^2 E)^{1/2} \quad (3)$$

The fracture energy (γ) was calculated according to the Griffith equation:

$$\gamma = K_{lc}^2 (1-\upsilon^2)/2E \quad (4)$$

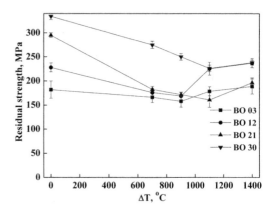

Figure 1. The residual strength of $(ZrB_2 + 3Y-ZrO_2)/BN$ composites after thermal shock at different ΔT.

where σ was the flexural strength, E was the Young's modulus, K_{IC} was the fracture toughness, α was the coefficient of thermal expansion (CTE) and υ was the Poisson's ratio. Higher R represented greater resistance to the initiation of fracture during rapid quenching and during steady-state heat flew down a steep temperature gradient. R'''' decided the resistance to catastrophic crack propagation of ceramics under a critical temperature difference (Wang et al. 2009).

3 RESULTS

3.1 Residual strength

The variation between the residual strength and the thermal shock temperature difference ΔT for $(ZrB_2 + ZrO_2)/BN$ composites was shown in Figure 1. The residual strength of $(ZrB_2 + ZrO_2)/BN$ composites decreased with thermal shock temperature difference (ΔT) increasing when ΔT was lower than 900°C. On the contrary, the residual strength of the composites increased slightly with ΔT up to 1100°C. The residual strength of composite BO03 and BO12 presented a modest variation with ΔT from 700°C to 1400°C, and did not show a sharp degradation. The critical thermal shock temperature (ΔT$_c$) of BO03 and BO12 are both higher than 1400°C. The residual strength of BO03 and BO12 were 189 MPa and 237 MPa at ΔT = 1400°C, respectively. Comparing with the flexure strength of BO03 and BO12 (182 MPa and 228 MPa, respectively), the ratio of the residual strength was higher than 100%. However, the residual strength of BO21 and BO30 decreased quickly at the thermal shock temperature region of 700°C to 1100°C. This result indicated that the critical thermal shock temperature of BO21 and BO30 were about 570°C and 1030°C, respectively.

Figure 2 showed the residual strength of $(ZrB_2 + 3Y-ZrO_2)/BN$ composites after multiply-cycle thermal shock with cycle times as 1, 5, 10, 20, and 40 at ΔT = 1100°C. The residual strength of composites

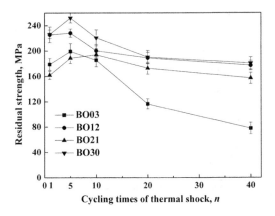

Figure 2. Residual strength of $(ZrB_2 + 3Y-ZrO_2)/BN$ composites after multiply-cycle thermal shock at $\Delta T = 1100°C$.

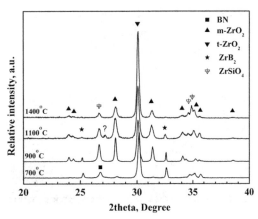

Figure 3. XRD patterns of surface of BO12 composite after thermal shock at different ΔT.

increased slightly after five-cycle thermal shock comparing with single thermal shock. The residual strength exhibited a modest degradation when the times of thermal shock was more than five, especially the residual strength of BO03 showed a sharp decrease after twenty-cycle thermal shock. It was worthy to note that the residual strength of BO12 after multiply-cycle thermal shock (except forty-cycle thermal shock) was higher than the residual strength after single thermal shock, even after forty-cycle thermal shock, the residual strength reached 157 MPa.

3.2 Phase

The XRD patterns of the surfaces of BO12 after thermal shock at the temperature differences of 700, 900, 1100 and 1400°C were shown in Figure 3. The results indicated the predominant phases were $t-ZrO_2$, ZrB_2, h-BN and a small amount of $m-ZrO_2$ after thermal shocked with $\Delta T = 700°C$, which were similar to the phases of the sample before the thermal shock. When the thermal shock temperature was higher than 900°C, the peaks of BN disappeared and new phase $ZrSiO_4$ was found in the XRD patterns. The presence of $ZrSiO_4$ was attributed to the reaction of ZrO_2 and SiO_2 as the additive in the composites. The amount of $m-ZrO_2$ increased after thermal shocked when ΔT was higher than 700°C, which is attributed to two reasons: (1) more spontaneous martensitic transformation of $t-ZrO_2$ to $m-ZrO_2$ occurred during cooling from the thermal shock temperature; (2) the oxidation of ZrB_2 forming ZrO_2 and B_2O_3 (Hu et al. 2010). Moreover, ZrB_2 phase decreased with increasing thermal shock temperature, which means that partial ZrB_2 was oxidized.

Figure 4 showed the XRD patterns of BO03 composite after multiply cycle thermal shock at $\Delta T = 1100°C$. The amount of $t-ZrO_2$ was very little, and even $t-ZrO_2$ disappeared when the sample was thermal shocked for more than five cycles. The amount of $ZrSiO_4$ increased with increasing the cycle number of thermal shock, due to the oxidation time at high

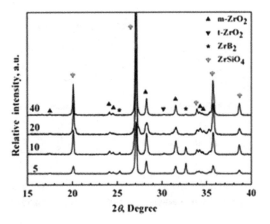

Figure 4. XRD patterns of surface of BO12 composite after multiply-cycle thermal shock at $\Delta T = 1100°C$.

temperature prolonging. Similarly, the ZrB_2 content decreased with the cycle number increasing, which is also attributed to the oxidation time increment.

3.3 Microstructure

The surface morphologies of BO03 composite after thermal shock at different ΔT were shown in Figure 5. Extensive cracks were found in the surface after thermal shock when $\Delta T < 900°C$. Moreover, when the thermal shock temperature was higher than 900°C, a thin glass layer was formed at the surface of BO03. EDS analysis revealed that this glass layer mainly composed of zirconium, silicon, oxygen and a small amount of boron. Previous studies on the oxidation of BN composite have shown that BN began to oxide to the boron oxide, and $ZrSiO_4$ was formed via the reaction between SiO_2 with ZrO_2 at more than 900°C (Mori et al. 1993; Yang et al. 2008), which were consistent with XRD results. The glass layer cracked after thermal shock at $\Delta T = 1100°C$, attributing to both the thermal stress during the thermal shock and the volume

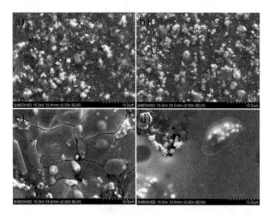

Figure 5. Surface morphologies of BO03 composite after thermal shock at different ΔT: a) 700°C; b) 900°C; c) 1100°C; d) 1400°C.

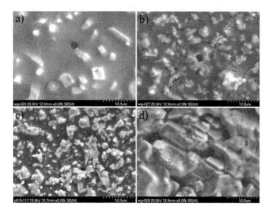

Figure 6. Surface micrographs of BO03 composite after multiply-cycle thermal shock at ΔT = 1100°C: a) 5; b) 10; c) 20; d) 40.

change caused by the oxidation of ZrB$_2$. when ΔT was increasing up to 1400°C, the glass layer became thick and some bubbles existed in the layer as shown in the Figure 5d, which might be caused by B$_2$O$_3$ evaporation. The glass layer did not crack after thermal shock at ΔT = 1400°C, this layer could be beneficial to the residual strength.

Figure 6 showed the surface microstructures of BO03 composite after multiply cycle thermal shock at ΔT = 1100°C. The surface morphology after five-cycle thermal shock was different from that after single thermal shock. The amount of glass phase increased and no crack existed in the glass layer. The white phase increased and formed the tetragonal grains with an average length of 5–10 μm. According to the EDS analysis, the white phase composed of zirconium, silicon and oxygen. The white phase was identified as ZrSiO$_4$ phase in contrast with the XRD pattern (as shown in Figure 4). After ten-cycle thermal shock, the more ZrSiO$_4$ phase formed, and the average size of this phase decreased. With thermal shock cycle further increases ZrSiO$_4$ phase became more and more, and the average size of grains increased obviously, whereas the glass layer and cracks disappeared. Especially, the average size of ZrSiO$_4$ grain was more than 10 μm after forty-cycle thermal shock. The dense glass layer is porous with some big holes as shown in Figure 6d).

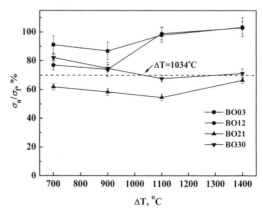

Figure 7. Residual strength ratio of (ZrB$_2$ + 3Y-ZrO$_2$)/BN composites after thermal shock at different ΔT.

4 DISCUSSIONS

4.1 Ratio of residual strength

The ratio of the residual strength of (ZrB$_2$ + 3Y-ZrO$_2$)/BN composites as a function of thermal shock temperature was shown in Figure 7. The results indicated that critical thermal shock temperature of BO21 was lower than 700°C, and there was no critical thermal shock temperature of BO03 and BO12 at which the strength decreases catastrophically, up to 1400°C. The thermal shock resistance of BO03 was most excellent among all the composites.

The ratios of the residual strength of BO03 and BO12 increased after thermal shock at ΔT = 900°C and the ratios of BO21 and BO30 also became higher when thermal shock temperature was higher than 1100°C, it was attributed to the surface crack healing by glass phase as shown in Figure 5. The borosilicate or silicon oxide glass phase had active influence on healing the surface cracks and decrease the sensitivity of cracking. However, the inflexion temperature in the residual strength curve of BO21 and BO30 were higher than that of BO03 and BO12. And the flexural strength of BO21 and BO30 were stronger than that of BO21 and BO30. It was implicated that there were two reasons about these results: (1) the crack propagation of BO21 and BO30 was accelerated rapidly after thermal shock; (2) the surface layer contained some loose ZrO$_2$ particles and big bubbles by the oxidation. The crack and defect promote crack initiation and growth during the thermal shock processing, so that more glass phase was needed to heal the crack.

Figure 8 showed the ratio of the residual strength after multiple-cycle thermal shocked at ΔT = 1100°C.

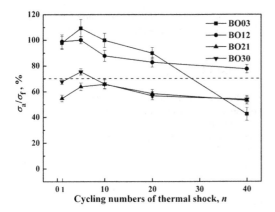

Figure 8. Residual strength ratio of (ZrB$_2$ + 3Y-ZrO$_2$)/BN composites after multiply-cycle thermal shock at ΔT = 1100°C.

The ratio of the residual strength increased and reached the maximum value after five-cycle thermal shock. For example, the ratio presented a modest degradation when the thermal shock cycles were less than thirty, and the ratio of residual strength of BO03 and BO12 were 109.3% and 100.1% after five-cycle thermal shock, respectively. The flexural strength increment was attributed to the glass layer, which healed surface cracks and decreased propagation of crack. The surface micrograph of BO03 after single thermal shock showed some differences from that after multiply-cycle thermal shock as shown in Figure 4c) and Figure 6. The glass layer on the surface and cracks were formed after single thermal shock. However, the glass layer became thick and the cracks disappeared after five-cycle thermal shock. The disappearance of cracks might be caused by the crack healing induced by oxidation, which is beneficial to the thermal shock resistance of the composites.

However, the residual strength of BO03 exhibited a fast degradation after forty-cycle thermal shock, the ratio is down to 42.5%. The degradation could also attribute to oxidation at the surface. When the samples under more five-cycle thermal shocked, the samples had sufficient time for boron oxide evaporation and to enable SiO$_2$ reacting with ZrO$_2$ to form ZrSiO$_4$ in the stagnant. The surface of composite was covered by ZrSiO$_4$ grains and some bigger bubbles caused by the evaporation of B$_2$O$_3$ in the oxidized layer, it was harmful to the thermal shock resistance of the composites. The influence of oxidation layer on thermal shock resistance was counterbalanced by the generation of new defects and the healing cracks, either in the oxide scale or at the interface between the oxide layer and bulk materials (Park et al. 1998).

4.2 Parameter of thermal shock resistance

The calculated thermal shock properties of composites were presented in Table 2. The ratio of residual strength of BO21 reached 60%, which was lowest

Table 2. The thermal shock parameters of (ZrB$_2$ + 3Y-ZrO$_2$)/BN composites.

	γ_f (Pa·m)	R(K)	R'''' (m × 10^{-6})	R_{st} (K·m$^{1/2}$)
BO03	67.02	279.04	269.49	5.29
BO12	69.90	352.94	182.87	5.51
BO21	62.08	394.74	113.19	4.85
BO30	75.14	408.11	118.55	5.13

among composites as shown in the Figure 7. The results were in accordance with the theoretical calculated thermal shock properties (R'''' and R_{st}), the mechanical of thermal shock resistance was mainly expressed to resist crack propagation. The surrounding second phase grains (ZrO$_2$ and/or ZrB$_2$) or glassy phase (SiO$_2$) were constrained in the BN platelets, large tensile stresses were developed perpendicular to the basal plane, resulting in separating BN platelets into layers along the basal plane direction. Thus the specimens had many microcracks within the BN-rich cell boundaries before the thermal shock treatment, so that crack propagations along BN cell boundaries became more favorable (Koh, Y.H. et al. 2004).

Furthermore, the experiment critical thermal shock temperature ΔT$_c$ value was significantly higher than the theoretical values. The discrepancy between the calculated and experimental thermal shock parameters indicated that R captured the initiation of thermal shock cracking and relied only upon common mechanical properties. Thermal shock resistance had a relation with the thermal conductivity of the materials, surface heat transfer and sample size, which reduced the thermal stress during the water quenching (Zimmermann et al. 2008; Zhang et al. 2009). All sorts of formula were proposed to calculate thermal shock parameter, such as the Eq. (1) could be modified by a stress reduction factor $f(\beta)$, so the Eq. (1) changed to Eq. (5). Moreover, several formulas of the tress reduction factor $f(\beta)$ had been proposed, such as Eq. (6) and Eq. (7) (Becher, P.F. et al. 1980, Fellner, M. & Supancic, P. 2002).

$$R = \frac{\sigma_f(1-\upsilon)}{E\alpha}f(\beta) \tag{5}$$

$$f(\beta) = 1 + \frac{4k}{ah} \tag{6}$$

$$f(\beta) = 1.5 + \frac{4k}{ah} - 0.5\exp(-\frac{5k}{ah}) \tag{7}$$

where, k was the thermal conductivity, a was the characteristic heat-transfer length, and h was the surface heat transfer coefficient. Water bath at room temperature was frequently used for thermal shock test because of its simplicity. However, the heat of sample made water converted into vapor, resulting in protective steam bubbles at the interface between water and specimen. This phenomenon led to a rapid reduction of the surface heat transfer coefficient (h)

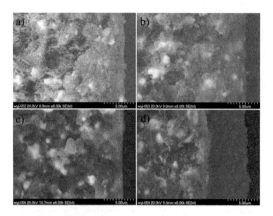

Figure 9. Micrographs of oxide layer of BO03 composite after thermal shock at different ΔT: a) 700°C; b) 900°C; c) 1100°C; d) 1400°C.

due to the random formation of bubbles surrounding the sample, which changed the heat transfer mechanism (Zhang et al. 2008). Unfortunately, Surface heat transfer coefficient h is usually unknown precisely. It was difficult to quantificationally evaluate the thermal stress. Even that, the influence of thermal shock parameters could be qualitatively analyzed by parameters such as k, a and h according to Eqs. (5)–(7). It was easily to generalize a conclusion that the theoretical calculations R was lower than experiment critical thermal shock temperature ΔT_c value because of the surface heat transfer coefficient reducing.

4.3 *The oxide layer*

The oxidation layer was observed when the thermal shock temperature was more than 1100°C. According to the XRD and EDS, the thick glass covered on the surface of composites composed of $ZrSiO_4$ and amorphous SiO_2. The surface heat coefficients of $ZrSiO_4$ and amorphous SiO_2 were much lower than that for ZrO_2 and BN, so the heat transfer coefficient of oxidation layer was lower than that of composites. Figure 9 showed the oxidation layer cross-section of BO03 at different thermal shock temperatures. The thickness of oxidation surface layer on the composite was about 5 μm at $\Delta T = 1400°C$. The oxide layer acted as the thermal barrier coating decelerated the temperature of samples cooling down during water quenching, which reduced the thermal stress and associated with the enhancement in the residual strength. Thus, residual strength of BO03 was improved after thermal shock at $\Delta T = 1100°C$ even if the oxidation layer had some cracks. However, the residual strength was improved and even the ratio of the residual strength was higher than 100% at $\Delta T = 1400°C$. This result was attributed to the formation of the oxidation layer on the surface after the composite quenched at $\Delta T = 1400°C$. The density oxidation layer healed cracks and decreased heat transfer coefficient. This result also conjectured that the ratio of residual strength was higher after five-cycle thermal shocked than one-cycle thermal shocked as shown in Figure 6 and Figure 9.

5 CONCLUSIONS

The thermal shock resistance of $(ZrB_2 + 3Y\text{-}ZrO_2)/BN$ composites had been studied by single and multiply-cycle thermal shocked, ranging from 700°C to 1400°C. The residual strength decreased and reached the minimum value at $\Delta T = 900–1100°C$ after single thermal shocked. The residual strength ratios of 30% ZrO_2/BN and $(10\% ZrB_2 + 20\% ZrO_2)/BN$ composites were higher than 100% at $\Delta T \geq 1100°C$. Furthermore, when the composites after multiply-cycle thermal shock, the residual strength increased firstly and reached their maximum values after five-cycle thermal shock, and then decreased with increasing thermal shock cycle. The residual strength ratio of $(10\% ZrB_2 + 20\% ZrO_2)/BN$ composite was 77.6%, whereas the residual strength of 30% ZrO_2/BN composites exhibited a fast degradation, down to 42.5%. The composites with high ZrO_2 content had a better thermal shock resistance than those with high ZrB_2 content, the $(10\% ZrB_2 + 20\% ZrO_2)/BN$ composites had excellent thermal shock resistance. The oxidation layer was beneficial to the residual strength by the healing-crack, and the compressive stress existed in the oxidation layer also improved the residual strength when the oxidation layer was density and composed of continuous glass phase. However, the surface oxidation layer was composed of some loose $ZrSiO_4$ and bubbles when sample was oxidized for a prolonged time, which was also harmful to the thermal shock resistance.

ACKNOWLEDGMENT

This work is partially supported by the National Natural Science Foundation of China (Nos.: 51602074, 51372050 and 51672060), Natural Science Foundation of Heilongjiang Province (Nos.: E2016026), and the Fundamental Research Funds for the Central Universities (Grant No. HIT. NSRIF. 2016 4).

REFERENCES

Aksel, C. & Warren, P.D. (2003). Thermal shock parameters [R, R′′′ and R′′′′] of magnesia-spinel composites. *Journal of the European Ceramic Society*. 23(2): 301–308.

Becher, P.F. Lewis, D. Cerman, K.R. & Gonzalez, A.C. (1980). Thermal shock resistance of Ceramics. *American Ceramic Society Bulletin*. 59: 542–545.

Buchheit, A.A. Hilmas, G.E. Fahrenholtz, W.G. & Deason, D.M. (2009). Thermal shock resistance of an AlN-BN-SiC ceramic. *Journal of the American Ceramic Society*. 92(6): 1358–1361.

Cheng, W. Wei, J. & Lin, Y.P. (2000). Mechanical and thermal shock properties of size graded MgO-PSZ refractory. *Journal of the European Ceramic Society*. 20(8): 1159–1167.

Eichler, J. & Lesniak, C. (2008). Boron nitride (BN) and BN composites for high-temperature applications. *Journal of the European Ceramic Society.* 28(5): 1105–1109.

Fellner, M. & Supancic, P. (2002). Thermal shock failure of brittle materials. *Key Engineering Materials.* 223: 97–106.

Hu, P. Wang, Z. & Sun, X. (2010). Effect of surface oxidation on thermal shock resistance of ZrB_2-SiC-G composite. *International Journal of Refractory Metals and Hard Materials.* 28(2): 280–285.

Jiang, T. Jin, Z.H. Yang, J.F. & Qiao, G.J. (2008). Mechanical property and R-curve behavior of the B_4C/BN ceramics composites. *Materials Science and Engineering: A.* 494(1): 203–216.

Jung, J. & Baik, S. (2007). Combustion synthesis of AlON-BN composites under low nitrogen pressure. *Journal of the American Ceramic Society.* 90(10): 3063–3069.

Koh, Y.H. Kim, H.W. Kim, H.E. & Halloran, J.W. (2004). Thermal shock resistance of fibrous monolithic Si_3N_4/BN ceramics. *Journal of the European Ceramic Society.* 24(8): 2339–2347.

Kusunose, T. Sekino, T. & Niihara, K. (2007). Contact damage of silicon carbide/boron nitride nanocomposites. *Journal of the American Ceramic Society.* 90(10): 3341–3344.

Li, H.B. Zheng, Y.T. Han, J.C. & Zhou, L.J. (2011). Microstructure, mechanical properties and thermal shock behavior of h-BN-AlN ceramic composites prepared by combustion synthesis. *Journal of Alloys & Compounds.* 509(5): 1661–1664.

Li, Y.L. Zhang, J. Qiao, G.J. & Jin, Z.H. (2005). Fabrication and properties of machinable $3Y$-ZrO_2/BN nanocomposites. *Materials Science and Engineering A.* 397(1–2): 35–40.

Li, Y.Y. Qiao, G.J. & Jin, Z. H. (2002). Machinable Al_2O_3/BN composite ceramics with strong mechanical properties. *Materials Research Bulletin.* 37(8): 1401–1409.

Lipp, A. Schwetz, K.A. & Hunold, K. (1989). Hexagonal boron nitride: Fabrication, properties and applications. *Journal of the European Ceramic Society.* 5(1): 3–9.

Mori, T. Yamamura, H. Kobayashi, H. & Mitamura, T. (1993). Formation mechanism of $ZrSiO_4$ powders. *Journal of Materials Science.* 28(18): 4970–4973.

Park, H. Kim, H.W. & Kim, H.E. (1998). Oxidation and strength retention of monolithic Si_3N_4 and nanocomposite Si_3N_4-SiC with Yb_2O_3 as a sintering aid. *Journal of the American Ceramic Society.* 81(8): 2130–2134.

Vikulin, V.V. Kelina, I.Y. Shatalin, A.S. & Rusanova, L.N. (2004). Advanced ceramic structural materials. *Refractories and Industrial Ceramics.* 45(6): 383–386.

Wang, R.G. Pan, W. Chen, J. Jiang, M.N. & Fang, M.H. (2002). Fabrication and characterization of machinable Si_3N_4/h-BN functionally graded materials. *Materials Research Bulletin.* 37(7): 1269–1277.

Wang, Z. Hong, C. Zhang, X. Sun, X. & Han, J. (2009). Microstructure and thermal shock behavior of ZrB_2-SiC-graphite composite. *Materials Chemistry & Physics.* 113(1): 338–341.

Wu, H.T. & Zhang, W.G. (2010). Fabrication and properties of ZrB_2-SiC-BN machinable ceramics. *Journal of the European Ceramic Society.* 30(4): 1035–1042.

Yang, Z.H. Jia, D.C. Zhou, Y. Meng, Q.C. Shi, P.Y. & Song, C.B. (2008). Thermal shock resistance of in situ formed SiC-BN composites. *Materials Chemistry & Physics.* 107(s2–3): 476–479.

Zhang, X.H. Zhang, R.B. Chen, G.Q. & Han, W.B. (2008). Microstructure, mechanical properties and thermal shock resistance of hot-pressed ZrO_2(3Y)-BN composites. *Materials Science and Engineering A.* 497(1–2): 195–199.

Zimmermann, J.W. Hilmas, G.E. & Fahrenholtz, W.G. (2008). Thermal shock resistance of ZrB_2 and ZrB_2-30% SiC. *Materials Chemistry & Physics.* 112(1): 140–145.

Zhang, X.H. Wang, Z. Hu, P. Han, W.B. & Hong, C.Q. (2009). Mechanical properties and thermal shock resistance of ZrB_2-SiC ceramic toughened with graphite flake and SiC whiskers. *Scripta Materialia.* 61(8): 809–812.

Zhang, X.H. Xu, L. Du, S.Y. Han, W.B. Han, J.C. & Liu, C.Y. (2008). Thermal shock behavior of SiC-whisker-reinforced diboride ultrahigh-temperature ceramics. *Scripta Materialia.* 59(1): 55–58.

Electronic Engineering – Wang (ed)
© 2018 Taylor & Francis Group, London, ISBN 978-1-138-60260-1

Evaluation model for traffic pollution control using multi-attribute group decision-making based on pure linguistic information

H. Liu
Postdoctoral Research Center of Computer Science and Technology,
Harbin University of Science and Technology, Harbin, China
School of Software, Harbin University of Science and Technology, Harbin, China

S. Zhang
School of Computer Science and Technology, Harbin Engineering University, Harbin, China
School of Informatics and technology, Heilongjiang University, Harbin, China

V.V. Krasnoproshin
Faculty of Applied Mathematics and Computer Science, Belarusian State University, Minsk, Belarus

C.X. Zhang, B. Zhang, B. Yu, H.W. Xuan & Y.D. Jiang
School of Software, Harbin University of Science and Technology, Harbin, China

ABSTRACT: In order to effectively evaluate the advantages and disadvantages of uncertain traffic pollution control programs, an index system influencing the choice of programs is constructed. A multi-attribute decision-making model using linguistic variables is presented to describe the uncertainties. All preferences of decision maker to different programs are considered in the model. The uncertainty of traffic pollution control programs is covered by linguistic variables. Five programs for traffic pollution control are taken as examples to verify the validity of the model and algorithm in a detailed case analysis. Results from the model proposed are compared to those from the models with no subjective preferences of decision makers. The traffic pollution control decision-making solution is more scientific and rational.

Keywords: Traffic pollution control; Uncertain linguistic variable; Evaluation model; Pure linguistic information

1 INTRODUCTION

Exhaust from motor vehicles is the main component of urban air pollution. It is related to planning, designing, management and control of traffic. There are many control programs, such as signal control, speed control, one-way traffic and bus priority. Therefore, evaluating the traffic pollution control programs is an urgent problem to be solved for traffic management departments (Moltchanov et al. 2015).

Zhang et al. (2012) studied the traffic flow allocation model considering traffic pollution emissions; Ye et al. (2016) studied the traffic capability of urban road sections; Lv et al. (2006) studied the multi-objective double-layer programming model for urban expressway ramp and pollution control; Dijkema et al. (2008) reveals the relationship between highway speed and traffic exhaust. AHP method, principal component analysis method, survey on experts and other traditional methods are generally taken. In most of these studies, it is needed to assess the program to give a quantitative assessment of information directly. However, when the experts are restricted due to a number of subjective and objective factors, their preferences on programs affect ranking and optimization results. So, it can be seen from the above, traffic pollution control program evaluation with the uncertainty information is more realistic.

Because of complexity of objective things, uncertainty and fuzziness of human mind, the attribute value of the problem is evaluated by the qualitative measure of uncertain linguistic variables, and the problem of linguistic multiple attribute decision making (LMADM) has been widely studied (Höhle & Rodabaugh 2012; Liu et al. 2016; Wang et al. 2016; Wei 2009; Xu 2015; Xu & Liao 2015; Zheng et al. 2015). On the basis of constructing the operation rules for uncertain linguistic variables, a multi-attribute decision-making model with pure linguistic attributes is constructed. Obtain the comprehensive attribute

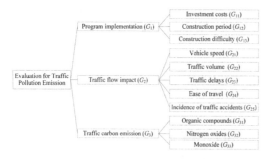

Figure 1. Index system of evaluation for traffic pollution.

value of each program, and sort and choose the best decision program, and then select the best traffic pollution emission control program,

A case study is presented comparing the result sequence of five programs in case with subjective preference and without subjective preference to verify the validity of the model and algorithm.

2 EVALUATION INDEXES

The selection of appropriate evaluation indexes is the core of evaluation of traffic pollution control programs. The comprehensive evaluation set influences the traffic pollution discharge control program. And it is determined from the viewpoints of program implementation, traffic flow impact and traffic carbon discharge, based on the research results of the literatures.

The index system of evaluation for traffic pollution discharge is shown in Figure 1. The program implementation is set for the collection of investment costs, the construction period and construction difficulty; traffic flow impact is set for the collection of vehicle speed, traffic volume, traffic delays, ease of travel and the incidence of traffic accidents; traffic carbon discharge is set for the collection of organic compounds, nitrogen oxides and monoxide.

3 MULTI-ATTRIBUTE GROUP DECISION-MAKING MODEL USING PURE LINGUISTIC

For a certain city, there are m control programs of traffic pollution discharge in set $X = \{X_1, X_2, \ldots, X_m\}$, which involves n attributes in sets $G = \{G_1, G_2, \ldots, G_n\}$. Considering the complexity of objects and the fuzziness of human thinking, linguistic variables are used measuring attribute $G_j \in G$ of any program $X_i \in X$. The decision matrix $\tilde{R} = (\tilde{r}_{ij})_{m \times n} = (r_{ij}^L, r_{ij}^U)_{m \times n}$ is made up with uncertain linguistic variable. Suppose that each decision-maker has subjective preference for any program $X_i \in X$. Similarly, the subjective preference is given by using the linguistic variable $\tilde{\theta}_i = (\theta^L, \theta^R)$, $\tilde{S}$ is a set of uncertain language variables and $\tilde{\theta}_i, \tilde{r}_{ij} \in \tilde{S}$ are linguistic variables. The multi-attribute group decision-making method is given as below for traffic pollution discharge control using pure linguistic.

3.1 Formal definitions

The decision-maker can pre-set the linguistic evaluation scale $S = \{s_i | i = -t, \ldots, t\}$, when dealing with uncertain things. And $\tilde{S} = \{s_a | a \in [-q, q], (q \geq t)\}$ is an extended scale defined based on S to facilitate the calculation and to avoid loss of decision information. In general, decision-maker takes original term $s_i \in S$ to evaluate decision-making programs, then utilizes extended term $s_i \in \tilde{S} - S$ to carry out the operations and rank all programs (Zhang, Lv & Zietsman 2012).

Let $\tilde{S} = [s_\alpha, s_\beta]$ be uncertain linguistic variable, $s_\alpha, s_\beta \in \tilde{S}$ denote its upper and lower limits respectively. For any three linguistic variables $\tilde{S} = [s_\alpha, s_\beta]$, $\tilde{S}_1 = [s_{\alpha 1}, s_{\beta 1}]$ and $\tilde{S}_2 = [s_{\alpha 2}, s_{\beta 2}]$, the operation rules are defined as follows:

$$\tilde{S}_1 \oplus \tilde{S}_2 = [s_{\alpha 1}, s_{\beta 1}] \oplus [s_{\alpha 2}, s_{\beta 2}] = [s_{\alpha 1 + \alpha 2}, s_{\beta 1 + \beta 2}] \quad (1)$$

$$\lambda \tilde{S} = \lambda [s_\alpha, s_\beta] = [s_{\lambda \alpha}, s_{\lambda \beta}], \quad \lambda \in [0,1] \quad (2)$$

Definition 1: Let $len(\tilde{S}_1) = (\beta_1 - \alpha_1)$ and $len(\tilde{S}_2) = (\beta_2 - \alpha_2)$ be the length of two linguistic variables $\tilde{S}_1 = [s_{\alpha 1}, s_{\beta 1}]$ and $\tilde{S}_2 = [s_{\alpha 2}, s_{\beta 2}]$, possibility of $\tilde{S}_1 \geq \tilde{S}_2$ is defined as:

$$p(\tilde{S}_1 \geq \tilde{S}_2) = \frac{\max[0, len(\tilde{S}_1) + len(\tilde{S}_2) - \max(\beta_2 - \alpha_2, 0)]}{len(\tilde{S}_1) + len(\tilde{S}_2)}.$$

Definition 2: The distance between two linguistic variables $\tilde{S}_1 = [s_{\alpha 1}, s_{\beta 1}]$ and $\tilde{S}_2 = [s_{\alpha 2}, s_{\beta 2}]$, possibility of $\tilde{S}_1 \geq \tilde{S}_2$ is defined as:

$$d(\tilde{S}_1, \tilde{S}_2) = \frac{1}{2}(|\beta_1 - \beta_2| + |\alpha_1 + \alpha_2|).$$

Definition 3: If $WAO : (\tilde{S})^n \to \tilde{S}$ and $WAO_{S_\lambda}(\tilde{S}_1, \tilde{S}_2, \cdots, \tilde{S}_n) = \sum_{j}^{n} w \tilde{S}_j$, then the function WAO is defined as the weighted average operator of n-dimensional uncertain linguistic variables, where $w = (w_1, w_2, \cdots, w_n)^T$ is a weighted weight vector such that $\sum_{j=1}^{n} w_j = 1 \, (w_j \geq 0)$.

3.2 Decision process with pure linguistic variables

The problem of multi-attribute decision-making for traffic pollution control has been studied to improve the applicability of this model for the uncertain factors. In the solution process, values of pure linguistic variables on attributes of programs are given by decision-maker based on their subjective preferences. Considering the unknown attributes weights, decision-making programs are sorted and selected aiming for the minimum deviation between subjective and objective preference values of the decision-maker. Therefore, main steps of problem are listed as follows.

Step 1: Let the decision-maker measure the attribute $G_j \in G$ for the program $X_i \in X$ and construct the linguistic decision matrix $\tilde{R} = (\tilde{r}_{ij})_{m \times n} = (r_{ij}^L, r_{ij}^U)_{m \times n}$, and set the subjective preference of program $X_i \in X$ as $\tilde{\theta}_i = (\theta^L, \theta^R)$.

Step 2: The optimal weight vector w is obtained by minimizing the deviation of the subjective and objective values of the decision maker. The main procedure is listed as follows.

1) According to the determination of the attribute weight value, the multi-objective optimization model $\min D_i(w) = \sum_{j=1}^{n} d^2(\tilde{r}_{ij}, \tilde{\theta}_i) w_j^2 = 1 \ (w_j \geq 0)$ is established; where $d(\tilde{r}_{ij}, \tilde{\theta}_i)$ represents the deviation between objective preference value $\tilde{r}_{ij}$ and the subjective preference value $\tilde{\theta}_i$ of decision-maker on attribute G_j of program X_i.

2) Similarly, since each program is fair and there is not any preference relation, the above LMADM problem are transformed into a single-objective optimization problem $\min N_i(w) = \sum_{j=1}^{n} d^2(\tilde{r}_{ij}, \tilde{r}_j^+) w_j^2$.

3) According to the idea of combinatorial method, the objective function $\min H(w) = aD(w) + bN$ $(w) = \sum_{i=1}^{m} \sum_{j=1}^{n} [(ad^2(\tilde{r}_{ij}, \tilde{\theta}_i) + bd^2(\tilde{r}_{ij}, \tilde{r}_j^+)) w_j^2]$ is constructed by minimizing the sum of the total deviations of all the decision-making programs; where a and b are the degrees of preference on two different information of decision-maker with $a + b = 1$, and $a, b \geq 0$.

4) Construct the Lagrangian function $L(w, \lambda) =$ $\sum_{i=1}^{m} \sum_{j=1}^{n} [(ad^2(\tilde{r}_{ij}, \tilde{\theta}_i) + bd^2(\tilde{r}_{ij}, \tilde{r}_j^+)) w_j^2] + 2\lambda$ $\left(\sum_{j=1}^{n} w_j - 1\right)$. For solving this problem, be Find the partial derivative $\frac{\partial L(w, \lambda)}{\partial w_j} = 0$ and $\frac{\partial L(w, \lambda)}{\lambda w_j} = 0$ taking w_j and λ, and $w_j = \left(\sum_{j=1}^{n} \left(\sum_{i=1}^{m} (ad^2(\tilde{r}_{ij},$ $\tilde{\theta}_i) + bd^2(\tilde{r}_{ij}, \tilde{r}_j^+))\right)^{-1}\right)^{-1} \cdot \left(\sum_{i=1}^{m} (ad^2(\tilde{r}_{ij}, \tilde{\theta}_i) +$ $bd^2(\tilde{r}_{ij}, \tilde{r}_j^+))\right)^{-1}$ is received.

Step 3: Taking the *WAO* operator, the comprehensive attribute value of each program $\tilde{z}_i(w) =$ $WAO_w(\tilde{r}_{i1}, \tilde{r}_{i2}, \cdots, \tilde{r}_{in}) = w_1 \tilde{r}_{i1} \oplus w_2 \tilde{r}_{i2} \oplus \cdots \oplus w_n \tilde{r}_{in}$ is obtained for the uncertain linguistic decision matrix $\tilde{R} = (\tilde{r}_{ij})_{m \times n}$. Obviously, the greater $\tilde{z}_i(w)$ of any program X_i is, the higher its ranking is.

Step 4: Comparing $\tilde{z}_i(w)$ and $\tilde{z}_j(w)$ of any two programs X_i and X_j, the complementary judgment matrix $P = (p_{ij})_{m \times n}$ is constructed; where $p_{ij} = p(\tilde{z}_i(w) \geq \tilde{z}_j(w))$ satisfying $(\forall i \neq j)p_{ij} + p_{ji} = 1$ $(p_{ij} \geq 0)$ and $(\forall i = j)p_{ij} = 0.5$.

Step 5: Rank the corresponding programs according to value of $\omega_i = \frac{1}{m(m-1)}(\sum_{j=1}^{n} p_{ij} + \frac{m}{2} - 1)$, and ranking vector $\omega = (\omega_1, \omega_2, \cdots, \omega_m)^T$ of the complementary judgment matrix P is obtained.

4 CASE ANALYSIS

For the traffic pollution control problem in a certain city, there are 5 programs $X_1 \sim X_5$, 11 evaluation attributes $G_1 \sim G_2$ and set linguistic assessment scales $S = \{s_{-4} = extremely\ poor, s_{-3} = very\ poor, s_{-2} = poor, s_{-1} = little\ poor, s_0 = normal, s_1 = slightly\ better, s_2 = good, s_3 = very\ good, s_4 = extremely\ good\}$. The resulting uncertain linguistic decision matrix is shown below with data collected by using Oracle. The programs' ranking results are respectively calculated to verify the effectivity of model and algorithm, under the two scenarios whether the decision-maker has subjective preferences on five programs.

$$
\begin{array}{c}
X_1 \\
X_2 \\
X_3 \\
X_4 \\
X_5
\end{array}
\begin{array}{ccccccccccc}
G_1 & G_2 & G_3 & G_4 & G_5 & G_6 & G_7 & G_8 & G_9 & G_{10} & G_{11}
\end{array}
$$

$$
\left[
\begin{array}{ccccccccccc}
[s_2, s_3] & [s_3, s_4] & [s_1, s_2] & [s_1, s_2] & [s_2, s_3] & [s_3, s_4] & [s_1, s_2] & [s_1, s_2] & [s_{-1}, s_2] & [s_{-2}, s_2] & [s_0, s_2] \\
[s_1, s_2] & [s_{-1}, s_3] & [s_3, s_4] & [s_2, s_3] & [s_1, s_2] & [s_{-4}, s_2] & [s_3, s_3] & [s_{-2}, s_4] & [s_1, s_2] & [s_{-3}, s_4] & [s_2, s_3] \\
[s_2, s_4] & [s_1, s_3] & [s_1, s_2] & [s_0, s_1] & [s_{-2}, s_2] & [s_1, s_4] & [s_2, s_3] & [s_1, s_2] & [s_2, s_3] & [s_2, s_3] & [s_{-4}, s_2] \\
[s_2, s_3] & [s_1, s_2] & [s_2, s_3] & [s_2, s_3] & [s_2, s_4] & [s_3, s_4] & [s_1, s_3] & [s_3, s_4] & [s_0, s_3] & [s_1, s_4] & [s_{-1}, s_4] \\
[s_3, s_4] & [s_0, s_3] & [s_1, s_4] & [s_3, s_4] & [s_{-1}, s_2] & [s_{-3}, s_1] & [s_2, s_4] & [s_1, s_2] & [s_{-1}, s_3] & [s_{-1}, s_2] & [s_2, s_4]
\end{array}
\right]
$$

In the case when decision-maker has subjective preferences on five programs, the values of subjective preference of five alternatives are $\tilde{\theta}_1 = (s_2, s_3), \tilde{\theta}_2 = (s_{-2}, s_0), \tilde{\theta}_3 = (s_{-3}, s_{-2}), \tilde{\theta}_4 = (s_{-1}, s_1)$ and $\tilde{\theta}_5 = (s_3, s_4)$. Then the calculation process progresses in the model defined in Section 3.

Step 1: Let $a = b = \frac{1}{2}$, the attribute weight vector is calculated as $w = (0.0894\ 0.0990\ 0.0982\ 0.0900\ 0.1186\ 0.8228\ 0.0960\ 0.0779\ 0.2383\ 0.1980\ 0.0978)^T$.

Step 2: The comprehensive attribute values of five alternatives X_i are calculated as: $\tilde{z}_1(w) = [s_{5.58},$

$s_{6.10}], \tilde{z}_2(w) = [s_{-2.74}, s_{5.18}], \quad \tilde{z}_3(w) = [s_{1.71}, s_{6.42}],$
$\tilde{z}_4(w) = [s_{3.79}, s_{7.29}]$ and $\tilde{z}_5(w) = [s_{1.92}, s_{4.51}]$.

Step 3: Construct the complementary judgment matrices as follows by comparing the $\tilde{z}_i(w)$ of any two programs.

$$P = \begin{bmatrix} 0.5000 & 1 & 0.8394 & 0.5721 & 1 \\ 0 & 0.5000 & 0.7086 & 0.1217 & 0.4948 \\ 0.1606 & 0.2914 & 0.5000 & 0.3203 & 0.7487 \\ 0.4279 & 0.8783 & 0.6797 & 0.5000 & 0.9275 \\ 0 & 0.5052 & 0.2513 & 0.0725 & 0.5000 \end{bmatrix}$$

Step 4: The ranking vector $\omega = (0.2706\ 0.0913\ 0.1011\ 0.1707\ 0.0665)^T$ of the complementary judgment matrix P is calculated. Taking ω_i to rank $\tilde{z}_i(w)$ in descending order, the result sequence is $X_1 > X_4 > X_3 > X_2 > X_5$.

And in the case when decision-maker has no subjective preferences on five programs, the values of subjective preference are $\tilde{\theta}_i = 0$. Likewise, the result sequence of values of $\tilde{z}_i(w)$ in descending order is $X_4 > X_3 > X_5 > X_2 > X_1$.

5 CONCLUSION

A multi-attribute group decision-making model using pure linguistic information is proposed to evaluate programs of traffic pollution control. In the model the evaluation index system was constructed. This model takes into account not only the uncertainties of the decision-making process of traffic pollution emission control, but also the preference of the different programs when the subjective and objective factors are restricted. A detailed case shows that using this model is simple and easy to be operate, the best program can be quickly and effectively analyzed out.

ACKNOWLEDGEMENTS

This work is sponsored by University Nursing Program for Young Scholars with Creative Talents in Heilongjiang Province (UNPYSCT-2016037), Natural Science Foundation of Heilongjiang Province of China (F2015041), Science and Technology Research Project of the Education Department of Heilongjiang Province (12541150), Harbin Science and Technology Innovation Talent Research Special Funds (2016RAQXJ039), Higher Education Research Project in "The thirteenth Five-Year Plan" of Heilongjiang Higher Education Association (16Q081) and Education and Teaching Research Project of Harbin University of Science and Technology (220160014). And this project was supported by Hei Long Jiang Postdoctoral Foundation and China Postdoctoral Science Foundation.

REFERENCES

Dijkema, M. Zee, S. & Brunek, B et al. (2008). Air quality effects of an urban highway speed limit reduction. *Atmospheric Enironment.* 42: 9098–9105.

Höhle, U. & Rodabaugh, S. (2012). *Mathematics of fuzzy sets: logic, topology, and measure theory.* Springer Science & Business Media.

Liu, X. Ju, Y. & Yang, S. (2016). Some generalized interval-valued hesitant uncertain linguistic aggregation operators and their applications to multiple attribute group decision making. *Soft Computing.* 20(2): 495–510.

Lv, Z. Fan, B. & Liu, J et al. (2006). Bi-level Multi-objective Programming Model for the Ramp Control and Pollution Control on Urban Expressway Networks. *Control & Decision.* 21(1): 64–52.

Moltchanov, S. Levy, I. & Etzion, Y et al. (2015). On the feasibility of measuring urban air pollution by wireless distributed sensor networks. *Science of the Total Environment.* 502: 537–547.

Wang, J. Wang, J. & Zhang, H. (2016). Multi-criteria group decision-making approach based on 2-tuple linguistic aggregation operators with multi-hesitant fuzzy linguistic information. *International Journal of Fuzzy Systems.* 18(1): 81–97.

Wei, G. (2009). Uncertain linguistic hybrid geometric mean operator and its application to group decision making under uncertain linguistic environment. *International Journal of Uncertainty, Fuzziness and Knowledge-Based Systems.* 17(02): 251–267.

Xu, Z. (2015). *Uncertain multi-attribute decision making: Methods and applications.* Springer.

Xu, Z. & Liao, H. (2014). Intuitionistic fuzzy analytic hierarchy process. *IEEE Transactions on Fuzzy Systems.* 22(4): 749–761.

Ye, P. Wu, B. & Fan, W. (2016). Modified Betweenness-Based Measure for Traffic Flow Prediction of Urban Roads. *Transportation Research Record Journal of the Transportation Research Board.* 2563(19).

Zhang, Y. Lv, J. & Zietsman, J. (2012). A convex approach to traffic assignment problem under air quality constraints for planning purposes. *Advances in Transportation Studies.* 12: 85–96.

Zheng, X. Gu, C. & Qin, D. (2015). Dam's risk identification under interval-valued intuitionistic fuzzy environment. *Civil Engineering and Environmental Systems.* 32(4): 351–363.

Electronic Engineering – Wang (ed)
© 2018 Taylor & Francis Group, London, ISBN 978-1-138-60260-1

A bid evaluation model of a hydropower project based on the theory of intuitionistic fuzzy sets

H. Liu
Postdoctoral Research Center of Computer Science and Technology,
Harbin University of Science and Technology, Harbin, China
School of Software, Harbin University of Science and Technology, Harbin, China

S. Zhang
School of Computer Science and Technology, Harbin Engineering University, Harbin, China
School of Informatics and Technology, Heilongjiang University, Harbin, China

C.X. Zhang, Y.P. Liu, B. Yu & B. Zhang
School of Software, Harbin University of Science and Technology, Harbin, China

V.V. Krasnoproshin
Faculty of Applied Mathematics and Computer Science, Belarusian State University, Minsk, Belarus

H.W. Xuan & J.F. Gao
School of Software, Harbin University of Science and Technology, Harbin, China

ABSTRACT: Project bidding plays a decisive role in the fulfilment of a hydropower project. Bid evaluation involves many factors, and it is difficult to evaluate directly. In this paper, an intuitionistic fuzzy linear evaluation model is established, based on the theory of intuitionistic fuzzy sets. Indexes such as quoted price, construction design and business conditions are selected as evaluation factors of the index matrix. Degrees of membership, non-membership and hesitancy are all considered. The priority order of the bidding companies can be obtained by use of this scoring function and the TOPSIS method. The validity and practicability of this model were verified by a case analysis.

Keywords: intuitionistic fuzzy set; hydropower project bidding; group decision-making; bid evaluation

1 INTRODUCTION

Bidding is a common, scientific and rational way of contracting a project. Choosing the right contractor through the form of bidding is a key link in project construction and plays a decisive role in the successful realisation of a project construction plan. The evaluation work is the core of the whole bidding exercise, and selecting the best bidder effectively determines the success or failure of the construction project (Kun & Maoshan, 2004).

Many scholars have conducted research on bid evaluation methods. Sheng et al. (2008) established a fuzzy comprehensive evaluation of water conservancy project evaluation methods; Shemshadi et al. (2011) introduced the entropy weight theory into the decision-making associated with bid evaluation; Zheng et al. (2015) established a hydropower project evaluation model based on composite elements and information entropy. Although the above-mentioned work has improved the accuracy of bid evaluation to a certain extent, most of the evaluations need experts to generate an exact score for each evaluation index based on personal experiences. In these methods, only the index of membership is considered. In Intuitionistic Fuzzy Set (IFS) theory, the degrees of membership, non-membership and hesitancy are all taken into account. IFS is more expressive and scientific than traditional methods of dealing with fuzzy and uncertain information. It has been widely used in many fields, such as feasibility analysis, electronic testing, building engineering, and system optimisation, and has produced good results.

IFS theory can address hydropower project bidding and make the evaluation result more rational. In this paper, appropriate evaluation factors are selected according to IFS theory, and a corresponding bid evaluation model of a hydropower project is established.

Table 1. The evaluation index of a bid evaluation system.

Primary index	Secondary indexes	Index features
Quoted price (C_1)	Total bid price (C_{11})	The total bid offer that bidder is willing to take for completing the project.
	Tender offer rationality (C_{12})	Rationality of the composition of bidding price.
Construction design (C_2)	Main construction methods (C_{21})	The reasonableness of the main construction methods adopted by the bidder.
	Organisation and technical personnel quality (C_{22})	Reasonableness of bidder's organisation structure and professional quality of bid technical personnel.
	Construction schedule and schedule guarantee measures (C_{23})	The reasonableness and reliability of the schedule arrangement and guarantee measures of the bidder.
	Quality assurance system and quality assurance measures (C_{24})	Rationality and reliability of assurance measures, and bidders' schedule.
	Safety, environmental protection and civilised construction measures (C_{25})	Reasonableness of bidders' safety, environmental protection and civilised construction measures arrangements.
Business conditions (C_3)	Construction performance and experience (C_{31})	Amount of similar projects completed by bidders in recent years.
	Financial status (C_{32})	Financial situation of bidders in recent years.
	Enterprise credit (C_{33})	Bidders' enterprise credit.

A detailed case study demonstrates the whole evaluation process, and the conclusion summarises this work and its main contributions.

2 EVALUATION INDEX SYSTEM

The construction of a hydropower project is very complex, and many factors influence project construction. In this paper, by reference to the relevant research (Sharma & Thakur, 2015; Chen & Zou, 2015), an evaluation index system is built, as shown in Table 1, that considers all factors.

3 EVALUATION MODEL ESTABLISHMENT

3.1 Intuitionistic fuzzy sets

Let X be a given set; IFS belonging to X is $A = \{\langle x, \mu_A(x), v_A(x)\rangle | x \in X\}$, where $\mu_A(x)$ represents the membership function of x to A, and $v_A(x)$ is the non-membership function of x to A.

$$\mu_A(x): X \to [0, 1], x \in \mu_A(x) \to \mu_A \in [0, 1]$$

$$v_A(x): X \to [0, 1], x \in v_A(x) \to v_A \in [0, 1]$$

and $0 \le \mu_A(x) + v_A(x) \le 1$ holds for all elements $|x \in X$ that belong to A. $\pi_A(x) (x \in X)$ is an intuitionistic index, indicating the degree of uncertainty or hesitancy about a problem.

$$\pi_A(x) = 1 - \mu_A(x) - v_A(x) \tag{1}$$

In addition, the Intuitionistic Fuzzy Number (IFN) is expressed by $\alpha = (\mu_\alpha, v_\alpha)$, where $\mu_\alpha \in [0, 1]$, $v_\alpha \in [0, 1]$, and $\mu_\alpha + v_\alpha \le 1$. Obviously $\alpha^+ = (1, 0)$, and is called the maximum IFN; $\alpha^- = (1, 0)$ is called the minimum IFN. For any fuzzy number, the formula of the score function can be defined as:

$$s(\alpha) = \mu_\alpha - v_\alpha \tag{2}$$

where $s(\alpha)$ represents the score of α; obviously $s(\alpha) \in [0, 1]$. The greater the value of $s(\alpha)$, the higher the satisfaction of the decision maker. However, the scoring function operates with significant limitations when used in a group decision-making process, and the score function is corrected according to this situation:

$$s(\alpha_i) = (\mu_i - v_i) + (\mu_i - v_i)\pi_i$$
$$= (\mu_i - v_i)(1 + \pi_i) \tag{3}$$

In this function, the proportion of those who tend to vote in favour of voting is assigned to $\mu_i\pi_i$, and the proportion of voted to vote is $v_i\pi_i$ (Joshi & Kumar, 2016; Xu, 2015).

3.2 Index attributes set

In the evaluation system of a multi-objective group decision-making problem, the evaluation object set is determined as $A = \{A_1, A_2, ..., A_N\}$, and there are two levels of evaluation indexes under this evaluation object set. Set $C = \{C_1, C_2, ..., C_N\}$ as the primary index attribute set, and set $C_i = \{C_{i1}, C_{i2}, ..., C_{im}\}$ as the secondary index attribute set. Let $W = \{\omega_1, \omega_2, ..., \omega_n\}$ denote the weight of the first index attribute set and $W_i = \{\omega_{i1}, \omega_{i2}, ..., \omega_{in}\}$ denote the weight of the second index attribute set, where $\sum_{i=1}^{n} w_i = 1$ and $\sum_{j=1}^{n} w_{ij} = 1$. In the multi-attribute analysis of index C_i, μ_{ij} is denoted as the membership degree of the evaluation object in relation to the evaluation index attribute C_{ij}; v_{ij} is denoted as the non-membership degree of the evaluation object attribute C_{ij}; and $\alpha_{ij} = (\mu_{ij}, v_{ij})$, $\pi_{ij} = 1 - \mu_{ij} - v_{ij}$. The larger the intuition index π_{ij}, the higher the hesitancy margin for satisfaction of the evaluator (Höhle & Rodabaugh, 2012; Xu & Liao, 2014).

3.3 Weights solving model with multiple attributes

Similarly to the determination of IFN, the weight of each factor in the evaluation system can be expressed by ρ_{ij} and τ_{ij}, which mean the degree of the determination to the importance of all the attributes C_{ij}. ρ_{ij} and τ_{ij} are also known as the membership degree and non-membership degree of C_{ij} in the fuzzy concept of "important". It is necessary to satisfy the conditions $0 \leq \rho_{ij}$, $\tau_{ij} \leq 1$ and $0 \leq \rho_{ij} + \tau_{ij} \leq 1$. $\xi_{ij} = 1 - \rho_{ij} - \tau_{ij}$ represents the intuitionistic index, and the greater it is, the greater the degree of uncertainty of importance of the decision index C_{ij} to the decision-making objective. In practical application, the dependency interval of assignment degree from experts is $[\omega_{ij}^l, \omega_{ij}^u] = [\rho_{ij}, \rho_{ij} + \xi_{ij}]$, where $\omega_{ij}^l = \rho_{ij}$ and $\omega_{ij}^u = \rho_{ij} + \xi_{ij}$. The weight ω_{ij} should satisfy $\omega_{ij}^l \leq \omega_{ij} \leq \omega_{ij}^u$ and $\sum_{j=1}^{m} \omega_{ij} = 1$.

According to the minimisation principle of the weighted average of the intuitionistic index, a linear solution model can be constructed as follows (Zhang et al., 2013):

$$\min d = \sum_{j=1}^{m} \sum_{a=1}^{n} \pi_{ij}^{A_a} \omega_{ij} \tag{4}$$

The optimal solution vector of the model can be solved by Wolfram Mathematica software, and $W_i = \{\omega_{1j}, \omega_{2j}, \ldots, w_{ij}\}$ can be obtained.

3.4 TOPSIS method

The Technique for Order of Preference by Similarity to Ideal Solution (TOPSIS) method is a sorting approach based on the degree of closeness between evaluation object and ideal solution. The method is used to sort alternatives, calculating distance between an object with optimal solution and the worst solution, that is, the value of scoring function is used as the evaluation scale; this method is considered for excellence to be close to a positive ideal point and far away from negative ideal point (Vahdani et al., 2015).

The formula is expressed as follows, calculating the distance between scoring functions $S(\alpha)$ and $S(\beta)$ of the IFNs α and β:

$$d(S(\alpha), S(\beta)) = \frac{1}{2}|S(\alpha) - S(\beta)|$$
$$= \frac{1}{2}|(\mu_\alpha - v_\alpha)(1 + \pi_\alpha) - (\mu_\beta - v\beta)(1 + \pi_\beta)| \tag{5}$$

From the definition of IFN, the maximal value of IFN is $(1, 0)$, and $A^+ = \{\alpha_1^+, \alpha_2^+, \ldots, \alpha_n^+\}^T$ is defined as the intuitionistic fuzzy positive ideal point, where $\alpha_i^+ = (1, 0)$, $i = (1, 2, \ldots, n)$, and the score vector is $S(A^+) = (1, 1, \ldots, 1)_{1 \times n}^T$. Similarly, $\alpha^- = (1, 0)$ is the least IFN, and its score vector is $S(A^-) = (-1, -1, \ldots, -1)_{1 \times n}^T$. The formulas are expressed as follows, calculating the distance between the score vector of each target A_α and the positive and negative ideal points A^+ and A^+:

$$d(S(A_\alpha), S(A^+)) = \sum_{i=1}^{n} \omega_i d(S(\alpha_i) - S(\alpha_i^+))$$
$$= \frac{1}{2} \sum_{i=1}^{n} \omega_i |(\mu_i - v_i)(1 + \pi_i) - 1| \tag{6}$$

Table 2. Statistical results of the evaluation of target index.

Evaluated company	Business conditions (C_3)		
	C_{31}	C_{32}	C_{33}
A_1	(0.62,0.17)	(0.76,0.12)	(0.51,0.37)
A_2	(0.52,0.21)	(0.74,0.09)	(0.63,0.25)
A_3	(0.44,0.33)	(0.69,0.16)	(0.62,0.21)
A_4	(0.48,0.28)	(0.69,0.21)	(0.59,0.23)

$$d(S(A_\alpha), S(A^-)) = \sum_{i=1}^{n} \omega_i d(S(\alpha_i) - S(\alpha_i^-))$$
$$= \frac{1}{2} \sum_{i=1}^{n} \omega_i |(\mu_i - v_i)(1 + \pi_i) + 1| \tag{7}$$

Thus, an evaluation index $R(A_\alpha)$ reflecting the satisfactory degree of the decision object is obtained as follows:

$$R(A_\alpha) = \frac{d(S(A_\alpha), S(A^-))}{d(S(A_\alpha), S(A^+)) + d(S(A_\alpha), S(A^-))} \tag{8}$$

The greater the value of $R(A_\alpha)$, the more satisfactory the decision object A_α.

4 CASE ANALYSIS

A hydropower project was put out to public tender, and four companies, A_1, A_2, A_3 and A_4, passed the prequalification in the bidding. The multi-attribute evaluation algorithm based on intuitionistic fuzzy sets was used to evaluate the bids of these four companies. An expert evaluation group was set up, and the evaluation index was assigned by the expert group. Using Mathematica and Oracle 11 g software, the decision makers proceeded as follows.

Taking the primary index "Business conditions (C_3)" as an example, the results of averaging the experts' scores are shown in Table 2.

Evaluation values f indexes C_{31}, C_{32} and C_{33} of the indicators are assigned as follows:

$$\beta_{31} = (0.35,0.16), \beta_{31} = (0.32,0.24), \beta_{33} = (0.28,0.31).$$

The dependency intervals of weight according to β_{ij} are calculated as follows:

$$\omega_{31} = [0.35, 0.84], \omega_{31} = [0.32, 0.76], \omega_{33} = [0.28, 0.69].$$

Using Equation 1, the following can be obtained:

for A_1, $\pi_{31}^{A_1} = 0.21$, $\pi_{32}^{A_1} = 0.12$, $\pi_{33}^{A_1} = 0.12$;
for A_2, $\pi_{31}^{A_2} = 0.27$, $\pi_{32}^{A_2} = 0.17$, $\pi_{33}^{A_2} = 0.12$;
for A_3, $\pi_{31}^{A_3} = 0.23$, $\pi_{32}^{A_3} = 0.15$, $\pi_{33}^{A_3} = 0.17$;
for A_4, $\pi_{31}^{A_4} = 0.24$, $\pi_{32}^{A_4} = 0.10$, $\pi_{33}^{A_4} = 0.18$.

Table 3. Primary index values and overall evaluations.

Evaluated company	C_1	C_2	C_3	Overall evaluation
A_1	(0.75,0.16)	(0.60,0.23)	(0.64,0.21)	0.762
A_2	(0.73,0.16)	(0.50,0.35)	(0.63,0.18)	0.717
A_3	(0.53,0.28)	(0.56,0.28)	(0.58,0.23)	0.666
A_4	(0.42,0.29)	(0.74,0.14)	(0.59,0.24)	0.712

Using Equation 4, the following linear model can be obtained:

$$\min d = \sum_{j=1}^{m} \sum_{a=1}^{n} \pi_{ij}^{A_a} w_{ij} = 0.95\omega_{31} + 0.54\omega_{32} + 0.59\omega_{33}$$

$s.t.\ 0.35 < \omega_{31} < 0.84,\ 0.32 < \omega_{32} < 0.76,$

$0.28 < \omega_{33} < 0.69$

$\omega_{31} + \omega_{32} + \omega_{33} = 1$

The optimal weight can be calculated via Mathematica as $\omega_3 = (0.35, 0.37, 0.28)$.

The secondary indexes $\alpha_{ij} = (\mu_{ij}, \nu_{ij})$ and the optimal weight ω_{ij} are weight-averaged to calculate the primary index values as $\alpha_i = (\mu_i, \nu_i)$, where $\mu_i = \sum_{j=1}^{m} \mu_{ij}\omega_{ij}$ and $\nu_i = \sum_{j=1}^{m} \nu_{ij}\omega_{ij}$. The results are shown in Table 3. The above steps are repeated to get the secondary index weight, and the index scores and evaluation results of companies A_1, A_2, A_3 and A_4 are calculated. The optimal weight is $\omega = (0.38,0.4,0.22)$. Using the TOPSIS method, the overall evaluations of companies A_1, A_2, A_3 and A_4 are calculated as shown in Table 3. Overall, the order of priority of the companies for the bid is: $A_1 > A_2 > A_4 > A_3$. Therefore, this project should select company A_1 to win the bid.

5 CONCLUSION

Hydropower project bid evaluation is a complex system engineering problem, involving many considerations. Selecting the best bidder quickly and effectively affects the success of the construction project. In this paper, a two-level evaluation index system is established. The membership, non-membership and hesitancy of different evaluation objects are considered. A formal evaluation model is established based on intuitionistic fuzzy set theory. The bidding preference is determined with the aid of the score function and the TOPSIS method, which greatly improves the accuracy of the evaluation. A detailed case study shows that using this model, the best bidder can be quickly and effectively assessed to provide the basis for bid decision-making.

ACKNOWLEDGEMENTS

This work is sponsored by the University Nursing Programme for Young Scholars with Creative Talents in Heilongjiang Province (UNPYSCT-2016037), the Natural Science Foundation of Heilongjiang Province of China (F2015041), the Science and Technology Research Project of the Education Department of Heilongjiang Province (12541150), Harbin Science and Technology Innovation Talent Research Special Funds (2016RAQXJ039), the Higher Education Research Project in "The thirteenth Five-Year Plan" of Heilongjiang Higher Education Association (16Q081), and the Education and Teaching Research Project of Harbin University of Science and Technology (220160014). This project was supported by the Hei Long Jiang Postdoctoral Foundation and the China Postdoctoral Science Foundation.

REFERENCES

Chen, Z. & Zou, Q. (2015). Risk analysis and evaluation of hydropower EPC project cost based on entropy weight. *Water Resources and Power, 33*(2), 168–171.

Höhle, U. & Rodabaugh, S. (2012). *Mathematics of fuzzy sets: Logic, topology, and measure theory.* New York, NY: Springer Science & Business Media.

Joshi, D. & Kumar, S. (2016). Interval-valued intuitionistic hesitant fuzzy Choquet integral based TOPSIS method for multi-criteria group decision making. *European Journal of Operational Research, 248*(1), 183–191.

Kun, P. & Maoshan, Q. (2004). Study on application of Fuzzy-AHP in risk evaluation and bidding decision-making of Duber Khwar hydropower project. *Journal of Hydroelectric Engineering, 23*(3), 44–50.

Sharma, A. & Thakur, N. (2015). Resource potential and development of small hydro power projects in Jammu and Kashmir in the western Himalayan region: India. *Renewable and Sustainable Energy Reviews, 52*, 1354–1368.

Shemshadi, A., Shirazi, H., Toreihi, M. & Tarokh, M.J. (2011). A fuzzy VIKOR method for supplier selection based on entropy measure for objective weighting. *Expert Systems with Applications, 38*(10), 12160–12167.

Sheng, S., Mao, J. & Su, C. (2008). Application of fuzzy comprehensive evaluation method in bidding assessment of water conservancy works. *Yangtze River, 39*(3), 104–106.

Vahdani, B., Salimi, M. & Charkhchian, M. (2015). A new FMEA method by integrating fuzzy belief structure and TOPSIS to improve risk evaluation process. *International Journal of Advanced Manufacturing Technology, 77*(1–4), 357–368.

Xu, Z. (2015). *Uncertain multi-attribute decision making: Methods and applications.* Heidelberg, Germany: Springer.

Xu, Z. & Liao, H. (2014). Intuitionistic fuzzy analytic hierarchy process. *IEEE Transactions on Fuzzy Systems, 22*(4), 749–761.

Zhang, X., Jin, F. & Liu, P. (2013). A grey relational projection method for multi-attribute decision making based on intuitionistic trapezoidal fuzzy number. *Applied Mathematical Modelling, 37*(5), 3467–3477.

Zheng, X., Gu, C. & Qin, D. (2015). Dam's risk identification under interval-valued intuitionistic fuzzy environment. *Civil Engineering and Environmental Systems, 32*(4), 351–363.

Realisation of a virtual campus roaming system based on the Unity 3D game engine

J.H. Dong, J.W. Dong & C. Liu
Higher Educational Key Laboratory for Signal & Information Processing of Heilongjiang Province,
School of Measurement-Control Technology and Communication Engineering,
Harbin University of Science and Technology, Harbin, Heilongjiang Province, China

ABSTRACT: The digital campus is an essential component of high-level colleges and universities. As an important part of a digital campus, a virtual campus roaming system is developed in this paper taking the West Campus of Harbin University of Science and Technology (HUST) as an example. The procedures of 3D model creation for campus building, model mapping and model exporting using Autodesk 3ds Max are first illustrated. Then 3D terrain creation for the campus environment and terrain mapping are developed using the Unity 3D game engine. By importing models into the terrain and roaming via the first person controller, the system can realise the display and interactive roaming of the 3D campus map.

Keywords: Virtual campus; roaming system; 3D modelling; Unity 3D

1 INTRODUCTION

The use of the Internet has now penetrated into all aspects of many people's lives, and many colleges have been promoting the construction of digital campus information. It can be said that the construction of a digital campus is a significant trend in the information development of institutions of higher learning in the information age (Liu et al., 2014). An intelligent digital campus not only provides students with a variety of useful real-time information, but also facilitates the institution's management of students and campus facilities, improving the social impact and competitiveness of the university (Shi, 2014).

A virtual campus roaming system, which can display the campus scene information by way of a three-dimensional view, is a fundamental part of the digital campus. Taking the West Campus of Harbin University of Science and Technology (HUST) as an example, the design procedures and methods of the development of a virtual campus roaming system are analysed in detail.

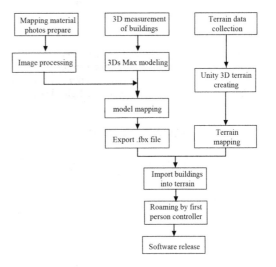

Figure 1. Diagram of development process of the virtual campus roaming system.

2 THE DEVELOPMENT PROCESS OF THE VIRTUAL CAMPUS ROAMING SYSTEM

The development process of the virtual campus roaming system is shown in Figure 1. The development process is divided into two parts: the production of the model and the production of the terrain. First, three-dimensional data of the buildings was measured, and a 3D model of the buildings on the campus was produced using Autodesk 3ds Max software. Then photos of campus buildings were taken and the images processed to obtain the model mapping materials. After model mapping, the model is exported as an FBX file. The production of the terrain in Unity 3D can

Figure 2. Photo of the main gate after compression.

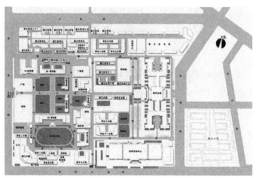

Figure 4. Plan figure of West Campus of HUST.

Figure 3. Photo of windows after cropping and fuzzy processing.

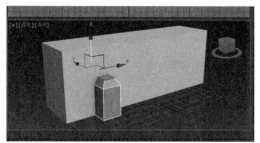

Figure 5. Shape the small box.

be accomplished by creating a terrain and mapping the terrain using the images of the campus environment. Finally, the FBX file is imported into Unity 3D to layout the buildings and render the light shadow.

3 BASIC MATERIAL PREPARATION

Real images are the raw material necessary for 3D model mapping for campus buildings. However, the direct use of high-definition photos will generate a large amount of data, causing the operation of the roaming software to judder. Before model mapping, the original photos must be processed by means such as compression and fuzzy processing. The main gate photo after compression (1024 × 680 pixels) is shown in Figure 2; the photo of a building's windows after cropping and fuzzy processing is shown in Figure 3; the campus plan figure based on field measurement is shown in Figure 4.

4 3DS MAX MODELLING

There are many buildings on campus. In the roaming system, each of them needs a corresponding 3D model based on appropriate proportions derived from field measurement data. In this paper, the procedures for 3D modelling using 3ds Max are described in detail, taking the main gate as an example.

4.1 *Creating a model*

To create a model using 3ds Max, three steps are needed as follows:

Step 1: According to the data from the field measurement of the main gate, a standard *primitive* is created.
Step 2: A small box of the size of the main gate's doorway is created and placed in the corresponding position. A *modifier* is added to the box; select *Edit mesh – Edit poly* menu items, then use the *Outline* and *Insert* tools on the right-hand side to shape the box, as shown in Figure 5.
Step 3: Combine the small box with the standard primitive. Select the primitives, use the *Compound Objects – Boolean* functions, and then select the small box. A doorway has been created in the main gate, as shown in Figure 6. Repeat step 3 to shape the main gate, as shown in Figure 7.

4.2 *Mapping the model*

In order to make the scenes in the roaming system look vivid and real, real photos of the buildings are used in mapping the 3D model so that the buildings' appearance is improved. In this paper, photos of the

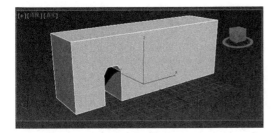

Figure 6. Create a doorway.

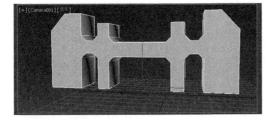

Figure 7. Main gate.

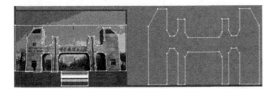

Figure 8. Front elevation mapping.

Figure 9. Main gate after model mapping.

front, back and sides of the main gate are employed using the *UVW Map* method.

Step 1: Apply *Unwrap UVW* command to the model, open *UV Coordinate Modifiers*, select the front elevation photo of the main gate as material, select the front elevation of the model, and use *Projection* to set the projecting direction, as shown in Figure 8. Repeat step 1 to map the other sides of the model, as shown in Figure 9.

Step 2: Export the model from 3ds Max as an FBX file. When selecting the *Export* menu item, check

Figure 10. Design sketch in Unity 3D.

Embedded Media to ensure the mapping images are included in the FBX file.

5 ROAMING AND RELEASE

Open Unity 3D and select *File – New Scene* menu item to create a scene. Import the FBX file to place the built model into the scene. Create terrain for the campus environment; *Import Package* can be used to obtain trees. The design sketch in Unity 3D is shown in Figure 10.

To roam on campus, free perspective is essential. *Character Controller* is imported as a resource via the *Assets – Import Package – Character Controller* menu item. In the resource package, there are two types of characters: *First Person Controller* and *Third Person Controller*. Because the first person perspective gives a greater sense of presence, the vision is more open, and the *First Person Controller* is employed to roam the demonstration. To drag the controller to a place where you want to start roaming, click the *Play* button at the top of the main interface to preview.

Using Unity 3D's multi-platform release, the roaming system can be released on both PC and Android mobile phone. To release a scene on the Android platform, the Android Software Development Kit (SDK) is needed, which can be downloaded from the official Android website. Using *Edit – Preferences – External Tools*, input the path of the Android SDK in the *Android SDK Location* item. To build an APK file, in *File – Building Settings – Platform*, check the Android item, then press *Building* button.

6 CONCLUSION

A virtual campus roaming system is not only a vivid simulation of the campus landscape, helpful for image building and the campus culture of the college, but is also an auxiliary tool for campus layout planning. The design method for the virtual campus roaming system using Unity 3D described in this paper can also be widely used in the development of virtual museums, real estate showcases, 3D games, and so on.

ACKNOWLEDGEMENTS

This project is financially supported by the Natural Science Foundation for Returned Overseas Scholars of Heilongjiang Province of China (No. LC201427) and the Heilongjiang Provincial College Students Innovation Program (No. 201510214015).

REFERENCES

Liu, H., Zhang, Z.M. & Zhou, Y. (2014). Design and implementation of 3D virtual campus in Shandong Jianzhu University. *Journal of Shandong Jianzhu University, 29,* 280–285.

Shi, M. (2014). Virtual scene construction and roaming. *Journal of System Simulation, 26,* 1969–1979.

Electronic Engineering – Wang (ed)
© 2018 Taylor & Francis Group, London, ISBN 978-1-138-60260-1

An efficient high-order masking scheme for the S-Box of AES using composite field arithmetic

J.X. Jiang, Y.Y. Zhao & J. Hou
School of Applied Sciences, Harbin University of Science and Technology, Harbin, China

X.X. Feng
School of Computer Sciences and Technology, Harbin University of Science and Technology, Harbin, China

H. Huang
School of Software, Harbin University of Science and Technology, Harbin, China

ABSTRACT: Most of the existing high-order masking schemes are based on either a lookup table or addition chains. This paper proposes an efficient new type high-order masking scheme for the S-Box using composite field arithmetic. In this scheme, the non-linear inverse function over Galois Field $GF(2^8)$ can be easily evaluated by mapping it into $GF(2^2)$, which significantly reduces the masking complexity. In order to verify the proposed scheme, various order masking schemes for AES S-Box were modelled in Verilog and implemented with EDA tools. The simulation results showed that this scheme has low hardware complexity compared with existing schemes. Furthermore, the proposed masking scheme has obvious advantages, such as provable security and supporting high-order masking.

Keywords: high-order masking scheme; AES; S-Box; composite field arithmetic

1 INTRODUCTION

Since the concept of Differential Power Attacks (DPA) was proposed by Kocher (1999), various countermeasures against DPA had been taken considerable attention. Most of them focused on the countermeasures against First-Order DPA (FODPA), which were known to be secure and practicable (Herbst et al., 2006; Kim et al., 2010). More recently, in order to counteract High-Order DPA (HODPA), high-order masking schemes have been proposed (Schramm & Paar, 2006; Rivain & Prouff, 2010). However, these high-order masking schemes provide higher security at the cost of significant time and space overheads, which seriously affect the application in practice.

Otherwise, to the best of the authors' knowledge, most of the existing masking schemes for AES S-Boxes were studied based on either Lookup Tables (LUT) (Coron, 2014) or addition chains (Carlet et al., 2012). Therefore, there was still a lack of research on high-order masking schemes for other implementation types of S-Boxes, such as composite field-based S-Box (Satoh et al., 2001). Instead of masking the intermediate values over $GF(2^8)$, this scheme masked the intermediate values over $GF(2^4)$, thus significantly reducing the number of multiplications required during an inverse operation. However, up to now, it is still a challenge to design a high-order masking scheme which can guarantee certain security with an affordable overhead. In this paper, a high-order masking scheme for S-Boxes using composite field arithmetic is proposed. Due to the adoption of a composite field arithmetic, an inverse operation in S-Box can be carried out efficiently with combinational logic; thus this scheme does not require any ROM space. The security of the proposed scheme has been proved theoretically through the method proposed in Blömer et al. (2004). Compared with existing similar schemes, the proposed scheme has lower masking complexity without any loss of security. Furthermore, the proposed scheme is not only suitable for AES S-Boxes, but also for inverse operations with affine transformation.

The remainder of the paper is organised as follows: Section 2 presents related work; Section 3 describes the composite field-based AES and its high-order masking scheme; and Section 4 concludes this paper.

2 PREVIOUS WORK

Composite field arithmetic can be used to reduce the computation cost of inverse operations in S-Boxes, which leads to compact hardware implementation. Based on this observation, various masking schemes founded on composite field arithmetic were discovered in the literature review of this study. Akkar and

121

Giraud (2001) introduced a new scheme, called transform masking, which calculated an inverse operation by transforming from Boolean masks and multiplicative masks to each other. However, it was susceptible to zero-value attack when applying it to protect DES and AES. In order to thwart zero-value attacks, Oswald et al. (2005) developed a scheme by combining the additive masks with multiplicative masks. Blömer et al. (2004) claimed that various choices of field polynomials would lead to different efficient masking schemes in the base of literature (Akkar & Giraud, 2001). In order to speed up the computation, Oswald et al. (2005) and Kim et al. (2011) replaced the inverse operation over composite field with a series precomputed LUTs. The former was based on composite field and the latter was based on an addition chain. After that, Ahn and Choi (2016) improved the performance of the former schemes. Up to now, the above mentioned schemes were the fastest schemes amongst the existing masking schemes for AES. However, they all required much more ROM space which was used to store the precomputed values, and except for the scheme (Kim et al., 2011), were limited to first-order. There was also a lack of research on high-order masking over composite fields.

As mentioned above, high-order masking schemes are provably secure, but have higher overheads both in software and hardware implementation. In this paper, a new high-order masking scheme was proposed to reduce the overhead caused by the masking processing. This scheme was based on the idea of reducing the number of multiplications by using improved arithmetic over $GF(2^4)$, and aims to offer an optimised trade-off between security and performance. In order to illustrate the proposed masking scheme, the composite field-based AES, and its high-order masking scheme, were proposed sequentially.

3 COMPOSITE FIELD-BASED AES AND ITS HIGH-ORDER MASKING SCHEME

3.1 Composite field-based AES algorithm

The AES is a block cipher composed of an SPN architecture. For a 128-bit AES of 10 rounds, each round consisted of four different layers. These layers were the SubBytes (S-Box), ShiftRows, MixColumns, and AddRoundKey. Amongst them, the S-Box was the only non-linear function.

3.2 S-Box evaluation using composite field

AES S-Box was defined by a multiplicative inverse of $b = a^{-1}$ followed by an affine operation over $GF(2^8)$. The diagram of the S-Box evaluation using a composite field is shown in Figure 1. In order to reduce the computation complexity, the inversion operation over $GF(2^8)$ was mapped into $GF(2^2)$. Since the inversion is a linear operation over $GF(2^2)$, it is easy to apply a masking scheme to it.

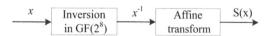

Figure 1. Diagram of the S-Box evaluation using composite field.

Figure 2. Diagram of a high-order masking scheme for AES.

3.3 High-order masking scheme for the AES

For high-order masking schemes, all input and output values in the operation must be masked. In order to meet this requirement and to make a compact hardware implementation, some of the correction terms were reused. Due to the reuse of these correction terms, the number of multiplications were significantly reduced. The diagram of a high-order masking scheme is shown in Figure 2. The XOR values of input and masks, as well as the masks, were mapped from composite field $GF(2^8)$ to $GF(2^4)$. Then all the arithmetic operations are done over $GF(2^4)$ and remapped from $GF(2^4)$ to $GF(2^8)$. Since the mapping and the operation over $GF(2^4)$ are linear operations which are very easy to mask, all the blocks in this masking scheme for AES were linear operations.

The detailed arithmetic operations over $GF(2^4)$ were analysed. The high-order masking scheme used masked every intermediate value and reused the correction terms as often as possible to reduce the area required in hardware implementation. The key point of this scheme was to obtain the inverse value over $GF(2^8)$ efficiently. It is equivalent to obtain the isomorphic elements over $GF(2^4)$. Firstly, a value of d with $d+1$ addition masks was computed.

$$\begin{aligned}
& d + m_{h0} + \cdots + m_{hd} \\
&= \underbrace{(a_h + m_{h0} + \cdots + m_{hd})^2 P}_{m_1} + \underbrace{(m_{h0} + \cdots + m_{hd})^2 P_0}_{c_1} \\
&+ \underbrace{(a_h + a_l + m_{h0} + \cdots + m_{hd})(m_{h0} + \cdots + m_{hd})}_{c_2} \\
&+ \underbrace{(a_l + m_{h0} + \cdots + m_{hd})^2}_{m_3} \\
&+ \underbrace{(m_{h0} + \cdots + m_{hd})^2}_{c_3} + \underbrace{m_{h0} + \cdots + m_{hd}}_{c_4} \\
&+ \underbrace{(a_h + m_{h0} + \cdots + m_{hd})(a_l + m_{h0} + \cdots + m_{hd})}_{m_2}
\end{aligned} \quad (1)$$

The masked d was computed over $GF(2^2)$ to get its inverse value with the same $d+1$ addition masks $d^{-1} + m_{h0} + \cdots + m_{hd}$.

Then the masked 4-bit values over $GF(2^4)$ was evaluated respectively:

$$a_h' + m_{h0} + \cdots + m_{hd}$$
$$= \underbrace{(a_h + m_{h0} + \cdots + m_{hd})(d^{-1} + m_{h0} + \cdots + m_{hd})}_{m_4}$$
$$+ \underbrace{(a_h + d^{-1} + m_{h0} + \cdots + m_{hd})(m_{h0} + \cdots + m_{hd})}_{c_5} \quad (2)$$
$$+ \underbrace{m_{h0} + \cdots + m_{hd}}_{c_4}$$

$$a_l' + m_{l0} + \cdots m_{ld}$$
$$= \underbrace{(a_h + m_{h0} + \cdots + m_{hd})(d^{-1} + m_{h0} + \cdots + m_{hd})}_{m_4}$$
$$+ \underbrace{(a_l + m_{h0} + \cdots + m_{hd})(d^{-1} + m_{h0} + \cdots + m_{hd})}_{m_5} \quad (3)$$
$$+ \underbrace{(a_h + a_l + m_{h0} + \cdots + m_{hd})(m_{h0} + \cdots + m_{hd})}_{c_2}$$
$$+ \underbrace{(m_{h0} + \cdots + m_{hd})^2}_{c_3} + \underbrace{m_{h0} + \cdots + m_{hd}}_{c_4}$$

The terms which were signed as m_1 to m_6 point to the masked value. The terms which were signed as c_1 to c_7 point to the correction terms, referred to in Oswald (2005). From Equation 1 to 3 it required, in total, five multiplications, two constant multiplications and one square for the entire process of the AES S-Box. As it can be seen, compared with the scheme in Oswald (2005), it reduces the number of operations significantly by reusing the correction terms.

3.4 Security analysis of our masking scheme

The security of the proposed scheme was proved by considering the distribution probability of the masked data and masks. The analysis methods in Blömer et al. (2004) & Oswald (2005) were adopted to analyse the distribution probability of the intermediate values.

Lemma 1. Let $a \in GF(2^n)$ be arbitrary. Let $m \in GF(2^n)$ be uniformly distributed over $GF(2^n)$ and independent of a. Then, $a + m$ is uniformly distributed regardless of a. Thus, the distribution of $a + m$ is independent of a.

Lemma 2. Let $a, b \in GF(2^n)$ be arbitrary. Let $m_a, m_b \in GF(2^n)$ be independently and uniformly distributed over $GF(2^n)$. Then, the probability distribution of $(a + m_a) \times (b + m_b)$ is:

$$\Pr((a+m_a) \times (b+m_b) = i) = \begin{cases} \dfrac{2^{n+1}-1}{2^{2n}}, & \text{if } i=0 \\ \dfrac{2^n-1}{2^{2n}}, & \text{if } i \neq 0 \end{cases} \quad (4)$$

Thus, the distribution of $(a + m_a) \times (b + m_b)$ is independent of a and b.

Lemma 3. Let $a \in GF(2^n)$ be arbitrary. Let $m_a, m_b \in GF(2^n)$ be independently and uniformly distributed

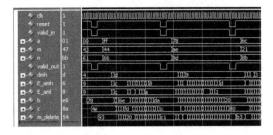

Figure 3. Simulation results of the second-order masking scheme.

over $GF(2^n)$. Then the probability distribution of $(a + m_a) \times m_b$ is:

$$\Pr((a+m_a) \times m_b = i) = \begin{cases} \dfrac{2^{n+1}-1}{2^{2n}}, & \text{if } i=0 \\ \dfrac{2^n-1}{2^{2n}}, & \text{if } i \neq 0 \end{cases} \quad (5)$$

Thus, the distribution of $(a + m_a) \times m_b$ is independent of a.

Lemma 4. Let $a \in GF(2^n)$ be arbitrary and $p \in GF(2^n)$ a constant. Let $m_a \in GF(2^n)$ be independently and uniformly distributed over $GF(2^n)$. Thus, the distribution of $(a + m_a)^2$ and $(a + m_a)^2 \times p$ is independent of a.

Lemma 5. Let $a_i \in GF(2^n)$ be arbitrary and $m_i \in GF(2^n)$ be independent of all a_i and uniformly distributed over $GF(2^n)$. Then the distribution of $\sum_i a_i + \sum_i m_i$ is independent of a_i.

This shows that the distribution probability of each intermediate value is completely unrelated to sensitive information in this high-order masking scheme for AES; thus, its security was proved.

3.5 Experiments and comparisons

In order to verify the proposed scheme, various order masking schemes for AES S-Box were modelled in Verilog and implemented with EDA tools. Figure 3 shows the simulation results of a second-order masking scheme. The value a refers to the input of the S-Box, and m and n refer to the various mask values, respectively. The term c refers to the output of the S-Box which is masked with m and n. The value dmh, E_amh and E_aml refer to the calculation result of Equations 1, 2 and 3. In order to verify the correctness of the simulation results, the simulation results were compared with the lookup table of the S-Box. The comparison results indicated that this study's high-order masking scheme was completely correct.

Table 1 shows the hardware resource comparisons of distinct masking schemes by Akkar and Giraud (2001), Blömer et al. (2004), MS-IAIK (2005), and this study. The number of different operations of multiplication, constant multiplication and square over $GF(2^4)$ of each masking scheme is given. As it can

123

Table 1. Comparisons amongst a distinct masking scheme.

	Multi	Constant multi	Square
S-Akkar [9]	18	6	4
S-Blömer [10]	12	1	2
MS-IAIK [11]	9	2	2
This study	5	2	1

Table 2. Comparisons of different AES S-Box implementation.

Method	Total combinational functions	Dedicated logic registers	Fmax
Unprotected	97	70	128.57
Oswald [11]	309	234	109.2
First-order	(218%)	(234%)	(84.9%)
This work	205	162	127.67
First-order	(111%)	(131%)	(99.29%)
This work	270	236	125.1
Second-order	(178%)	(237%)	(97.32%)

be seen, the number of all of the arithmetic operations used in this paper is the smallest. Compared to the other three masking schemes, this masking scheme required the least number of operations and achieves the most efficient area hardware implementation.

Table 2 lists the area and speed of different AES S-Box implementations: the unprotected, first-order; and the second-order masking scheme. The number of total combinational functions, dedicated logic registers and the maximum operation frequency were given. The percentages below the figures indicate the ratio from the unprotected AES S-Box. The implementation results show that the second-order scheme of this study is about twice that of the unprotected masking scheme, and about 25% less than the first-order masking scheme found in Oswald (2005) at the cost of 3% reduction of the maximum frequency. Compared with the scheme in Oswald, this high-order masking scheme has much lower masking overheads. Furthermore, the proposed scheme provides a practicable solution with considerably low area and speed overhead.

4 CONCLUSION

This paper proposes an efficient higher-order masking scheme based on composite field arithmetic for S-Boxes' of block cipher. Different from existing schemes, the proposed scheme provides a new way to mask the S-Boxes, and can be applied to any S-Boxes which are composed of inverse operations and affine transformations. The detailed performance of the implementation results is given and the security of this scheme was proved theoretically. Future work by the authors will focus on the research of efficient hardware architecture and its VLSI hardware implementation.

ACKNOWLEDGEMENT

This work was sponsored by the National Natural Science Foundation of China (Grant Nos. 61604050) and Grant Nos. 51672062).

REFERENCES

Ahn, S. & Choi, D. (2016). An improved masking scheme for S-Box software implementations. *Information Security Applications.*

Akkar, M.L. & Giraud, C. (2001). An implementation of DES and AES, secure against some attacks. *Cryptographic Hardware and Embedded Systems – CHES 2001, Third International Workshop 2162*, 309–318.

Blömer, J., Guajardo, J. & Krummel, V. (2004). Provably secure masking of AES. *Selected Areas in Cryptography.*

Carlet, C., Goubin, L., Prouff, E., Quisquater, M. & Rivain, M. (2012). Higher-order masking schemes for S-Boxes. *International Conference on FAST Software Encryption 549*, 366–384.

Coron, J.S. (2014). Higher order masking of look-up tables. *Advances in Cryptology.*

Herbst, C., Oswald, E. & Mangard, S. (2006). An AES smart card implementation resistant to power analysis attacks. *International Conference on Applied Cryptography and Network Security 3989*, 239–252.

Kim, H.S., Hong, S. & Lim, J. (2011). A fast and provably secure higher-order masking of AES S-Box. *Lecture Notes in Computer Science, 6917*, 95–107.

Kim, H.S., Kim, T.H., Han, D.G. & Hong, S. (2010). Efficient masking methods appropriate for the block ciphers ARIA and AES. *ETRI Journal, 32*, 370–379.

Oswald, E. & Kai, S. (2005). An efficient masking scheme for AES software implementations. *Information Security Applications, International Workshop 3786*, 292–305.

Oswald, E., Mangard, S., Pramstaller, N. & Rijmen, V. (2005). A side-channel analysis resistant description of the AES S-Box. *Lecture Notes in Computer Science, 3557*, 413–423.

Paul, C., Jaffe, J., Jun. & Benjamin. (1999). Differential power analysis. *Advances in Cryptology – CRYPTO' 99.*

Rivain, M. & Prouff, E. (2010). Provably secure higher-order masking of AES. *International Workshop on Cryptographic Hardware and Embedded Systems 6225*, 413–427.

Satoh, A., Morioka, S., Takano, K. & Munetoh, S. (2001). A compact rijndael hardware architecture with S-Box optimization. *Advances in Cryptology – ASIACRYPT 2001.*

Schramm, K. & Paar, C. (2006). Higher order masking of the AES. *Topics in Cryptology – CT-RSA 2006, The Cryptographers'Track at the RSA Conference 2006, San Jose, CA, USA, February 13–17, 2006, Proceedings 3860*, 08–225.

Experimental study on the effect of hydroxyl grinding aids on the properties of cement

B. Yang
Qinghai College of Architectural Technology, Xining, China

X.F. Wang, X.F. He, C.X. Yang & K.M. Wong
TongJi University, Shanghai, China

ABSTRACT: The effect of different dosage, hydroxyl group number, carbon chain length and grinding time on the properties of cement were studied. The results showed that in the same carbon chain length, with functional groups increased, the cement strength gradually reduced. The hydroxyl grinding aid had no obvious effect on the early strength of cement, but the intermediate and later strength of cement were significantly improved. The grinding aid dosage of 0.02% or more hydroxyl groups had no obvious effect on increasing the development of later strength. The higher the content of the single hydroxyl grinding aid, the more the specific surface area, the greater the adding value of water consumption of normal consistency.

Keywords: grinding aids; cement; properties

1 RAW MATERIALS

Grinding aids can change the surface properties and hydration rate of cement particles by physical or chemical action, and will affect the basic properties of the cement.

1.1 Cement

Using the rotary kiln cement clinker from Anhui Wuhu Conch Cement Company Limited, the size of the solid block was 10–40 mm.

Table 1. Chemical composition of clinker.

Name	Loss	SiO_2	Al_2O_3	Fe_2O_3	Total (%)
Content (%)	0.84	65.04	4.78	3.94	99.24
Name	CaO	MgO	SO_x	F-CaO	
Content (%)	22.28	2.00	0.36	1.39	

Table 2. The mineral composition of clinker.

Name	KH	H	P	C_3S
Content (%)	0.894	2.56	1.21	50.94
Name	C_2S	C_3A	C_4AF	–
Content (%)	25.53	5.98	11.98	–

1.2 Gypsum

From the Xiamen MEIYI cement factory, the size of the gypsum was 10–40 mm.

1.3 Sand

A good grading of sand was selected with a maximum particle size of not more than 5 mm.

Table 3. Chemical composition of gypsum.

Name	Loss	SiO_2	Al_2O_3	Fe_2O_3
Content (%)	19.23	2.79	1.27	0.33
Name	CaO	MgO	SO_3	Total
Content (%)	32.89	0.68	38.55	95.74

Table 4. Technical index of sand.

Fineness modulus	Mud content (%)	Solid-mud content (%)
2.83	4.5	2.5
Density (kg/m^3)	Particle diameter (mm)	Bulk density (kg/m^3)
2650	<5.0	1574

Table 5. Sand grain composition.

Sieve diameter (mm)	Sieve residue ratio (%)	Cumulative sieve residue ratio (%)
4.75	3.6	3.6
2.36	11.4	15.0
1.18	10.0	25.0
0.60	14.4	39.4
0.30	31.0	70.4
0.15	15.6	86.0
<0.15	14.0	100

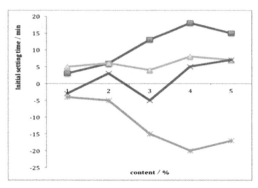

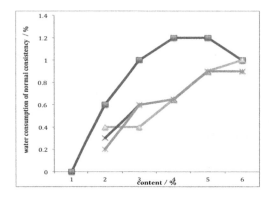

Figure 1. Effect of hydroxyl organic matter on the water consumption of standard consistency cement.

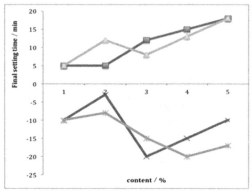

Figure 2. Effect of hydroxyl group on cement setting time.

1.4 Grinding aid component

Several organic compounds that appeared most often in the hydroxyl grinding aids were: AR products of ethylene glycol; propylene glycol; diethylene glycol; and glycerine. Four hydroxy compounds from Sinopharm Chemical Reagent Company Limited were selected as grinding aids.

2 TEST METHOD

2.1 Requirement of normal consistency

Ding aid influenced standard cement water consistency mainly from the following aspects: (1) to change cement particle morphology, effect friction between the cement particles bite force; (2) cement particles become finer from changes in the number of voids between particles which are filled with water; and (3) the surface properties of cement particles, are changed by the adsorption of cement particles with water.

The variation of water consumption of normal consistency cement is related to the content of the organic compounds and its molecular structure. Overall, the increase in the amount of water consumption of normal consistency cement increases with increasing hydroxyl group, and the change law of the specific surface area is similar. This is because the hydroxyl group makes cement particles smaller, and to improve the cement specific surface area, an increase in corresponding water consumption is needed in order to achieve standard consistency.

2.2 Effect of hydroxyl group on cement setting time

From Figure 2, it can be seen that diethylene glycol makes the cement setting time slightly shorter, while other hydroxyl compounds make the cement setting time slightly longer. Two glycol and glycerol will shorten the setting time of cement; ethylene glycol and glycol bin will prolong the setting time of cement.

2.3 Effect of hydroxyl group on the strength of cement

The data in Table 6 showed that the strength of cement mortar containing a 0.01% hydroxyl group was lower than that of the standard group. With the increase in strength, the difference between the grinding aids and the blank group was increased by 0.01%, while the hydroxyl group was not conducive to the development of cement strength, especially in the later stage.

In addition to polyethylene glycol, the cement mortar strength of the admixture of 0.02% grinding aid was higher than that of the standard. So, when the content reached a certain degree, the hydroxyl group promotes the strength of the cement. The early

Table 6. Effect of different amounts of hydroxyl groups on the strength of cement.

Grinding aid component	Adding content /%	Flexural strength / MPa			Compressive strength / MPa		
		3d	7d	28d	3d	7d	28d
Blank	a	4.8	5.6	7.1	21.5	28.0	24.2
	b	4.0	4.9	6.0	16.7	20.3	30.5
	c	3.4	4.5	5.8	13.2	19.2	29.4
Methanol	0.01	4.4	5.6	7.0	20.1	25.7	21.8
	0.02	4.4	4.9	6.1	17.5	20.3	32.0
	0.04	3.7	4.4	6.5	14.3	18.9	30.1
Ethanol	0.01	4.5	5.6	7.3	20.2	26.8	23.4
	0.02	4.2	5.1	6.9	17.2	21.3	31.4
	0.04	3.6	4.4	6.4	13.0	19.0	31.9
Glycol	0.01	4.7	5.3	6.9	20.1	27.1	23.2
	0.02	4.1	5.0	6.5	16.2	22.2	31.4
	0.04	3.6	4.6	6.4	12.8	18.8	33.3
Isopropanol	0.01	4.4	5.3	7.0	19.2	25.9	23.5
	0.02	4.1	5.1	6.4	17.5	23.0	34.4
	0.04	3.6	4.4	6.0	13.5	19.9	34.9
Propylene glycol	0.01	4.2	5.2	6.7	18.9	27.0	22.8
	0.02	3.8	5.3	6.9	16.8	22.9	33.9
	0.04	3.3	4.0	6.3	12.8	18.4	33.8
Glycerol	0.01	4.9	5.6	7.0	20.6	28.1	22.7
	0.02	4.5	5.0	6.5	16.7	23.0	32.3
	0.04	3.7	4.4	6.3	12.9	18.0	31.8
N-butyl alcohol	0.01	4.0	5.2	6.8	17.5	24.9	21.9
	0.02	4.0	5.2	6.4	15.0	22.6	35.7
	0.04	3.2	4.0	6.5	12.9	17.7	34.2

strength of the cement was not significantly affected by the hydroxyl group, and improved middle and late strength obviously. The strength of 28d was increased by 3–5 MPa and enhanced by 10–18%. Thus, n-butyl alcohol > isopropanol > propylene glycol. Early strength of cement mortar with 0.04% grinding aid was generally lower than the reference group, and the late strength was generally higher than the reference group. Combined with the content of 0.01% and 0.02% it was not difficult to find that the hydroxyl containing grinding aid did not significantly improve the early strength, and even caused low strength in the reference group; but the late strength is basically stable. It can be concluded that the hydroxyl group grinding aid can improve the flexural and compressive strength of cement mortar at the later stage, but the effect of early strength is not obvious.

3 CONCLUSIONS

The optimum content of the hydroxyl group was 0.02% to approximately 0.04%. The influence of water content on standard consistency cement and cement setting time was related to the dosage of the hydroxyl group. From the analysis of the molecular structure of the grinding aids, the effect of the cement grinding aid and the strengthening effect was negative. While the carbon chain length was 3, the grinding aid effect was the worst, but the strength was the best. The same length of carbon chain, with the number of functional groups increased, made the cement grinding aid effect better, while the strength reduced accordingly. The test results showed that ethylene glycol, propylene glycol and glycerol performed better in the grinding aid enhancing effect. Hydroxyl saturated content in the range of 0.01 to approximately 0.02%, had no obvious effect, while the content of 0.02% or more hydroxyl groups in the grinding aid increased late strength. Therefore the optimum dosage of hydroxyl in grinding aids is 0.02 to approximately 0.04%.

REFERENCES

Ren, C. (2010). Study on the effect of cement grinding aids and composite grinding aids. *Guangdong Building Materials, 8*, 35.

Zhang, Z., Zhu, J. & Fu, X. (2011). Study on single component cement grinding aids effect. *21st Century Building Materials Create Living, 12*, 86–89.

Zhu, K.C., Huaicheng. & Xia, Z.. (2010). Development status and trend of cement grinding aids at home and abroad. *China Cement, 3*, 64–66.

Electronic Engineering – Wang (ed)
© 2018 Taylor & Francis Group, London, ISBN 978-1-138-60260-1

An input delay approach in guaranteed cost sampled-data control

L.Y. Fan, J.N. Zhang & J.J. Song
Department of Applied Mathematics, Harbin University of Science and Technology, Harbin, China

ABSTRACT: In this paper, the problem of robust guaranteed cost sampled-data control was studied for a linear parametric uncertain system. The system was transformed into a continuous system with time-delay, through a input delay approach. With a linear matrix inequality (LMIs) method and the Lyapounov stability theory, a robust guaranteed cost sampled-data controller was derived to guarantee the asymptotical stability of the closed-loop system and the quadratic performance index less a given bound for all admissible uncertainties. The sufficient conditions for the existence of state-feedback controller were given in the form of LMIs. A convex optimization problem is formulated to get the optimal state-feedback controller which can minimize the proposed method. The simulation example indicates that illustrated the approach is effective.

Keywords: Input delay approach; Guaranteed cost sampled-data control; Linear matrix inequalities (LMIs)

1 INTRODUCTION

In recent years, the problem of guaranteed cost control on uncertain linear systems is one of the most active subjects in control filed. Considerable attention has been focused on the stability analysis of uncertain time-delay systems. As we all know, time-delay as well as parameter uncertainties is frequently a cause of instability and performance degradation and occurs in many dynamic systems (Fridman 2010). It also designs a control system which is not only stable but can also guarantee an adequate level of performance (Fridman 2006). The design method on a guarantee cost was first proposed by Chang and Peng, where the objective is to design the controller so that the performance index is no larger than a given bound for all admissible uncertainties (Fan & Wu 2011). Based on the idea, many significant results have been obtained for the continuous-time case (Hu et al. 2003) and the discrete-time case (Yu & Gao 2001). However, there have been few results (Liu et al. 2008) in the literature for the robust guaranteed cost sampled-data control based on LMI technique (Hao & Zhang 2011). It is still open and remains challenging. It motivates the research of the paper.

The objective of the paper is to design a robust guaranteed cost sampled-data controller for an uncertain linear system with time-varying delay. By using input delay approach, sufficient conditions are proposed in terms of LMIs, which guarantee the asymptotically stable and performance index less than a certain bound for all admissible uncertainties. At the same time, the design of optimal guaranteed cost controller is proposed. A simulation example is given to show that the method is effective.

2 PROBLEM FORMULATION

Consider the following class of uncertain nonlinear time-varying delay system:

$$\begin{cases} \dot{x}(t) = A(t)x(t) + A_d(t)x(t - \tau(t)) \\ \qquad + B(t)u(t) + f(t, x(t)) & (1) \\ x(t) = \varphi(t) \quad t \in [-h, 0] \end{cases}$$

where $x(t) \in \mathrm{R}^n$ and $u(t) \in \mathrm{R}^m$ denote the state and control input vector respectively. A, B and A_d are known real constant matrices with appropriate dimensions, ΔA, ΔB and ΔA_d are real value unknown matrices representing time-varying parameter uncertainties. In the paper, the admissible parameter uncertainties are assumed to be of the following form:

$$[\Delta A \ \Delta B \ \Delta A_d] = MF(t)[E_1 \ E_2 \ E_3]$$

where M, E_1 and E_2 are real constant matrices and $F(t) \in \mathrm{R}^{i \times j}$ is an unknown time-varying matrix function satisfying $F^{\mathrm{T}}(t)F(t) \le I$, $\forall t$.

It is assumed that all the elements of $F(t)$ are Lebesgue measurable. $\tau(t)$ is time-varying delay. $\varphi(t)$ is a given continuous vector value initial function.

For state-feedback sampled-data control with zero-order holder, the controller takes the following form:

$$u(t) = u_d(t_k), \quad t_k \le t < t_{k+1}$$

where $\lim_{k \to \infty} t_k = \infty$ and u_d is a discrete-time control signal.

Assume that $t_{k+1} - t_k \le h \ \forall k \ge 0$. So, $\tau(t) \in (0, h]$ and $\dot{\tau}(t) = 1$ for $t \ne t_k$.

129

We define the continuous quadratic cost function as follows:

$$J = \int_0^\infty \left[x^\mathrm{T}(t)Qx(t) + u^\mathrm{T}(t)Ru(t) \right] \tag{2}$$

where $Q \in \mathrm{R}^{n \times n}, R \in \mathrm{R}^{m \times m}$ are known symmetric and positive definite matrices.

We consider a state-feedback control law of the form:

$$u(t) = Kx(t_k), \quad t_k \leq t < t_{k+1}$$

We represent the digital control law as a delayed control as follows:

$$u(t) = u_d(t_k) = u_d(t - \tau(t))$$
$$\tau(t) = t - t_k, \quad t_k \leq t < t_{k+1} \tag{3}$$

under control law (3), the closed-loop system is expressed by

$$\begin{cases} \dot{x}(t) = A(t)x(t) + \left[A_d(t) + B(t)K \right] x(t - \tau(t)) \\ \qquad + f(t, x(t)) \\ x(t) = \varphi(t) \quad t \in [-h, 0] \end{cases} \tag{4}$$

Definition 1. J^* is said to be a guaranteed cost of the uncertain system in (1) and u^* is a robust guaranteed cost sampled-data control law if there exist state-feedback control law u^* and constant $J^* > 0$, so that the closed-loop system in (4) is asymptotically stable, and cost function for all admissible uncertainties satisfies $J \leq J^*$.

Then, the objective of the paper is to design a state-feedback controller that satisfies the stability of a closed-loop system and guarantees an adequate level of performance under the parametric uncertainties and time-varying delay with an input delay approach.

3 MAIN RESULTS

In this section, by employing input delay approach, we first present a sufficient condition for the existence of state-feedback guaranteed cost sampled-data controller, and then give a representation of the robust guaranteed cost sampled-data controller in terms of LMIs. The following lemma is used in the proof of our main results.

Lemma 1: Let Y, M, N be given matrices of compatible dimension, Y is a symmetric matrix, then for any $F(t)$

satisfying $F^\mathrm{T}(t)F(t) \leq I \quad \forall t$

$$Y + MF(t)N + N^\mathrm{T}F^\mathrm{T}(t)M^\mathrm{T} < 0$$

if and only if there exists a constant $\varepsilon > 0$ such that

$$Y + \varepsilon^{-1}MM^\mathrm{T} + \varepsilon N^\mathrm{T}N < 0$$

The following theorem gives sufficient condition for state-feedback guaranteed cost controller.

Theorem 1 If there exist symmetric and positive definite matrices P, $S \in \mathrm{R}^{n \times n}$ such that for all admissible uncertainties, the following matrix inequalities hold:

$$\begin{bmatrix} \Xi_1 & P\left[A_d(t) + B(t)K \right] \\ * & \Xi_2 \end{bmatrix} < 0$$

where

$$\Xi_1 = PA(t) + A^\mathrm{T}(t)P + \left(h - \tau(t) \right)S + \alpha^{-1}P^2 + \alpha H + Q$$
$$\Xi_2 = -\tau(t)S + K^\mathrm{T}RK .$$

then

(i) system (4) is asymptotically stable;
(ii) $u(t) = Kx(t - \tau(t))$

is guaranteed cost sampled-data controller, and the cost function

$$J \leq \varphi^T(0)P\varphi(0) + \int_{-h}^0 \varphi^T(r)S\varphi(r)dr$$

In the following part, we shall establish the sufficient condition for the existence of guaranteed cost controller in terms of matrix inequalities.

Theorem 2 If there exist constants $\varepsilon > 0$, and symmetric and positive definite matrices X, S and a matrix W such that the following matrix inequalities hold:

$$\begin{bmatrix} \Omega & A_dX + BW & M & XE_1^\mathrm{T} & 0 \\ * & -\tau(t)XS & 0 & W^\mathrm{T}E_2^\mathrm{T} + XE_3^\mathrm{T} & W^\mathrm{T} \\ * & * & -\varepsilon I & 0 & 0 \\ * & * & * & -\varepsilon^{-1}I & 0 \\ * & * & * & * & -Z \end{bmatrix} < 0 \tag{5}$$

Then

(i) $u^*(t) = WX^{-1}x(t - \tau(t)) \tag{6}$

is a guaranteed cost sampled-data controller of system (1);
(ii) the cost function of system (8) is as follows:

$$J^* = \varphi^\mathrm{T}(0)P\varphi(0) + \int_{-h}^0 \varphi^\mathrm{T}(r)S\varphi(r)\mathrm{d}r \tag{7}$$

The following theorem developed this optimization problem.

Theorem 3 Consider the uncertain nonlinear system (1) under sampled measurements with cost function (2), if the following optimization problem:

$$\min_{\varepsilon, \beta, W, X, D, \Phi} \beta + \mathrm{tr}(\Phi) \tag{8}$$

s. t. (i) inequality (5) hold;

(ii) $\begin{bmatrix} -\beta & \varphi^T(0) \\ \varphi(0) & -X \end{bmatrix} < 0$;

(iii) $\begin{bmatrix} -\Phi & N^T \\ N & -D \end{bmatrix} < 0$.

where $tr(\Phi)$ represents the trace of Φ, $D = S^{-1}$ and N satisfy

$$\int_{-h}^{0} x^T(r)x(r)\mathrm{d}r = NN^T$$

has a feasible solution $\varepsilon, \alpha, X, W, D, \Psi$. Then the controller u^* in (6) is an optimal state-feedback guaranteed cost sampled-data controller which ensures the minimization of the guaranteed cost (7).

4　NUMERICAL EXAMPLE

In this section, we use an example to illustrate the effectiveness of the designed robust guaranteed cost samped-data controller with input delay.

Consider the uncertain nonlinear time-varying delay system (1), and the parameters are given by

$$A = \begin{bmatrix} -0.8 & -0.1 & -0.3 & 0.1 \\ 0.6 & -0.5 & 0.2 & 1 \\ 0.1 & 0.3 & 0.2 & -0.2 \\ 0.1 & 0 & 0 & 0.1 \end{bmatrix},$$

$$B = \begin{bmatrix} 0 \\ 1.1 \\ 0.8 \\ 0.4 \end{bmatrix}, A_d = \begin{bmatrix} -0.3 & -0.2 & 0.4 & 0.3 \\ -0.2 & 0.3 & 0.5 & -0.2 \\ 0 & -0.1 & 0 & 0 \\ 0.3 & 0 & -0.1 & 0.1 \end{bmatrix},$$

$$Q = \begin{bmatrix} 1 & 0 & 0 & 0 \\ 0 & 1 & 0 & 0 \\ 0 & 0 & 1 & 0 \\ 0 & 0 & 0 & 1 \end{bmatrix}, H = \begin{bmatrix} 0.5 & 0 & 0 & 0 \\ 0 & 0.1 & 0 & 0 \\ 0 & 0 & 0.3 & 0 \\ 0 & 0 & 0 & 0.1 \end{bmatrix},$$

$$M = \begin{bmatrix} -0.6 \\ 0.2 \\ 0.2 \\ 0.1 \end{bmatrix}$$

$x(0) = [0.6796 \; 2 \; 0 \; 0]^T$,

$E_1 = [0.8 \; 0.2 \; 0.3 \; 0]$, $E_2 = 0.11$

$E_3 = [0.2 \; 0 \; 0.8 \; 0.2]$, $R = 1$, $h = 2.7$,

$\tau(t) = 2.5$

By using feasp solver of Matlab LMI Control Toolbox to solve the LMIs (5), we obtain the desired

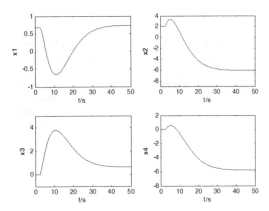

Figure 1. State response curve of four-order uncertain nonlinear time-delay system.

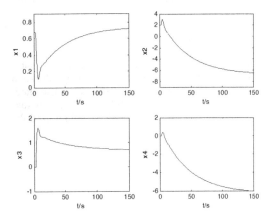

Figure 2. State response curve of uncertain time-delay system.

guaranteed cost controller and corresponding cost function of system (4) as follows:

$u^* = [-0.0376 \; -0.0771 \; -0.4312 \; -0.0130]$
$\cdot x(t - \tau(t))$

$J^* = 27.0314$

By using mincx solver of Matlab LMI Control Toolbox to solve (8), we obtain the optimal guaranteed cost controller and optimal guaranteed cost of the uncertain close-loop system (4) as follows:

$J^*_{opt} = 16.1219$

$u^*_{opt} = [1.3195 \; -0.0771 \; -0.4312 \; -0.0130]$
$\cdot x(t - \tau(t))$

The state response curve of uncertain time-delay system is as follows in Fig. 2.

From the above simulation results, we conclude that the performance index of the optimal guaranteed cost controller is obviously smaller than that of general guaranteed cost controller.

5 CONCLUSION

In the paper, the problem of designing robust guaranteed cost sampled-data controller for uncertain nonlinear systems with time-varying delay has been investigated. The parameter uncertainties are assumed to be time-varying but norm-bounded. Based on the Lyapunov stability theory and LMIs, an input delay approach has been developed to design a state-feedback controller, which guarantee not only the asymptotical stability but also a bound of quadratic performance level for the closed-loop system for all admissible parametric uncertainties. Sufficient conditions have been formulated in the form of LMIs. Furthermore, a convex optimization problem has been introduced to select the optimal robust guaranteed cost sampled-data controller. The simulation result has confirmed the effectiveness of the proposed method.

ACKNOWLEDGMENT

This work was supported by science and technology studies foundation of the Heilongjiang Education Committee of 2014 (12541161). The author also gratefully acknowledges the helpful comments and suggestions of the reviewers, which have improved the presentation.

REFERENCES

Fridman, E. (2010). A refined input delay approach to sampled-data control. *Autonatica*, 2 (46): 421–427.

Fridman, E. (2006). A new Lyapunov technique for robust control of systems with uncertain non-small delays. *IMA Journal of Math. Contr. & Information*. 2 (23): 165–179.

Fan, Liying. & Wu, Junfeng. (2011). Guaranteed cost sampled-data control: an input delay approach. *The 6th International Forum*. 8.

Hao, Ningmei. & Zhang, Yong. (2011). Guaranteed cost control for a class of uncertain nonlinear systems with time-delay. *Science Technology and Engineering*. 8 (11): 1728–1732.

Hu, L. Lam, J. Cao, Y. & Shao, H. (2003). A LMI approach to robust H_2 sampled-data control for linear uncertain systems. *IEEE Transactions on Systems, Man and Cybernetics, Part B*. 1 (33): 149–155.

L, Yu. & F, Gao. (2001). Optimal guaranteed cost control of discrete-time uncertain systems with both state and input delays. *Journal of the Franklin Insitute*. 1 (338): 101–110.

Liu, Fuchun. Yao, Yu. He, Fenghua. & Ji, Denggao. (2008). Robust Guaranteed Cost Control for Sampled-data Systems with Parametric Uncertainties. *Proceedings of the 27th Chinese Control Conference*.

Electronic Engineering – Wang (ed)
© 2018 Taylor & Francis Group, London, ISBN 978-1-138-60260-1

Design of a multilevel cache datapath for coarse-grained reconfigurable array

L. Wang
College of Electrical Engineering, Zhejiang University, China
State Key Laboratory of Cryptology, Beijing, China

X. Wang, J.L. Zhu & X.G. Guan
State Key Laboratory of Cryptology, Beijing, China

H. Huang
Institute of Microelectronics, Tsinghua University, Beijing, China

H.B. Shen
College of Electrical Engineering, Zhejiang University, China

H. Yuan
State Key Laboratory of Cryptology, Beijing, China

ABSTRACT: Multiple Coarse-Grained Reconfigurable Arrays (CGRA) can be dynamically reconfigured and parallelly operated, which leads to an efficient solution for data-intensive applications with high throughout and flexibility in a specific domain. In order to further improve performance of multiple CGRAs, there are two practical ways: fully exploring the high-level parallelism; and reducing the executing time of dynamic reconfiguration. In this paper, a multilevel cache datapath for CGRA is proposed in order to improve the parallel efficiency and reduce the dynamic reconfiguration time, especially during pipeline iterations. A pipelining approach for parallelly mapping an algorithm with loop kernels onto the CGRA is also proposed. The experimental results showed that this approach achieved about 1.48 to approximately 2.04 times improvement in throughput and reduced about 1.7 times the reconfiguration time compared with existing schemes.

Keywords: coarse-grained reconfigurable array; multilevel caches; reconfigurable datapath

1 INTRODUCTION

Today, the Coarse-Grained Reconfigurable Array (CGRA), which can perform computations in a specially designed Reconfigurable Processing Units (RPU) to greatly enhance performance and reduce power without loss of flexibility, are becoming an attractive topic in the field of computing architecture. They combine the post-fabrication programmability of processors with the spatial computational style that are most commonly employed in hardware designs (Page, 1996). There are two key characteristics that distinguish these architectures from others: firstly, they can be customised to solve any problem after device fabrication; and secondly, they exploit a large degree of spatially customised computation in order to perform their computation (DeHon & Wawrzynek, 1999). Most reconfigurable processors have the following characteristics (Bobda, 2007; Vassiliadis & Soudris, 2007).The scheme is composed of two parts: the Reconfigurable Datapath (RD); and the

Reconfigurable Controller (RC). The RD consists of reconfigurable PE arrays.

The data flow and control flow are separated to avoid potential communication bottlenecks and centralisation.

Furthermore, the CGRA also has several characteristics that set them apart from other reconfigurable processors, such as fine-grained FPGA. Firstly, CGRA is customisable domain-specific, generally having a high level of flexibility performance and power efficiency in one domain or a set of domains; secondly, it has the advantages of dynamic and partial reconfiguration and, the dynamic and partial reconfiguration can be very fast, since configuration codes are very short.

The current applications of CGRA in specific domains are summarised in Table 1, which includes multimedia, software radio and cryptography.

According to the analysis of a variety of CGRAs, traditionally, the mesh array is the core component of the entire reconfigurable architecture and mainly responds to data flow calculations. However, the cache datapath from previous research work was found to be ineffective, caused by the low degree of parallelism and throughput. Since a multilevel cache datapath has the potential to yield very high parallelism and throughput, even on loops and iterations containing control flow, it is practicable to create an efficient data memory scheme with higher access bandwidth and less latency by adopting high-level parallel and pipelined architecture.

In this paper, the parallel optimisation of applications, by considering the critical loop mapping under the CGRA's resource constraints, was focused on. The architecture of the studied CGRA is shown in Figure 1. The RD contains the mash array and data memory. A novel approach to parallelise loops by multilevel caches is proposed in this study. A genetic algorithm was introduced to schedule tiled loops with memory-aware object functions. Data locality and communication costs were optimised to avoid accessing external memory frequently during the parallel processing. Besides providing a bandwidth to the RD, another important function of the memory subsystem was to support the dynamical reconfiguration. In this case, the reconfiguration process was managed by the RC. Moreover, main data and configuration memories were also allocated off the memory subsystem (Liu & Baas, 2013). For this reason, it is very important to create an independent cache to store or restore the main data memory. In this way, the architecture minimises the negative effects of data bottlenecks or acceleration of data transfers.

This study introduces a new CGRA scheme along with a number of optimisations; more specifically a multilevel cache datapath. This scheme is generic and was not customised to optimally compute any class of pipeline with iteration (e.g. encryption algorithm) more efficiently than others.

The lifetime of operators was rigorously analysed, and a formula for mapping a kernel (loop) onto the CGRA, which has a multilevel cache datapath in a pipelining way, was developed.

CGRA supporting encryption algorithms, to demonstrate the scalability of the introduced technique, were designed. This scheme was found to be 1.48 to 2.04 times faster than others.

2 THE DATAPATH OF CGRA

The proposed CGRA architecture consisted of a reconfigurable controller and a reconfigurable datapath. The reconfigurable controller is the main component for controlling and configuring the CGRA; it is dedicatedly designed to obtain high configuration bandwidth and high-speed dynamic reconfiguration. The reconfigurable datapath consisted of reconfigurable cell arrays and data memory, and as the computing engine of the CGRA was used to implement operation level parallelisation for high energy efficiency.

Table 1. The applications of CGRA in specific domains.

CGRA	Specific domain	Algorithm
DRRA (Farahini et al., 2013)	Communication	Correlation pool in UMTS
BIIRC (Atak & Atalar, 2013)		Turbo decoder in UTMS
FLEXDET (Chen et al., 2012)		Multi-mode MIMO detector in SDR
XPP-III (Baumgarte et al., 2003)	Multimedia	H.264 decoding in multimedia
ARDES (Mei et al., 2005)		
REMUS (Liu et al., 2014; Wang et al., 2013)		H.264, MPEG2, AVS and GPS
REPORC (Wang & Liu, 2015)	Cryptography	AES, DES, SHACAL-1, SMS4, ZUC

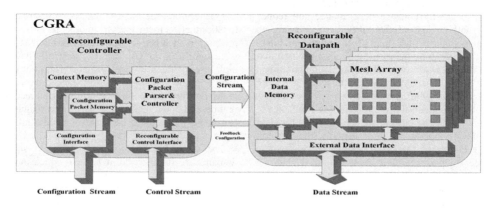

Figure 1. The architecture of the proposed coarse-grained reconfigurable array scheme.

2.1 Mesh arrays in the RCD

The mesh array with 4-level caches is shown in Figure 2. The main part of the mesh array is an array of heterogeneous Processing Elements (PEs) and the interconnecting elements between them. Each PE in the mesh array consisted of an ALU-like element, which computed different operations separately according to the configuration units, and registered chains for storing temporary data. The PE was developed similarly to the other CGRAs, including the RC granularity selection and operation sets of the ALUs. In this study's design, the basic RC granularity was 8-bit and the ALU operations included some basic 8-bit logic and arithmetic operations. The adjacent 4 PEs in the same row was jointed as a Reconfigurable Cell Group (RCG), which acted as the 32-bit operation element and included additional 32-bit operations. The adjacent PE rows were connected by a permutation based network, which was reconfigured as the arbitrary topology of the application DFGs. There were 4-level caches in this study: the level-0 caches provided direct connections between adjacent PE rows; the level-1 caches provided the high-speed and high-bandwidth shared memory between the mesh arrays; the level-2 caches were the internal buffers for the capacity constrained level-1 caches; and the level-3 caches were for the external data of the CGRA.

2.2 Level-0 cache

The architecture of the level-0 cache is shown in Figure 3.

The level-0 cache is the register in the mesh arrays, including the register between the adjacent PE rows and the shift register between the mesh arrays. The diagram of pipeline and feedback iteration mapping on level-0 caches is shown in Figure 4. The registers of the level-0 cache can be configured as direct connections between different computing elements, which can transfer the data streams in a short physical distance by the most efficient approach. It is usually used for storing the spatial locality data, which repeatedly appears in the single iteration. When this data is mapped onto the registers between the adjacent PE rows, it is efficient as an ASIC approach, but as the PE rows are limited, some data had to be mapped between different RC arrays. In order to transfer this data, the shift registers were placed between the adjacent arrays. The configuration registers for the level-0 cache were the constant when the single reconfiguration was performed. With the level-0 cache, the RC arrays could be implemented as a pipeline or as a parallel application.

2.3 Level-1 cache

The level-1 caches are Register Files (RFs) shared by the RC arrays. The RFs are multiport, high-bandwidth and high-speed components for the data transfer, sharing and synchronisation. Each RC array has dedicated read-ports and write-ports that can access the RFs in the same clock cycle.

Level-1 cache can store the data until the dedicated address port and write enable ports are selected. Since the level-1 cache is constrained by the capacity, it is usually used to store the short-lived data generated by RC arrays. The data stored in the level-1 cache can be used to implement the pipeline and feedback applications.

2.4 Level-2 cache

The level-2 cache is on-chip memory shared by the RC arrays. The on-chip memory is the standard dual-port memory, and in the same cycle; only one port of the RC array can write or read the memory. The last row of the mesh array can write the memory, and the first row of the mesh array can read the memory. The level-2 caches are shared by the RC arrays, but only one RC array can read or write the level-2 cache in the same cycle. Furthermore, it can also be used to store and recover the level-1 cache.

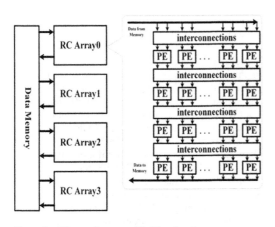

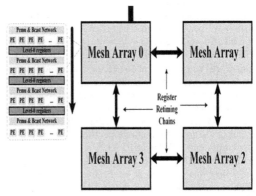

Figure 2. The mesh array with 4-level caches.   Figure 3. The level-0 cache.

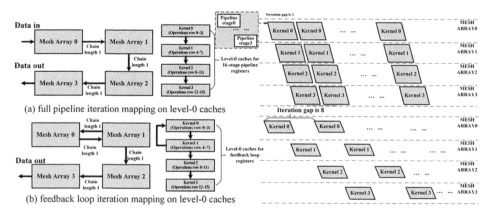

Figure 4. Pipeline and feedback iteration mapping on a level-0 cache.

2.5 Level-3 cache

The level-3 caches are the memory outside the RC arrays. The inside data of the RC array are exchanged with the outside data of the RC array through the FIFO interface. The level-3 caches are buffered by the FIFO. To read the level-3 cache, the controller outside the RC arrays transfers the required data to the FIFO, and the RC array reads the buffer of the FIFO. In order to write the level-3 cache, the RC array writes the data to the FIFO, and the controller outside the RC arrays transfers the required data in the buffer FIFO to the level-3 cache. The status of the RC array will be driven by the status of FIFO.

3 MAPPING ALGORITHM

In Section 2, the 4-level cache architecture was introduced. This architecture is suitable for different requirements of data storage, transfer and exchange depending on the spatial and temporal features in the iterations of the applications. To access the level-0 cache, the multiplexers of selected registers will be set by the configuration registers. The configuration context is included in the RC array context. To access the level-1 cache, the dedicated controller of the RC array will be configured. The controller will generate a read and write sequence depending on the configuration context. The read and write operation of level-1 cache will be executed in one clock cycle, and there is no need to wait for a response. To access the level-2 cache, more clock cycles and the write and read valid-flag in the control instructions are needed. The access of level-3 cache will be the begin point or the end point of the array computing operations depending on the empty and full flag of FIFO.

From the view of the temporal features, the mapping flow mainly depends on the lifetime of the data. Generally, the short-lived data will be mapped onto the lower level caches; while the long-lived data will be mapped onto the higher-level caches. While from the view of spatial features, the spatial locality data should be mapped onto the registers with direct connection between the PEs. If the computing recourse in the architecture is rich enough, all the internal data in the single iteration DFG can be used immediately after generated, and the lifetime of the spatial locality data is less than one. In the actual computing architecture, the computing recourse is always limited, and the generated data needs to be stored for a few cycles, which will increase the lifetime of the data. For the CGRA architecture in this study, the computing resource was rich enough for most iterations, and the data with the spatial locality features could be mapped onto the level-0 caches as direct connections between the PEs (which are mainly for parallel operation levels and pipeline realisation). When the operations within the single iteration DFG are mapped onto a mesh array and the corresponding level-0 cache, the loop is essentially turned into a pipeline DFG. With the level-1 cache, the generated loop kernel can be mapped to generate the feedback or longer pipeline DFG on the mesh arrays.

4 HARDWARE IMPLEMENTATION AND COMPARISONS

A multilevel cache datapath for CGRA is presented in this paper. The level-0 cache and level-1 cache improved the parallel computing capability of the array and support for pipeline architecture efficiency. The level-2 increased the flexibility of data storage and scheduling. With this architecture, CGRA can implement high-speed block cipher, hash and stream cipher algorithms. In this paper, four different algorithms, AES, SM3, SHA-1 and ZUC, were implemented and compared with other architectures. The comparison results are given in Table 2. Compared with many-core GPP arrays (Wang et al., 2014), the proposed CGRA achieved a 3.14 times throughout improvement for AES. Compared with REPROC (Wang & Liu, 2015),

Table 2. Results of multiple algorithms on different architectures.

CGRAs		GPP array [14]	REPROC [12]	Study's CGRA
	Technology (nm)	65	65	65
	Speed (MHz)	1210	400	250
	Area (mm^2)	6.63	4.28	12.89
Throughout (Gbps)	AES	1.02	2.16	3.2
	SM3	n/a	n/a	0.4
	SHA-1	n/a	1.55	1.6
	ZUC	n/a	0.49	1

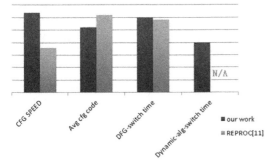

Figure 5. Comparison of configuration and switch performance of REPROC and proposed coarse-grained reconfigurable array scheme.

the proposed reconfigurable architecture achieved 1.48 times throughout for AES and 2.04 times throughout for ZUC. Besides, this CGRA can implement a SM3 cryptographic hash algorithm which has complex iterative structures.

The level-2 and level-3 caches can achieve the algorithm for real-time data storage and recovery rapidly, and increase the speed of dynamic algorithm switch. For example, CGRA used in this paper can switch from AES to ZUC dynamically in 20 clock periods. The use of hierarchical context organisation schemes reduces the scale of configuration code and improves the switching speed of DFG. Compared with REPROC[12], the proposed CGRA scheme in this study achieved a 1.7 times improvement in configuration speed and a 14.5% reduction in average configuration codes. Besides this, CGRA can achieve dynamic algorithm switch on a nanosecond level. The comparison results are shown in Figure 5.

5 CONCLUSION

This paper proposes a multilevel cache datapath implementation scheme for CGRAs. This scheme is generic and not customised to optimally compute any class of pipeline with iteration more efficiently than others. The lifetime of operators were analysed, and a formula for mapping a kernel (loop) onto the CGRA with a multilevel cache datapath in a pipelining way, was developed. During pipeline interleaving, the values were loaded every clock cycle from the retiming chain or shift registers rather than previously loaded and stored in RFs. Therefore, the designer can schedule a separate load operation for each application. This optimisation can improve the performance in terms of number of cycles and executed instructions, especially for the load and store operations. Therefore, it had a direct influence on the energy consumed during the processing of the loop. Experiments showed that the proposed architecture has obvious advantages over state-of-the-art architectures in literatures.

REFERENCES

Atak, O. & Atalar, A. (2013). BilRC: An execution triggered coarse grained reconfigurable architecture. *IEEE Transactions on Very Large Scale Integration (VLSI) Systems*, 21(7), 1285–1298.

Baumgarte, V., Ehlers, G., May, F., Nückel, A., Vorbach, M. & Weinhardt, M. (2003). PACT XPP – A self-reconfigurable data processing architecture. *The Journal of Supercomputing*, 26(2), 167–184.

Bobda, C. (2007). *Introduction to reconfigurable computing: Architectures, algorithms, and applications*. Springer Science & Business Media.

Chen, X., Minwegen, A., Hassan, Y., Kammler, D., Li, S., Kempf. & Ascheid, G. (2012, April). FLEXDET: Flexible, efficient multi-mode MIMO detection using reconfigurable ASIP. In *Field Programmable Custom Computing Machines (FCCM), 2012 IEEE 20th Annual International Symposium on* (pp. 69–76). IEEE.

DeHon, A. & Wawrzynek, J. (1999, June). Reconfigurable computing: What, why, and implications for design automation. In *Proceedings of the 36th annual ACM/IEEE Design Automation Conference* (pp. 610–615). ACM.

Farahini, N., Li, S., Tajammul, M.A., Shami, M.A., Chen, G., Hemani, A. & Ye, W. (2013, May). 39.9 Gops/watt multi-mode CGRA accelerator for a multi-standard base station. In *Circuits and Systems (ISCAS), 2013 IEEE International Symposium on* (pp. 1448–1451). IEEE.

Liu, B. & Baas, B.M. (2013). Parallel AES encryption engines for many-core processor arrays. *IEEE Transactions on Computers*, 62(3), 536–547.

Liu, L., Chen, Y., Wang, D., Yin, S., Wang, X., Wang, L. & Wei, S. (2014). Implementation of multi-standard video decoder on a heterogeneous coarse-grained reconfigurable processor. *Science China Information Sciences*, 57(8), 1–14.

Mei, B., Veredas, F.J. & Masschelein, B. (2005, August). Mapping an H. 264/AVC decoder onto the ADRES reconfigurable architecture. In *Field Programmable Logic and Applications, 2005. International Conference* on (pp. 622–625). IEEE.

Page, I. (1996). Constructing hardware-software systems from a single description. *Journal of VLSI signal processing systems for signal, image and video technology*, 12(1), 87–107.

Vassiliadis, S. & Soudris, D. (Eds.). (2007). *Fine-and coarse-grain reconfigurable computing* (Vol. 16). New York, NY: Springer.

Wang, B. & Liu, L. (2015, February). REPROC: A dynamically reconfigurable architecture for symmetric cryptography. *In Proceedings of the 2015 ACM/SIGDA International Symposium on Field-Programmable Gate Arrays* (pp. 269–269). ACM.

Wang, Y., Liu, L., Yin, S., Zhu, M., Cao, P., Yang, J. & Wei, S. (2014). On-chip memory hierarchy in one coarse-grained reconfigurable architecture to compress memory space and to reduce reconfiguration time and data-reference time. *IEEE Transactions on Very Large Scale Integration (VLSI) Systems*, 22(5), 983–994.

Wang, X., Nguyen, H.K., Cao, P., Qi, Z. & Liu, H. (2013, February). Mapping method of coarse-grained dynamically reconfigurable computing system-on-chip of RE-MUS-II. *In Proceedings of the 10th Workshop on Optimizations for DSP and Embedded Systems* (pp. 45–45). ACM.

Research on path planning for large-scale surface 3D stereo vision measurement

Y.J. Qiao & Y.Q. Fan
School of Mechanical Engineering, Harbin University of Science and Technology, China

ABSTRACT: Measurement path planning is required in three-dimensional measurement tasks of large-scale complex surfaces, and in this study, a path planning algorithm with a region constraint is proposed. In this study, according to the principle of stereo vision measurement, the regional constraints in path planning are were analyszed and determined and then a multi-layer road network model established; this expressed the regional conditions accurately and established the mathematical model of path planning for stereo vision measurement. On the basis of the above, using path planning algorithm under regional constraints, a path finding search through a two-layer path of regions and inter-regions was carried out to realize stereo vision measurement path planning. Finally, the effectiveness and feasibility of the path planning algorithm under the regional constraint was verified by comparing with the exhaustive algorithm.

Keywords: three-dimensional measurement; stereo vision; regional constraint; path planning

1 INTRODUCTION

With the development of the manufacturing industry, a large-scale, high precision and high speed measurement technique for measurement tasks is put forward in this paper. Conventional measuring methods, such as coordinate measuring machines and theodolites cannot meet current requirements. Based on this background, binocular vision measurement technology can be developed rapidly, and can complete the above mentioned measurement requirements well [1,2,3,4]. However, in the measurement of large-size complex surface parts, due to the large size of the part, it is necessary to perform local measurements several times, and it is necessary to carry out visual measurement network planning [5,6,7]. After completing the network planning, it is necessary to realise the measurement path planning based on the measurement network planning results, and the stereoscopic vision planning of the whole 3D measurement is completed.

Currently, the shortest loop planning is given by the user to access the required nodes information. A shortest loop is searched according to the cost minimization principle. For engineering problems, the shortest loop planning problem is usually limited by regional constraints. This requires the algorithm to plan out the shortest loop that can satisfy both the regional constraint and the cost minimization principle. But the current path finding algorithm cannot solve such problems.

In this paper, the path planning of a visual measurement network is the shortest loop planning problem. Binocular vision measurement requires the sensor to

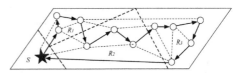

Figure 1. Region constraint.

start from the zero position, and then in turn run to the specified location for image acquisition. At the same time, the binocular vision sensor is required to pass through the viewpoint region with the same posture, and then enter the next posture region; the final return to the initial point. To solve this problem, a mathematical model of a planning problem is established in this paper. A path planning algorithm, based on region constraint, is proposed which can be used to solve the problem of small-scale and shortest loop planning under regional constraints.

An example of a path plan for a region constraint is shown in Figure 1, where S is the Planning Starting Point. As the planning goal is the loop path, and S is also the Planning Terminal Point, the Planning Terminal Point is indicated by the pentagram. R_1, R_2, and R_3 are three Constraining Regions, and the boundary of the region is indicated by a bold dotted line. The thick solid line in Figure 1 indicates the shortest loop path under the region constraint.

1.1 Road network model

In order to ensure that the regional constraints are effectively dealt with in the path planning, a multi-layer network model is proposed to rationally express the constraints contained in the path planning task. The constraint regions are mapped to a different network, and the constraint boundary is extended to the isolation surface, thus forming a multi-layer road network model. As shown in Figure 2, the Origin Point S is mapped to the Network Layer C_0, the constraint region R_1 is mapped to the Road Network Layer C_1, the constraint area R_2 is mapped to the road network layer C_2, and the constraint region R_3 is mapped to the road network layer C_3. Thus, the Isolation Surface B_0, B_1 and B_2 are produced. The thick solid lines in Figure 2 indicate the path finding process.

It can be seen from Figure 2 that the path in the network layer C_1 is through all nodes of the layer, then through the Isolation Surface B_1 into the road network layer C_2; this starts the next road network layer path finding. At this time, the path completes the traversal for all nodes of the network layer, and it can enter the next network layer. However, in the network layer C_2, the path tries to enter C_3 through the isolation plane B_2, but is blocked by the isolation plane B_2, since there are nodes not yet extended to the network layer C_2. From the above analysis, it can be seen that in this multi-layer road network model, the isolation surface plays the role of regional constraints, and the path finding direction is guided by a multi-layer road network. In the path planning process, only the path through all the nodes in the current network layer will let the isolation surface open, thus allowing the path through. The path finding path can only enter sequentially from low to high in the network layer, and cannot jump until returning to the road network layer C_1 again. A candidate loop can be obtained which satisfies the constraint condition.

1.2 Mathematical model

Simply put, the mathematical model of the optimal measurement path planning is to establish a basic logical relationship between the mathematical description of each measurement point and the path distance on a multi-layer road network, as shown in the following equation:

$$H = \min(\sum_{i=1}^{n} G_i + \sum_{e=0}^{n}\sum_{f=0}^{n} W_{ef} g_{ef})$$

$$s.t. \begin{cases} g_0: & G_i = G_{st} = \sum_{a=1}^{m_i}\sum_{b=1}^{m_i} w_{ab}^i x_{ab}^i - w_{ts}^i x_{ts}^i, x_{ts}^i = 1 \\ g_1: & \sum_{a=1}^{m_i} x_{ab}^i = 1, b = 1,2,...,m_i \\ g_2: & \sum_{b=1}^{m_i} x_{ab}^i = 1, a = 1,2,...,m_i \\ g_3: & x_{c_1c_2} + x_{c_2c_3} + ... + x_{c_kc_1} \leq k-1, c_1, c_2..., c_k = 1,2,...,m_i \\ & c_1 \neq c_2 \neq ... \neq c_k, k = 2,3,...,m_i-1 \\ g_4: & x_{ab} = 0 \text{ or } 1 \end{cases}$$

where i is the Number of the Constraint Region; G_i is the Local Path contained in the constraint region numbered i; W_{ef} denotes the Distance of the Communication Path between the Constraint Region e and the Constraint Region f; and G_{ef} indicates whether or

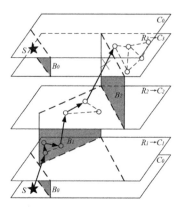

Figure 2. Sketch map of a multi-network model.

not the constraint region e, and the constraint region f are in communication. When the value equals 1, it indicates inter-region connectivity, and 0 means no inter-region connectivity. The w_{ab}^i represents the Path Cost between the Node i and the Node b in the constraint region, and the x_{ab}^i denotes whether the node i in the region a is connected to the node b. A value equal to 1 indicates inter-zone connectivity and 0 indicates no inter-zone connectivity. The n is the Total Number of Constraint Regions, m_i denotes that the Number of Nodes in the Constraint Region m is i, g_1 means that each node can only be reached once, g_2 means that each node can only start once, and g_3 guarantees that there will not be a leak node path through only the k nodes.

2 STEREO VISION MEASUREMENT PATH PLANNING ALGORITHM

2.1 Statement of problem

In optimal visual measurement path planning, in order to obtain the precise structural relationship between the visual sensors, the binocular vision sensors will be recalibrated whenever the binocular vision sensor posture changes, or it cannot guarantee access to the high precision point cloud segment. Because binocular vision sensor calibration process is complex, in the measurement process, the number of recalibrations should be minimized. The binocular vision sensor is required to enter the next posture position by all the viewpoint positions that having the same posture. The optimal constraint loop in the measurement loop planning has the same sensor position as the viewpoint position. This constraint should be met in the results of the optimal measurement loop plan, otherwise the planning result is considered invalid. The position and orientation of a binocular vision sensor in a visual measurement network can be abstracted to the basic attribute elements in the model: nodes and paths. Each viewpoint location is abstracted as a node in the path plan, the position of the sensor at the viewpoint position determines the constraint region to which the viewpoint belongs and maps to the corresponding network layer. Path is the length of the distance between nodes, which is described between the nodes and the spatial topology.

2.2 General idea of an algorithm

The algorithm of regional constraint path planning divides two search paths to search the solution space tree: inter-region path finding and path finding in the region. According to the order in which the constrained regions are projected to the network layer, the path finding between different regions is guided and controlled by the order of each constraint region. The backtracking method is used to explore the order, which can effectively implement the full pruning operation and improve the search efficiency in the solution space. Path finding in the region is based on the current order of path finding between regions to expand the results, to implement the path guidance and explore the local path. In this path finding layer, the branch-and-bound method is adopted to reduce branches in the process of node expansion to realize local shortest path planning in the region.

2.3 Algorithm implementation

2.3.1 Pathfinding between regions

Path finding between regions is extended between multi-layer road networks. The goal of path finding between regions is to find the order in which traversal constraints are followed during path discovery. When exploring new regions in the expansion, the results are planned by means of local path planning within the region, and the paths traced back between regions; the backtracking method is used to reduce the order tree and optimize the search.

2.3.1.1 Order expansion

The hierarchical order of the road network layer is where the located constraint region determines the order in which the path passes through. Figure 3 shows the constraint region sequential expansion process.

A thin solid line indicates a possible expansion, and a solid arrow indicates the direction of expansion. Starting from S, mapping is undertaken to the Network Layer C_0. The constraint region determines its hierarchy within the order tree according to its hierarchical correspondence at the network layer. The current order is expanded to the third level; S is at the root of the ordinal number, which is the first layer of the order tree. The region constraint R_2 is mapped to the network layer C_1, that is, the area constraint R_2 is located in the second layer of the order tree. Similarly, the region constraint R_1 is at the third level with the order number.

A thin solid line indicates a possible expansion, and a solid arrow indicates the direction of expansion. Start from S as a starting point and map to the network layer C_0. The constraint region determines its hierarchy within the order tree according to its hierarchical correspondence at the network layer. The current order is expanded to the third level; S is at the root of the ordinal number, which is the first layer of the order

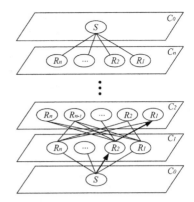

Figure 3. Process of sequence development.

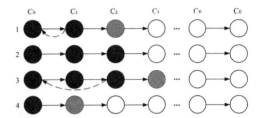

Figure 4. Process of path backtracking.

tree. The region constraint R_2 is mapped to the network layer C_1, i.e. the area constraint R_2 is located in the second layer of the order tree. Similarly, the region constraint R_1 is at the third level with the order number.

2.3.1.2 Path backtracking

It is necessary to use the node expansion results in each network layer to backtrace the path when entering the next network layer. The backtracking process is shown in Figure 4. The attribute C_i of the node member in the current network layer is used to find out the node with the least path cost. If the value is greater than the current optimal cost path, the sequence expansion is discarded, the node queue is cleared, and it is back to the starting point for reordering. If the value is smaller than the current optimal cost path, the network layer of the next regional constraint is entered to continue the order expansion. The above process is gradually repeated until the path finding terminates.

In the backtrack, if the path completes the path finding for the entire constrained region, the path cost is compared with the current optimal path cost. If the value is greater than the current optimal value, Queue Z is cleared and path planning is continued. If the value is smaller than the current value, the path is replaced with the current optimal value and the path information is saved.

2.3.1.3 Pathfinding termination

The termination of regional order expansion is also the termination of inter-regional path finding; the inter-region expansion order forms the order tree, which is the order of path through region constraints. The termination condition of inter-regional path finding is to complete the expansion of the whole sequence tree. If there is an optimal path, the path backtracking is started by the End Point S of the network layer C_0, and in turn, from C_n back to C_1, finally reaching the Origin Point S that is in the network layer C_0. The path obtained by the above process is the shortest loop path that satisfies the region constraint. This is the Hamiltonian path that satisfies the regional constraints.

2.3.2 Pathfinding in the region

The path finding within the region is extended in the C_i-layer road network. The goal of path finding within the region is to find the optimal path of network layer C_i by using branch-and-bound algorithms; the number of local shortest paths in the network layer C_i is the same as the total number of nodes.

2.3.2.1 Node extension

The node expansion of path finding in the region is different from the common node expansion; it does not extend a single shortest path through each node, but a shortest path with any node as the starting node. Moreover, the actual path cost is the shortest path distance value that the node arrives at the starting point S according to the current network layer order. As shown in Figure 5, it is the local path finding process of the node extension in the region. The white nodes in Figure 5 indicate that the node corresponding to the local optimal path has not been found in the network layer C_i, and the black nodes indicate that a local optimal path in its corresponding region has been found. Such a set of nodes forms a queue Z; the latter is shown in Equation 1:

$$Z_i = \{z_1^i, z_2^i, \cdots, z_j^i, \cdots, z_N^i\} \quad (1)$$

where z_j^i is the Node in the Network Layer; i is the Network Layer Number; j is the Number of Nodes in the Network Layer; and N is the Total Number of Nodes in the network layer C_i. In the initial case, the node queue is empty, and when the inter-region path finding changes, the node queue must be cleared to recalculate. Node z_j^i contains the node properties, as shown in Equation 2:

$$z_j^i = \{x_j, L_j, c_j, x_f\} \quad j = 1, 2, \cdots, N \quad (2)$$

In the Equaton 2, x_j is the Node Number, and L_j is the Node Number List of the Local Optimal Path corresponding to the Node z_j^i, C_i is the Cost of the Corresponding Optimal Path, x_f is the Number of the Nodes in the Parent Network Layer C_{i-1}, to which the end point in the optimal path is connected. The C_j is calculated as shown in Equation 3:

$$c_j = c + cf + c_j^{c_{i-1}} \quad (3)$$

where c is the Cost of the Path corresponding to the Linked List L_i in Node z_j^i attribute; C_f is the Path Cost of the End Point of the linked list L_i to a node within the parent network layer C_{i-1}; and $c_j^{c_{i-1}}$ is the Cost of a path corresponding to a node in the parent network layer C_{i-1}.

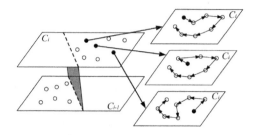

Figure 5. Sketch map of the process of node expansion.

Figure 6. Propeller blades' structure.

2.3.2.2 Expand termination

The termination condition of path finding in the region is that all the nodes of the extended network layer find the optimal path in the region corresponding to the current region order. That is, the nodes in the network layer are all added to queue Z.

3 EXPERIMENT AND COMPARATIVE VERIFICATION

3.1 Planning example

In the experiment for this paper, the measured model used was a propeller blade with a diameter of 5.8 m; the blade structure is shown in Figure 6.

Visual measurement network planning was carried out for one blade, and the planning results are shown in Figure 7. Position labels of different colours indicate different binocular vision sensor postures.

Based on the results of visual measurement network planning, path planning was carried out by using the path planning algorithm under the regional constraint. The planning results are shown in Figure 8. The red route represents the planned optimal path result. The locations of the different colour nodes indicate the different sensor postures; the location nodes of the same color form a constraint region.

3.2 Contrast verification

In order to test and verify the quality and efficiency of the path planning, the general exhaustive method was compared with the experimental algorithm proposed in this paper. The exhaustive method first records all

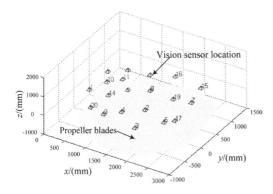

Figure 7. Layout result of the vision measurement network.

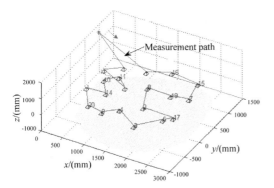

Figure 8. Path planning results.

Table 1. Path planning results.

	Proposed algorithm	Exhaustive algorithm
Path length of the plan	12099.5 mm	12099.5 mm
Time spent	67.58 s	≈5 hrs

of the loop paths through all nodes, and then verifies whether the constraint condition is satisfied and the cost value is recorded. Finally, the path that satisfies the constraint and the least cost is the final planning result. The experimental algorithm platform used was a notebook computer with a 2.27 GHz CPU frequency, 4 GB memory, and a Windows 7 operating system. The results are shown in Table 1.

As can be seen from Table 1, the proposed path planning algorithm and the general exhaustive algorithm under the regional constraints were all exact solution algorithms, and all achieved the exact shortest path value. This proved the correctness of the proposed algorithm, and its efficiency was found to be far better than the general exhaustive algorithm.

4 CONCLUSION

This paper analyzed the regional constraints in the stereo vision measurement path and established a corresponding road network model. On this basis, the mathematical model of visual measurement network planning was established. A region-constrained algorithm for measurement planning is proposed, which includes a two-layer path finding method in path finding for between regions and for inter-regions. The proposed algorithm effectively solves the problem of stereo vision measurement path planning, which is of great significance to improving the efficiency of stereo vision measurement systems.

REFERENCES

Geng, X., Qu, X., Jiang, W. and Zhang, F.M. Rapid measurement and modeling technologies of large pipes and their application. *Chinese Journal of Scientific Instrument*, Vol. 34, No. 2, pp. 338–343, 2013.

Qiao, Y.J., Wang, H.R. and Zhao, Y.J. Study on binocular vision measurement network layout for large curved surface parts. *Chinese Journal of Scientific Instrument*, Vol. 36, No. 4, pp. 913–918, 2015.

Yu, Z.Z., Yan, J.H. and Zhu, Y.H. Mobile robot path planning based on improved artificial potential field method. *Journal of Harbin Institute of Technology*, 43(01): 50–55, 2011.

Zhang, X.L., Yin, S.B. and Ren, Y.J. High-precision flexible visual measurement system based on global space control. *Infrared and Laser Engineering*. (09): 2805–2812, 2015.

Zhang, Y.W. and Zhang, X.P. Restoration of defocus blurred image in large scale 3D vision measurement. *Chinese Journal of Scientific Instrument*, Vol. 31, No. 12, pp. 2748–2753, 2010.

Zhang, Y.X., Wang, J.Q., Wang, S. and Zhang, X.P. Freight train gauge exceeding detection based on large scale 3D geometric vision measurement. *Chinese Journal of Scientific Instrument*, Vol. 33, No. 1, pp. 181–187, 2012.

Zhao, H., Liu, D.X. and Wang, W.L. The development of three-coordinate measuring machine (CMM) detection planning system based on CAD. *Chinese Journal of Scientific Instrument*. Vol. 30, No. 9, pp. 1846–1853, 2009.

Electronic Engineering – Wang (ed)
© 2018 Taylor & Francis Group, London, ISBN 978-1-138-60260-1

Research on the application of a chaotic theory for electronic information security and secrecy management

Y.Y. Lai
Guangdong Engineering Polytechnic, Guangzhou, Guangdong, China

ABSTRACT: Electronic information security is very important for countries and individuals. It is easy for electronic information in the process of transmission to be illegally intercepted and thus the security of information not guaranteed. The traditional electronic confidential information algorithm is very sensitive to initial condition and system parameters. When external disturbance appears, it will result in information receivers not being able to work. Therefore, in this study a form of chaotic electronic information system, based on a state observer, is put forward. Firstly, electronic information chaos characteristics were extracted in order to provide accurate data support for the state observer. Then a chaotic state observer was designed that eliminated the interference factors that impact on the synchronisation of the chaotic system. Synchronisation of the information transmitting and receiving device of the chaotic system was achieved. Experimental results showed that using this proposed algorithm can greatly improve the efficiency of decryption and has strong robustness.

Keywords: chaotic theory; electronic information security; security management; state observer

1 INTRODUCTION

With the rapid development in the technology industry, electronic information has gradually become one of the most essential means of human interaction (Gu & Yao, 2003). People today frequently exchange information electronically at all security levels and therefore, the security protection of electronic information shares a close relationship with a country's politics, economics, military and diplomatic affairs, as well as an individual's personal life. While electronic information makes modern life more convenient, there are also many underlying risks, such as illegal and stolen data, interception and copying. Therefore, electronic information security is irreplaceable in protecting the value of information and has been a research topic that scholars have focused on the most (Zhang, 2005).

Electronic information in a complex communication environment, can be illegally intercepted in the transmission process, and so information security cannot be guaranteed. The traditional algorithm used for electronic information security relies on a complex algorithm as the key process for control. Current electronic information communication systems transmit a huge amount of data, and fast decryption cannot be achieved, and so communication efficiency is reduced. In addition, with the rapid improvement in computer decryption technology, it is easier to crack the electronic information when intercepted, so using the traditional encryption algorithm has been unable to cope with the demands of the current electronic information security.

In order to avoid the above mentioned disadvantages, a new method of electronic information security, based on chaotic state observer, is proposed in this paper. Firstly, chaotic characteristics of electronic information were extracted in order to provide an accurate data basis for the design of the state observer. Then a state observer was designed which was aimed at the chaotic system of electronic information, and in the design process, the observer was designed to eliminate the interference caused by the error of the chaotic system; this was to realise the synchronisation of the electronic information transmitting device and the receiving device, and further to achieve electronic security confidence. The experiments conducted showed that using this algorithm can realise the synchronisation of a chaotic electronic information system, and the decryption efficiency of the signal receiving device can be greatly improved.

2 ANALYSIS MODELLING OF THE SECURITY AND SECRECY PRINCIPLE ON ELECTRONIC INFORMATION

In the process of using electronic information for communication, chaotic digital communication has a bright future due to its characteristics of high security, and easy control and synchronisation, for example. In electronic information communication systems, the signal transmitting device and the signal receiving device need to produce the same chaotic sequence (including errors). In order to be able to achieve this goal, the dynamic system is required to have the same initial condition, system parameters and initial time. The main role of the chaos digital information secrecy system is to generate a sequence, and so without real time control, a good job can be done in advance. The synchronisation process can be realised by using the phase locked system of the chaotic sequence of electronic information (Yang & Gao, 2004). The specific methods are as follows.

Set each binary symbol $b_k \in (-1, 1)$ to be able to use the length of the chaotic sequence being described, $n = 1 + (k-1)N, \ldots kN$. In this binary code, 0 corresponds to the Chaotic Sequence and is multiplied by -1, and then is sent. If it is 1 corresponding to the N chaotic sequence, then in accordance with the original transmission, the use of Equation 1 can be used for the process described above:

$$S^n = b^k \times x^n$$

$$n = 1 + (k-1)N, \ldots kN, k = 1, 2, 3 \ldots \qquad (1)$$

In the process of electronic information transmission, due to external interferences, some Noise $\{e^n\}$ will be produced, and the noise exists in the sequence of electronic information therefore, the signal received by the receiving device can be described with Equation 2:

$$r^n = s^n + e^n \qquad (2)$$

The signal receiving device can predicate $\{x(n)\}$, and the following electronic information secrecy system (Equation 3) can be obtained:

$$C_{rx}^{kN} = \sum_{n=1+(k-1)N}^{kN} r^n \times X^n \qquad (3)$$

According to Equation 3, it can be found that the electronic information of the k position is b^k for the N sequence of any length. If the impact of the external signal is small, and if it is $+1$, then the correlation coefficient is positive, and if it is -1, then the correlation coefficient is negative. Therefore, the following results can be obtained in Equation 4:

$$\begin{cases} b_{out}^k = 0, if \ C_{rx}^{kN} < 0 \\ b_{out}^k = 1, if \ C_{rx}^{kN} > 0 \end{cases} \qquad (4)$$

Using the method described above, the security and secrecy of electronic information can be realised. However, the security system of chaotic electronic information depends highly on initial conditions and the extreme sensitivity of system parameters. In the process of phase space position, modulation of the phase trajectory of chaotic electronic security systems exists in an unstable periodic trajectory. Its occurrence frequency relates closely to the initial conditions because a slight deviation will lead to quite different encoding scheme of chaotic sequences, and an inability to accurately decode.

In order to avoid these problems, this paper presents a state observer method based on chaotic electronic information security.

3 CHAOTIC ELECTRONIC INFORMATION SECURITY OPTIMISATION METHOD

The difficult point of a chaos system is how to realise its synchronisation. For electronic information in the transmission process, circuit resistance and capacitance have a small difference, but along with noise interference, it is of vital significance to make the synchronisation method have a certain robustness and security.

3.1 Extracting the chaotic feature of electronic information

The time sequence that the electronic information transmitted can reconstruct the phase space. Setting the Time Sequence can be described by $\{y_1, y_2, \ldots, y_P\}$, and phase space reconstruction can be realised by Equation 5 (Chen, 1992; Zhao, 2004):

$$Y = [t_1, t_2, \cdots, t_L] = \begin{bmatrix} y_1 & y_2 & \cdots & y_L \\ y_{1+v} & y_{2+v} & \cdots & y_{L+v} \\ \cdots & \cdots & \cdots & \cdots \\ y_{1+(n-1)v} & y_{2+(n-1)v} & \cdots & y_{N+(n-1)v} \end{bmatrix} \qquad (5)$$

In Equation 5, $L = P - (n-1)v$. The Time of the Delay of the electronic information reception can be described by v, the Input Information Vector of the electronic information system can be described by n, and $t_j = (y_j, y_{j+v}, \ldots, y_{j+(n-1)v})^T$ is the State Parameters of the electronic information system. In the process of feature extraction of information, the result of the time series is very important.

Equation 6 can be used to complete the process of singular value transformation of the time series:

$$Y = VEW^T \qquad (6)$$

In Equation 6, $V \in S^{n \times n}$ is used to describe the Result of the Orthogonal Transformation, and $\sum = diag(\sqrt{\mu_1}, \sqrt{\mu_2}, \ldots, \sqrt{\mu_n})$ is a Non-Null Vector feature in the above matrix.

Using Equation 7 to calculate the chaotic characteristic value of electronic information:

$$Y^T Y = VEW^T WE^T V^T = VEE^T V^T = V\Lambda V^T \qquad (7)$$

If $N \gg n$, in order to reduce the amount of computation of the information vector feature in the process of extraction, and to improve efficiency, a dimension reduction processing is needed, as follows:

(1) The matrix parameters of the chaotic characteristic component of the electronic information $\vec{Y}(m, p_j)$ can be obtained, and the corresponding characteristic value of $\mu_1, \mu_2, \ldots, \mu_m$ and the chaotic characteristic component matrix of the electronic information of

$Z = [z_1, z_2, \ldots, z_m]$, can also be obtained. Using Equation 8, the electronic information matrix parameters can be calculated:

$$\hat{T}_x = \sum_{j=1}^{d} q_j \frac{1}{p_j} \sum_{l=1}^{p_j} \left[\left(\vec{Y}_l^{(j)} - \vec{n}_j \right) \left(\vec{Y}_l^{(j)} - \vec{n}_j \right)^T \right] \qquad (8)$$

In Equation 8, q_j is the Prior Probability of the j-th Chaotic Characteristics of electronic information, p_j is the j-th Attribute Number of electronic information extraction, being used to describe the Average Value of set $\left\{ \vec{Y}_l^{(j)}, l = 1, 2, \ldots, p_j \right\}$ formed by the j-th electronic information attribute.

(2) Using Equation 9, the average value of the discrete degree matrix of the chaotic characteristics of the electronic information can be obtained:

$$T_c = \sum_{j=1}^{d} q_j \left(\vec{n}_j - \vec{n} \right) \left(\vec{n}_j - \vec{n} \right)^T \qquad (9)$$

(3) Equation 10 can be used to obtain the difference between the different types of electronic information's chaotic attributes:

$$K\left(\vec{Y}_k \right) = \frac{z_k^T T_c z_l}{\mu_k} \qquad (10)$$

In order to deal with the chaotic characteristics of the electronic information, Equation 11 can be used:

$$K\left(\vec{Y}_1 \right) \geq K\left(\vec{Y}_2 \right) \geq \cdots \geq K\left(\vec{Y}_m \right) \qquad (11)$$

(4) The chaotic characteristics of the extracted initial electronic information is reduced to the Dimension e, which can get the chaotic characteristic of the previous e large electronic information, and the Characteristic Matrix of the electronic information $X = [z_1, z_2, \ldots, z_e]$ is thus generated.

(5) The above mentioned electronic information is compressed and processed to obtain the following results in Equation 12:

$$\vec{Y}^* = X^T \vec{Y} \qquad (12)$$

According to the method described above, the chaotic characteristics of electronic information can be extracted effectively. The accurate data base is provided for the design of the electronic information state observer.

3.2 Realisation of the electronic information security algorithm

The electronic information chaotic system based on a state observer can overcome the deviation caused by the factors, such as noise interference, and can realise the synchronisation process better (Li & Qiu, 2005; Yu & Ma 2005).

To set up the electronic information chaotic system, Equation 13 can be used:

$$\begin{cases} \dot{x} = Ax(t) + Bf(y) \\ y(t) = Cx(t) \end{cases} \qquad (13)$$

In Equation 13, $x(t)$ is the State of the chaotic system, $y(t)$ is the Output of the chaotic system, $f(y)$ is the Nonlinear Feedback Input Vector of the chaotic system, and A, B, C are the constant matrix of the dimension.

Equation 13 can be converted to Equation 14 because of the interference of q (State Components in the system).

$$\begin{cases} \dot{x}_1 = A_1 x(t) + B_1 f(y) \\ \qquad \cdots \\ \dot{x}_{n-q-a} = A_{n-q-1} x(t) + B_{n-q-1} f(y) \\ \dot{x}_{n-q} = A_{n-q} x(t) + B_{n-q} f(y) \\ \dot{x}_{n-q+1} = A_{n-q+1} x(t) + B_{n-q+1} f(y) + d_1(t) \\ \qquad \cdots \\ \dot{x}_n = A_n x(t) + B_n f(y) + d_q(t) \\ y = Cx(t) \end{cases} \qquad (14)$$

In Equation 14,

$$A = \begin{bmatrix} A_1 \\ \cdots \\ A_n \end{bmatrix}, B = \begin{bmatrix} B_1 \\ \cdots \\ B_n \end{bmatrix}, d_i(t)(i = 1, \ldots q)$$

is the Interference of the chaotic system. The synchronous state observer is designed for the electronic information communication system which is disturbed by the interference.

Firstly, setting a new variable $z(t) \in R^q$, and $z(t) = A_{21} \bar{x}_1(t)$, then:

$$z(t) = \dot{\bar{x}}_2(t) - B_2 f(y) - d(t) - A_{22} \bar{x}(t) \qquad (15)$$

The new output variable of the chaotic system is:

$$\tilde{y} = -N_1 \bar{y}(t) = -[N_{11} N_{12}] \begin{bmatrix} y(t) \\ z(t) \end{bmatrix} \qquad (16)$$

A dynamic chaotic system of $(n - q)$ including $\bar{x}(t)$ factorial can be obtained:

$$\begin{cases} \dot{\bar{x}}_1(t) = [A_{11} - A_{12} M_2 - A_{12} N_{22} A_{21}] \bar{x}_1(t) - A_{12} N_{21} y(t) + B_1 f(y) \\ \tilde{y}(t) = M_1 \bar{x}_1(t) \end{cases} \qquad (17)$$

It is known that if the chaotic system (Equation 13) can be observed and the chaotic system (Equation 17) can be observed. The full order state observer of the chaotic system (Equation 17) can be described by Equation 18:

$$\dot{s}(t) = Fs(t) + B_1 f(y) + Hy(t) + L\tilde{y}(t) \qquad (18)$$

Figure 1. Original electronic image information.

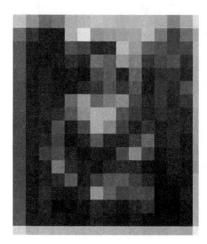

Figure 2. Results after the original image was decryption.

In Equation 18,

$F = A_{11} - A_{12}M_2 - A_{12}N_{22}A_{21} - LM_1, H = -A_{12}N_{21}$,

where L is used to describe the Gain Matrix of the observer.

The Interference Error in the electronic information chaotic system can be described by Equation 19:

$$e(t) = s(t) - \overline{x}_1(t) \quad (19)$$

The state observer of the electronic information system can be described by Equation 20:

$$\begin{cases} \hat{x}_1 = s(t) \\ \hat{x}_2 = -(M_2 + N_{22}A_{21})s(t) - N_{21}y(t) \end{cases} \quad (20)$$

According to the above information, the state observer of the chaotic system can obtain the estimation of the disturbance from:

$\hat{d}(t) = -(M_2 + N_{22}A_{21})\dot{s}(t) - N_{21}\dot{y}(t) - [A_{21} - A_{22}(M_2 + N_{22}A_{21})]s(t)$
$+ A_{22}N_{21}y(t) - B_2 f(y)$

Equation 13 was applied to an electronic information transmitting device where a transmitting electronic information chaotic system was shown. Equation 20 was applied to an electronic information receiving device It was shown that electronic information, was received thereby achieving the synchronisation of the chaotic system for electronic information. Because the design process of the state observer eliminated the influence of the interference factors in the chaotic system, it was demonstrated to have strong robustness.

4 EXPERIMENTAL RESULTS AND ANALYSIS

In order to verify the effectiveness of the algorithm, an experimental environment using MATLAB 7.0 was created.

Figure 3. Decryption results of the image using the traditional algorithm.

In the process of the experiment, the original image information which was transmitted by the electronic information transmitting device is shown in Figure 1.

The image information after encryption is shown in Figure 2.

Using the traditional algorithm to decrypt the image, the results obtained are shown in Figure 3.

Using this study's proposed algorithm to decrypt the image information, the results obtained by the experiment are shown in Figure 4.

From the experimental results, the accuracy of the information obtained by the proposed algorithm was higher than that of the traditional algorithm, which demonstrates the effectiveness of this algorithm.

In the course of this experiment, with by increasing the interference factors, the experimental results obtained by different algorithms are shown in Figures 5 & 6.

Using the above experimental results, the electronic information was decoded with the traditional

Figure 4. The results of image information decryption using the proposed algorithm.

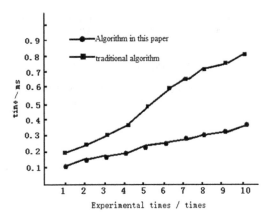

Figure 6. Decoding time of different algorithms.

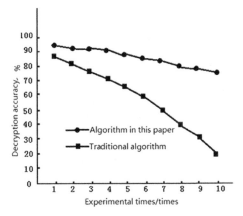

Figure 5. Decoding accuracy of different algorithms.

Table 1. Summary of experimental results of different algorithms.

Experimental times/times	interference factor	Decryption accuracy/% traditional algorithm	Algorithm in this paper	Decryption time/ms traditional algorithm	Algorithm in this paper
1	0.05	87	95	0.2	0.1
2	0.13	81	93	0.25	0.12
3	0.25	76	92	0.3	0.15
4	0.34	72	90	0.5	0.18
5	0.41	65	88	0.55	0.2
6	0.51	60	87	0.6	0.22
7	0.59	49	85	0.65	0.23
8	0.61	38	84	0.7	0.25
9	0.68	31	82	0.75	0.28
10	0.71	18	80	0.8	0.3

algorithm. With external interference factors gradually increased due to the extreme sensitivity of chaotic electronic information security systems, initial conditions and system parameters dependence was particularly high, and the resulting decryption accuracy and time was not guaranteed; making it difficult to undertake a decryption process. The proposed state observer algorithm design can realise the synchronisation of the chaotic system, and in the process of state observer design, better eliminates the interference factors. Therefore, decryption accuracy became higher and the decryption time was shorter, thus fully reflecting the superiority of this algorithm.

The data in the experiment process was analysed, and the results of the experiment are shown in Table 1.

5 CONCLUSIONS

According to the parameters of the initial state and the system of chaos in traditional extremely sensitive information security systems an external disturbance will lead to a drawback of the decryption algorithm; this is why this paper proposed a chaotic electronic information system based on a state observer. The electronic information extraction of chaotic characteristics provided accurate data support for the state observer, chaotic state observer design, and eliminated the interference factors caused by synchronisation of chaotic systems in the design process. The information transmission device and receiving device was able to realise synchronisation of the chaotic systems. Experimental results showed that the proposed algorithm can improve the efficiency of decryption, shorten the decryption time, and that the effect is satisfactory.

REFERENCES

Chen, G.S. (1992). *Mapping and chaos*. Beijin, China: National Defence Industry Press.
Gu, Q.L. & Yao, M. (2003). Research on digital image encryption based on logistic chaotic sequences. *Computer Engineering and Applications*, *23*, 114–116.

Li, P.F. & Qiu, S.S. (2005). Image encryption algorithm based on hybrid chaotic dynamic system. *Computer Application*, *25*, 543–545.

Yang, J. & Gao, J. (2004). Improved scheme of Logistic chaotic sequence encryption. *Computer Application, 23*, 58–61.

Yu, W.Z. & Ma, H.G. (2005). Image encryption method based on one dimensional chaotic map, *Computer Application*, *25*, 141–143.

Zhang, Q.C. (2005). *Bifurcation and chaos theory and its application*. Tianjin, China: Tianjin University Press.

Zhao, X.F. (2004). New digital image scrambling scheme. *Computer Application Research, 6*, 113–114.

Electronic Engineering – Wang (ed)
© 2018 Taylor & Francis Group, London, ISBN 978-1-138-60260-1

Research on feature selection algorithms based on random forest

D.J. Yao
School of Software, Harbin University of Science and Technology, Harbin, China

X.J. Zhan
College of Computer Science and Technology, Heilongjiang Institute of Technology, China

ABSTRACT: Feature selection can reduce the execution time of the classification algorithms, improve the classification accuracy of the classifier, simplify the classification model and improve the understandability of the classification model by removing irrelevant, redundant and noise features before training the classification model. Random forest is an ensemble machine learning method based on decision tree, which has been widely used in all kinds of classification, prediction, feature selection and outlier detection problems due to its higher classification performance. In this paper, we mainly study various feature selection methods based on random forest variables importance score. We also compare the performances of various algorithms on 5 UCI data sets.

Keywords: Feature selection; Random forest; Data mining

1 INTRODUCTION

In recent years, the rapid development and application of social network, image recognition, biological information technology has produced many types of high dimensional datasets. Before performing data mining on these high-dimensional datasets, it usually need execute feature selection. Feature selection can reduce the execution time of the classification algorithms, improve the classification accuracy of the classifier, simplify the classification model and improve the understandability of the classification model by removing irrelevant, redundant and noise features before training the classification model. With the advent of the era of big data, feature selection technology has become a hot topic in the field of machine learning, data mining, and pattern recognition and so on.

Random forest is an excellent integrated machine learning algorithm, which combines Bagging and random feature subspace method to realize the randomness and diversity among base classifiers. In the process of training random forest classifier, Bootstrap sampling technology are firstly used to randomly selected multiple samples subsets from the original sample space with replacement, and then decision tree models are trained on each samples subset of Bootstrap. In the process of training the decision tree model, at the time of each node splitting, random subspace method are used to randomly selects one attribute from an attribute subset for node splitting. Finally, the predicting results of multiple decision trees are combined for final the final prediction by the way of majority voting. A large number of studies show that, compared with other machine learning algorithms, random forest algorithm has many advantages (Verikas, Gelzinis & Bacauskiene 2011, Strobl, Boulesteix, Kneib, Augustin & Zeileis 2008, Liu & Li 2012): (1) can handle a variety of data types, including qualitative data (such as single nucleotide polymorphism SNPs) or quantitative data (such as microarray expression or proteomic data); (2) can provide variable importance measure which provides a easy to understand way for the analysis and research on relative importance of the features in datasets; (3) has higher classification accuracy compared to other classical classification algorithms; (4) has good robustness for noise data; (5) has ability of analyzing complex interactions among multiple features; (6) has a faster learning speed, and the computation time is suitable for large datasets with the number of input variables increasing. Over the past ten years, random forest has become a hot research topic in machine learning, data mining, pattern recognition, biomedical, bioinformatics and other application fields.

In this paper, we mainly study various feature selection methods based on random forest variables importance score. At last, experiments are performed to compare the performances of various feature selection algorithms on 5 UCI data sets.

2 RELATED WORKS

2.1 *Classification of feature selection algorithms*

In the field of machine learning and data mining, feature selection can be defined as a process of looking for the best subset of features: given a learning algorithm L, a dataset S with n features $X_1, X_2, X_3, \ldots, X_n$, class label Y, and being consistent with the distribution of D sample space, an optimal feature subset X_{opt} is a feature subset making a certain evaluation criteria $J = J(L, S)$ be optimal.

As for a dataset with n features, the search space can have 2^n-1 possible combinations, in practical applications; to meet the requirements of the minimum feature subset is a NP complete problem. Thus, various current feature selection methods mostly adopt the heuristic search algorithm to find the approximate optimal subset. Generally speaking, the process of feature selection generally includes 4 parts: the search strategy, the evaluation function, the stop criterion, and the verification process.

2.1.1 Classification by the search strategy

At present, the search strategies can be divided into 3 kinds: global search, random search and sequence search (Girish & Ferat 2014).

The global search strategy includes breadth first search, branch and bound search and directed search. The breadth first search breadth first traversal feature subspace, which belongs to the exhaustive search, the time complexity is O (2^n), the applicability of this type of algorithms is low. Branch and bound search introduces the branch and bound conditions based on exhaustive search: for some branch which may not find the better optimal solution than the current search, perform pruning, although less the time complexity than the exhaustive search but still very large.

The random search strategy combines feature selection with simulated annealing algorithm, genetic algorithm and random sampling algorithm in the process of feature selection, uses classification performance as evaluation criterion of feature subset, and usually can obtain an approximate optimal solution. However, the main problem of random search is that the experiment process is dependent on the random factors, and the experimental results are difficult to reproduce. In addition, the time complexity of the algorithm is still very large when the dimension of the feature space is high.

Sequential search strategy is the most widely used search strategy in the current feature selection algorithm, including: Sequence forward selection (SFS), Sequential backward selection (SBS), Increasing L minus R selection (LRS) and Sequence floating selection. Assuming that the original feature set is F, the selected feature subset is X, then SFS algorithm first set $X = \emptyset$, then each time select a feature $x \in F-X$, and add it into the feature subset X, so make evaluation function of the feature subset J (X) be optimal. SFS does not take into account the correlation between features; it can only add features and cannot remove the feature, which leads to the feature subset contains redundant features. On the contrary, SBS algorithm first set $X = F$, and then each time remove a feature $x \in X$, generate a new feature subset X making evaluation function of the feature subset J(X) be optimal.

Figure 1. The principle of feature selection based Filter.

SBS algorithm can only delete features and cannot add features, which may lead to the loss of important information. When using the same evaluation function, the SBS algorithm is better than the SFS algorithm in the actual computing performance and robustness. For LRS algorithm, when L > R, it starts from the empty set, then each time L features are added to the feature subset X firstly, and then R features are removed out from X, makes the evaluation function of the corresponding feature subset J(X) be optimal; when L < R, it starts from a full dataset, then moves out R features form the feature subset X firstly, and then add L features into X, makes the evaluation function of the corresponding feature subset be optimal. LRS algorithm combines the idea of SFS and SBS algorithm, which is faster than SBS with respective to speed and is better than SFS with respective to performance, but how to set the values of L and R is the key for LRS algorithms. Sequential floating selection algorithm is a development of the LRS algorithm, and the difference between them is that the value of L and R is not fixed in sequential floating selection but changed (floating). In the process of feature selection of each round, sequential floating selection can set the L and R values dynamically according to statistical characteristics of the features. Sequential floating selection combines with the characteristics of SFS, SBS and LRS algorithm, and make up for their shortcomings; however, similar to the LRS algorithm, sequential floating selection has a difficult how to set the L and R value in algorithm.

2.1.2 Classification by the evaluation method

According to the evaluation method adopted in feature selection process, feature selection algorithms can be divided into 3 kinds: Filter, Wrapper and Filter combined with Wrapper (Guyon & Elisseeff 2003).

Feature selection methods based on Filter use the variable ordering as a feature selection criterion, usually using some suitable criteria for scoring variables, through the preset threshold to remove some variables with smaller scores. This is actually a kind of filtering method, because before training the classification model, some relevant variables can be filter out. Filter methods are independent of the subsequent machine learning algorithm, through the analysis of the internal characteristics or statistical rules of the feature subset to measure the quality of the feature subset. The process is shown in Fig. 1.

The Filter method can quickly eliminate some noise and non-critical features, narrow the search scope for

Figure 2. The principle of feature selection based Wrapper.

the optimal feature subset, and can be used as a feature filter. However, the main problem is that it cannot guarantee to select the smaller optimal feature subset.

Being different from the Filter methods, Wrapper selection methods depend on the subsequent machine learning algorithm in the feature selection process. The selected feature subset are directly used to train classification model, and the classification performance of the model are used to evaluate the quality of the feature subset. The basic process is shown in Fig. 2.

Compared to feature selection algorithms based on Filter, feature selection algorithms based on Wrapper can usually choose a better feature subset, but its computational speed is usually slow, and the computational efficiency is not as good as the Filter methods.

According to the high dimensional datasets, such as microarray gene expression datasets, the feature dimension usually reach tens or even hundreds of thousands, including a large number of noise features and irrelevant features, Wrapper algorithms are not desirable to used in the original high-dimensional datasets with respective to computation time and model performance meanwhile Filter algorithms are difficult to identify the really important features. By combining the Filter and Wrapper feature selection algorithms, we can solve this problem well. The basic process of combining Filter and Wrapper is: firstly, using the Filter algorithm to pre-selection (filtering), quickly removes the noise features and irrelevant features to reduce the dimension of datasets; and then use the Wrapper algorithm to carefully select features, identify the most relevant feature subset and training model.

2.2 *Variable importance score of random forest*

In the process of training classifier, random forest algorithm computes variable importance score which have been widely used in a variety of feature selection applications. Random forest algorithm provides 4 variable importance scores for selection; next, we only introduce the most commonly used variable importance score based on permutation in this paper. The variable importance scores based on the permutation can be defined as the average reduction in the classification accuracy of the classification model before and after the occurrence of a slight disturbance in the out of bag data, which not only considering the influence of each variable separately but also considering the interaction between multiple variables, and being more widely, used in the literatures (Qi 2012).

Given Bootstrap sample set $D_b, b = 1, 2, \ldots, B$, B is the number of random samples, the variable importance scores of features X_j ($j = 1, \ldots, N$) are computed as the following:

1. Sampling B times from the original training dataset using Bootstrap random re-sampling, get B samples subset D_b;
2. Set $b = 1$;
3. Train decision tree model T_b on the sample set D_b, and the data out of bag data is labeled as L_b^{oob};
4. In the dataset L_b^{oob}, use T_b to classify the data, the number of the correct classification samples is labeled as R_b^{oob};
5. For feature $X_j, j = 1, \ldots, N$, randomly perturbs each sample in L_b^{oob} until its original relationship with the target variable is interrupted, and the perturbed dataset is labeled as L_{bj}^{oob};
6. In the dataset L_{bj}^{oob}, use T_b to classify the data, the number of the correct classification samples is labeled as R_{bj}^{oob}; If the feature X_j is related to target variable, then the classification performance of the classifier will be significantly reduced;
7. For $b = 2, \ldots, B$, repeat steps 3–6;
8. Computing the variable importance score of the feature X_j by the following formula:

$$\overline{D_j} = \frac{1}{B}\sum_{i=1}^{B}\left(R_b^{oob} - R_{bj}^{oob}\right) \qquad (1)$$

3 FEATURE SELECTION ALGORITHMS BASED ON RANDOM FOREST

3.1 *Algorithm designing*

In this paper, we focus on the combination of the random forest variable importance score and the sequential search strategy for feature selection. More concretely, we run random forest algorithm for computing variable importance scores of each features in training dataset firstly, then we sort all features by their variable importance score, next, we use sequence forward selection and sequence backward selection strategy respectively to select optimal feature subset. Finally, in all feature subsets produced in the process of feature selection, we select the feature subset with the highest classification accuracy as the optimal feature subset.

In order to ensure the reliability of the algorithm, when computing variable importance score, we run 10 times random forest algorithm and compute the average variable importance score of 10 times running as the basis for sorting all features in feature subsets. For the same purpose, we compute the average value of classification accuracy of 10 time running as the basis of evaluating the quality of feature subsets.

3.2 *Experiment and result analysis*

In order to study the performance of the proposed feature selection algorithm, we select 5 datasets for experiment from UCI Machine Learning Repository,

Table 1. The information of datasets used in experiments.

Name	Number of features	Number of samples
Wdbc	32	569
Wpbc	34	198
Musk	168	476
Ionosphere	34	351
German-org	25	666

Table 2. Experiment results on 5 UCI dataset.

	RF-SFS		RF-SBS	
Dataset	number of selected features	Classification accuracy	number of selected features	Classification accuracy
Wdbc	0.9676	29	0.9842	22
Wpbc	0.8024	28	0.8112	23
Musk	0.8159	22	0.8958	10
Ionosphere	0.9303	29	0.9332	26
German-org	0.7551	20	0.7696	20

and the basic information of those datasets is shown as Table 1.

We carried out experiments on those 5 datasets and run the feature selection algorithm based on random forest variable importance score and sequence forward selection (named RF-SFS) and the feature selection algorithm based on random forest variable importance score and sequence backward selection (named RF-SBS). The optimal feature subsets and the corresponsive classification accuracy are shown in table 2.

From table 2, one can see that the classification accuracy on the optimal feature subset of RF-SBS are higher than RF-SFS on all 5 dataset, and the size of the optimal feature subset is obviously smaller than the original feature set.

4 CONCLUSIONS

In this paper, we focus on the combination of the random forest variable importance score and the sequential search strategy for feature selection. Experiment showed that feature selection method based on random forest variable importance score and sequence selection strategy is effective.

ACKNOWLEDGMENTS

This work is sponsored by the Science and Technology Research Project of Education Department of Heilongjiang Province "Research on Feature selection and classification technology based on random forest" (No. 12541124).

REFERENCES

Girish, C. & Ferat, S. (2014). A survey on feature selection methods. *Computers and Electrical Engineering*. 40: 16–28.

Guyon, I. & Elisseeff, A. (2003). An introduction to variable and feature selection. *J Mach Learn Res*. 3: 1157–82P.

Liu, H. Q. & Li, J. Y. (2012). A comparative study on feature selection and classification methods using gene expression profiles and proteomic patterns. *Genome Informatics*. 13: 51–60.

Qi, Y. (2012). Random forest for Bioinformatics. *Ensemble Machine Learning*. 307–323.

Strobl, C. Boulesteix, A. Kneib, T. Augustin, T. & Zeileis, A. (2008). Conditional variable importance for random forests. *BMC Bioinformatics*. 9: 307.

Verikas, A. Gelzinis, A. & Bacauskiene, M. (2011). Mining data with random forests: A survey and results of new tests. *Pattern Recognition*. 44: 330–349.

Mobile-terminal sharing system for B-ultrasonic video

Q.H. Shang & F.Y. Liang
Faculty of Electrical and Electronic Engineering, Harbin University of Science and Technology, Harbin, China

ABSTRACT: The sharing system of a mobile terminal of the B-ultrasonic video is a medical image sharing system designed to overcome the existing disadvantage that sharing B-ultrasonic video is difficult. This system realises the collection of B-ultrasonic video and wireless transmission and the conversion from the B-ultrasonic video signal into a video link address through a video acquisition card based on a processor core of the ARM9 platform. A mobile-terminal sharing system for the B-ultrasonic video can combine a Picture Archiving and Communication System (PACS) with other medical imaging workstations. Patients can acquire and share B-ultrasonic video with different medical experts easily, which improves the accuracy of diagnosis for the patients.

Keywords: B-ultrasonic video; ARM9; video acquisition; video sharing; PACS

1 INTRODUCTION

With the rapid development of electronic technology the connection between electronics, communications and the Internet is getting closer. Video is widely used in industrial control, medicine, national defence and other fields because of its incomparable superiority, as it provides richer and more colourful information than language and characters. Among them, video is most widely applied in clinical medicine.

Not only does B-scan ultrasonography observe the movement and function of organs coherently and dynamically, it can also track the pathological changes and display the stereo changes without being restricted by the imaging stratification. In addition, B-scan ultrasonography combined with Doppler technology can identify the nature and extent of damage to the organ by monitoring the direction of blood flow. Therefore, B-scan ultrasonography is becoming an indispensable diagnostic method in modern clinical medicine.

There are deficiencies in video sharing related to the B-ultrasonic detector, although B-scan ultrasonography has an irreplaceable role in the field of medicine. If patients want a B-ultrasonic video to share it with other specialists to get more diagnostic information, a B-scan ultrasonic detector can only use its own CD to store and share the B-ultrasonic video with complex operation, inconvenient management and slow speed. It is not convenient for medical experts to receive video in this way for remote diagnosis.

The rapid development of the Internet and smart phones provides a good solution for the problems above. B-ultrasonic video will be created by the video sampling circuit and can be uploaded to the medical imaging website automatically (Jiang & Su, 2013; Fu et al., 2011). Furthermore, the video URL created by

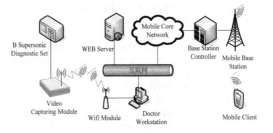

Figure 1. Schematic diagram of the overall system structure.

the software of a web server can be sent to the doctor's workstation. Patients can go to the video website directly and share the video.

2 OVERVIEW OF THE SYSTEM

The system consists of four parts, including a video capturing circuit, a doctor's workstation, a web server and a mobile phone client. Figure 1 shows the schematic diagram of the overall structure of the system.

The function of the video capturing circuit is to encode the B-ultrasonic video signal (PAL signal) coming from the B-ultrasonic detector into the form of a JPEG image, and then transmitting it to the doctor workstation through the WiFi communication module in the form of JPEG video streams.

The doctor workstation receives the video stream and converts it into a video in Audio Video Interleave (AVI) format, and the video will be uploaded to the web server via the Internet, which will form a two-dimensional code picture for patients to scan with their phones.

The purpose of the web server is to manage the B-ultrasound videos of each hospital and the software of the server can distribute hyperlink addresses of the B-ultrasonic videos. Patients can watch the video online when they visit the link.

The mobile phone client is the tool for patients to obtain the B-ultrasonic video. Patients can get the B-ultrasonic video link address by scanning the two-dimensional code on the doctor's workstation monitor with the mobile phone client (Anselmo & Madonia, 2016). At the same time, they can also share the link address with other medical experts to share the B-ultrasonic video.

3 DESIGN OF THE VIDEO CAPTURING CIRCUIT

The mobile terminal sharing system of the B-ultrasonic video requires that video capturing should be real-time, which is the reason why we chose video capture chip, image compression chip and the ARM processor platform as the basis for video capture. The method of hardware compression is adopted in this scheme, which has the advantages of fast compression and short development cycle.

The video capturing circuit is mainly composed of an ARM-based controlling module, a video decoding module, a video compression coding module and a WiFi communication module. The structure of the video capturing circuit is shown in Figure 2.

S3C2440A is used as the core of the ARM controlling module to realise the function of the controlling circuit initialisation and video data compression coding. This system uses one instance of K4S561632N to form the 16-bit Synchronous Dynamic Random-Access Memory (SDRAM), which is mainly used as the running space of the program and the data stack area.

The function of the video decoding module is to receive the analogue video signal coming from the B-supersonic diagnostic set, and convert it into a digital video signal conforming to a certain standard, which will be transmitted to the back circuit.

A video compression coding module is used to complete the compression and coding of JPEG data. The task of the WiFi communication module is to transmit the JPEG video stream to the doctor's workstation.

3.1 Design of the video decoding module

The circuit of the video decoding module is shown in Figure 3.

The core chip of the video decoding module is produced by Philips (Amsterdam, The Netherlands) and called SAA7113H, which can automatically recognise PAL or NTSC video signal formats and automatically detect the field frequency. The input pin AI22 of the SAA7113H chip connects to the analogue video

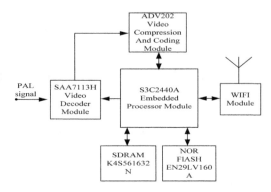

Figure 2. Structure of the video capture circuit.

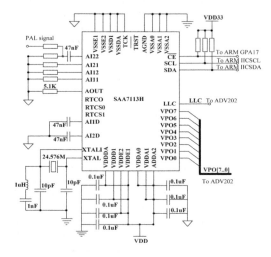

Figure 3. Circuit diagram of the video decoding module.

signal output from the B-supersonic diagnostic set; the other input pins are grounded. The output signal LLC of the video decoding circuit is the clock signal line of the ADV202, VPO7~VPO0 are signal lines of 8-bit image data. When the circuit is operated, S3C2440A will set the internal registers of SAA7113H through the I²C bus firstly, then it starts the analogue-to-digital (A/D) conversion and samples the VPO data and finally outputs the video signal according to CCIR 656 YUV 4:2:2.

3.2 Design of the video compression coding module

The circuit diagram of the video compression coding module is shown in Figure 4.

The module uses a video codec chip from ADI (Analog Devices, Inc., Norwood, MA) called ADV202 and its main function is to compress the decoding data coming from SAA7113H and complete the coding process of the JPEG image. The encoded JPEG pictures are sent to SDRAM and wait to be sent by the WiFi communication module. The clock signal LLC outputted by the video decoding module is used as the

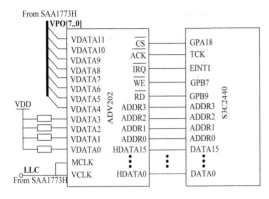

Figure 4. Circuit diagram of the video compression coding module.

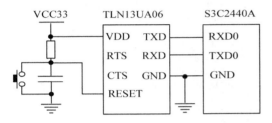

Figure 5. Connection diagram of the WiFi module.

clock signal of the system and the clock signal of the video data of the ADV202 directly. As the standard of CCIR656 YUV 4:2:2 requires that the synchronisation signal and image data must be transmitted at the same time, it is unnecessary to design hardware synchronisation. The address lines of S3C2440A are connected to address lines ADDR0~ADDR3 of ADV202, which are used to control the reading of the image data, HDATA15~HDATA0 is the data bus of ADV202.

3.3 Design of the WiFi communication module

This refers to two articles (Kang & Borriello, 2006; Hur et al., 2008). The schematic diagram of the circuit connection between the TLN13UA06 wireless WiFi module and the S3C2440A ARM processor is shown in Figure 5.

WiFi communication uses the TLN13UA06 module, which is a kind of embedded module based on the Universal Asynchronous Receiver/Transmitter (UART) interface with the WiFi standard of a wireless network. It has a built-in protocol stack of IEEE802.11 and TCP/IP in accordance with the wireless network protocol, and can realise the function of conversion of serial data from the user to the wireless network.

4 DESIGN OF THE SOFTWARE OF THE SYSTEM

The software includes three parts, namely, embedded software, the doctor's workstation software and the web server software.

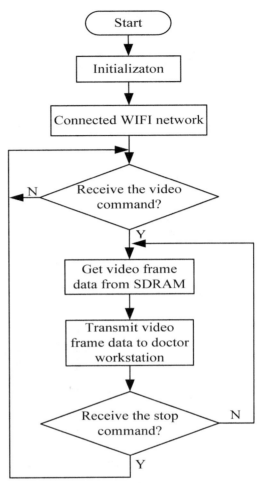

Figure 6. Flow chart of the application software.

4.1 The function of the embedded software

This system is initialised with the BootLoader developed by Tianxiang Technology Co. Ltd. Embedded software is mainly composed of a Linux kernel, YAFFS2 file system, device driver and application software. Among them, the device driver is written with the IOCTL function, which is used to initialise and control SAA7113H and ADV202 to obtain video data. The application software flow chart is shown in Figure 6.

4.2 The function of the doctor's workstation software

The software of the doctor's workstation has a visual interface. The software functions are to receive and convert the video stream, video upload and two-dimensional code generation. The software workflow diagram is shown in Figure 7.

Decoding the video stream is the process by which the workstation software receives the JPEG picture stream according to the IP address and port number

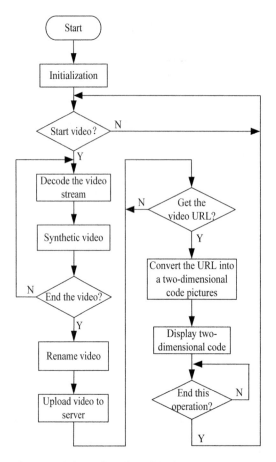

Figure 7. Software flow chart of the doctor workstation.

of the video capture module and parses the protocol. Video synthesis is the process by which the decoded video data is encapsulated in a container in the AVI format and a high-definition video in AVI format is stored on the local disk. The doctor's workstation software uploads the video according to the HTTP protocol and the local video address to the web server specified IP automatically when the doctor uploads the video, the doctor's workstation software will enter into the waiting process in the whole process of uploading video. If the server returns the video address, then the doctor's workstation software will process the URL for the two-dimensional code image with JIS and ISO standards and display it on the screen of the doctor's workstation.

4.3 The function of the web server

The web server is a server based on the Tomcat container; software on the doctor's workstation can access the program on the server through the HTTP protocol. In this system, the web server is mainly used to generate the web page with the B-ultrasonic video and return a link address to the doctor's workstation software. The software will process the link address into a two-dimensional code picture. The patient who scans the two-dimensional code with a mobile smart phone can conveniently obtain the link of the B-ultrasonic video.

5 CONCLUSIONS

The system improves the management of the B-ultrasonic video over the Internet on the basis of video capture. In addition, it simplifies the method for the patient to get and share B-ultrasonic video, while satisfying their need for high-quality B-ultrasonic video. It is possible for patients to share the URL link of B-ultrasonic video or two-dimensional code picture with QQ and WeChat friends to share B-ultrasonic video indirectly. Not only does this solve the difficulty of having access to the B-ultrasonic video for patients, but it also overcomes the problem of sharing large-capacity B-ultrasonic video, and it relieves the patient's pain of travelling back and forth to many hospitals, saving time and medical expenses.

REFERENCES

Anselmo, M. & Madonia, M. (2016). Two-dimensional comma-free and cylindric codes. *Theoretical Computer Science, 658*, 4–17.

Fu, W., Xu, Z., Liu, S., Wang, X. & Ke, H. (2011). The capture of moving object in video image. *Journal of Multimedia, 6*, 518–525.

Hur, J., Park, C., Shin, Y. & Yoon, H. (2008). An efficient proactive key distribution scheme for fast handoff in IEEE802.11 wireless networks. *Lecture Notes in Computer Science, 5200*, 86–102.

Jiang, H. & Su, X. (2013). The embedded image acquisition system based on the ARM. *Journal of Convergence Information Technology, 8*(9), 845–852.

Kang, J.H. & Borriello, G. (2006). Harvesting of location-specific information through WiFi networks. *Lecture Notes in Computer Science, 3987*, 46–53.

Electronic Engineering – Wang (ed)
© 2018 Taylor & Francis Group, London, ISBN 978-1-138-60260-1

A novel area-efficient Rotating S-boxes Masking (RSM) scheme for AES

J.X. Jiang, J. Hou & Y.Y. Zhao
School of Applied Sciences, Harbin University of Science and Technology, Harbin, China

X.X. Feng
School of Computer Sciences and Technology, Harbin University of Science and Technology, Harbin, China

H. Huang
School of Software, Harbin University of Science and Technology, Harbin, China

ABSTRACT: Rotating S-boxes Masking (RSM) scheme for AES was proposed by Nassar *et al.* in 2012, which was a type of Low-Entropy Masking Scheme (LEMS). Its security was proved against first-order and zero-offset second-order DPA attacks. This paper designs a novel area efficient RSM scheme for AES by analyzing the characteristics of the mask values of the RSM scheme. The main idea of this paper is that separating the sixteen masks into four groups and reusing the dedicated designed S-box in each group. The proposed masking scheme reduces the number of S-boxes required for the RSM scheme from sixteen to four (without considering the S-boxes in key expansion operation), which leads to a novel area-efficient RSM scheme. Compared with existing RSMs, this scheme dramatically reduces the hardware complexity without any loss of security.

Keywords: Rotating S-boxes Masking (RSM); LEMS; AES; S-box reusing

1 INTRODUCTION

The Advanced Encryption Standard (AES) is a standard symmetric encryption algorithm. It has a good nature to achieve speed, memory requirements in hardware and software. Thus AES is widely used in practical applications. Along with the Side-Channel Attacks (SCA) (Kocher 1996) technology appearing, especially the Differential Power Attacks (DPA) (Paul 1999) technology, the security of the encryption chip has been under a great threat (Nassar 2011).

During the last decades, there are two major countermeasures, named hiding and masking, which have been proposed against DPA (Mangard 2007). The masking scheme is more effective than the hiding scheme. Thus it is widely studied in existing literatures, and can be classified into three categories: Look-Up-Table (LUT) based schemes (Coron 2012), addition chain based schemes (Carlet 2012) and composite field based schemes (Satoh 2001). LUT-based schemes have the advantage of faster computation speed than other two schemes, however it requires a large memory area, which is not suitable for area constraint applications. In order to overcome this problem, a new type of Low-Entropy Masking Scheme (LEMS), named Rotating S-boxes Masking (RSM), was proposed for hardware implementations (Nassar 2012). After that, RSM was proposed for software targets in (Bhasin 2013). Compared to existing masking schemes, the RSM has much lower masking complexity, and its performance and complexity is close to an unprotected AES. Moreover, it can thwart first-order and zero offset second-order DPA. However, its masking complexity could be further reduced, if the property of the used masks is considered. This paper analysis the property of the masks used in RSM scheme, and gives the certain law. Thanks to the given law, the number of masked S-boxes can be reduced from sixteen to four by reusing the dedicatedly developed S-boxes. Therefore, the proposed scheme is called area-efficient RSM for AES. The security of the proposed scheme has been proved theoretically and experimentally. The function of algorithm is also verified by simulation.

The remainder of the paper is organized as follow. Section 2 presents related work. Section 3 describes the RSM scheme and S-box reuse masking scheme. Section 4 concludes this paper.

2 RELATED WORK

LUT-based masking scheme always adopts the table re-computation technique. In 1999 (Chari 1999), Chari *et al.* firstly proposed this scheme, the S-box was randomized into $T(x) = S(x \oplus r) \oplus s$, where x was an eight-bit unmasked data, r was input mask, s was output mask. This scheme only masked the first and last rounds and had been proved only against first-order DPA. Later, many high-order masking schemes were proposed, but most of the countermeasures brought a significant cost overhead in terms of memory, time or both. In order to make the masking schemes practicable, various LEMSs were proposed. In 2001 (Itoh 2001), Itoh et al. proposed the fixed value masking scheme to reduce the RAM and computation time for masked S-box, the main idea of this scheme was to use the same group masking values for each round of encryption, and each state byte also used the same masking byte, this idea could be expressed in this form $T(x) = S(x \oplus r) \oplus r$. The random masks and the recomputed S-boxes were stored in the ROM in advance. With this method, one of the pairs was selected randomly during the encryption process. It reduced not only the chip processor load, but also the space occupied by RAM. The disadvantage of the scheme was only against first-order CPA attacks. In 2012 (Nassar 2012), Nassar et al. proposed the RSM scheme, RSM was belong to the fixed value masking scheme, but unlike any previous masking schemes, all those S-boxes were different, the RSM scheme would have a detailed description in Section 3.1. Furthermore, the RSM has an advantage to counteract first-order and zero offset second-order DPA. In 2014 (Yamashita 2014), Yamashita et al proposed a variant RSM to reduce the complexity further by omitting the re-masking operation for the intermediate rounds, which was necessary for RSM. However, the output masking values were determined by the input masking values.

In summary, RSM was better than previous proposed LUT-based scheme in terms of security and performance. However, there was a lack of research on further reducing the masking overhead by using the property of the used masks. This paper proposes a novel area-efficient RSM scheme for AES, which aims at reducing the number of the S-boxes by reusing the developed S-boxed. In order to illustrate the proposed masking scheme, the RSM and S-box reusing masking scheme are revisited and introduction sequentially.

3 RSM SCHEME AND S-BOX REUSEING MASKING SCHEME

The AES is a block cipher composed of an SPN architecture. For AES-128, it has ten rounds and each round consists of four different layers: SubBytes (S-box), ShiftRows, MixColumns, and AddRoundKey. Among them, the Sbox is the only non-linear function, which is difficult to be masked. While the linear operations can be easily masked by a simple XOR operation with pre-computed constants applied at the end of each round. Therefore, the non-linear function (S-box) is mainly introduced in the following.

3.1 RSM scheme revisited

The RSM scheme adheres to the table re-computation technique, the S-boxes are addressed only by the masked data, thus it is a mono-path structure. It uses

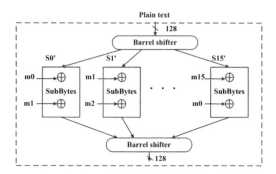

Figure 1. Diagram of RSM S-boxes.

the same number of S-boxes (namely sixteen) as an unprotected of the AES algorithm. And, all the masks are orthogonal array, such as:

{0x00, 0x0f, 0x36, 0x39, 0x53, 0x5c, 0x65, 0x6a, 0x95, 0x9a, 0xa3, 0xac, 0xc6, 0xc9, 0xf0, 0xff} Or

{0x03, 0x0c, 0x35, 0x3a, 0x50, 0x5f, 0x66, 0x69, 0x96, 0x99, 0xa0, 0xaf, 0xc5, 0xca, 0xf3, 0xfc}. The rotating S-boxes can be implemented in hardware by adding extra barrel shifters, which is shown in Figure 1.

3.2 S-box reuse masking scheme

As we can see, the used masks can be classified into four groups, and four masks in each group would be XORed to zero. Taking the masks {0x00, 0x0f, 0x36, 0x39, 0x53, 0x5c, 0x65, 0x6a, 0x95, 0x9a, 0xa3, 0xac, 0xc6, 0xc9, 0xf0, 0xff} as example, all of the sixteen masks can be classified into four groups, denote as M_j with $j \in \{1, 4\}$. And each group is shown as follow.

$$M_1 = [00, 0f, f0, ff] \quad (1)$$

$$M_2 = [36, 39, c6, c9] \quad (2)$$

$$M_3 = [53, 5c, a3, ac] \quad (3)$$

$$M_4 = [65, 6a, 95, 9a] \quad (4)$$

Therefore, each group can use same S-box, and the S-box is masked as follow.

$$S_m(x \oplus m_i) = S(x) \oplus m_{i+1 (\mod 4)} \quad (5)$$

where x is an eight-bit unmasked data. Each group can arbitrarily choose one masked S-box as reused S-box. For example, in first group M_1, the input mask '00' and output mask '0f' can be selected to develop the first reused S-box, which is expressed as S-box_m11_m12. Similarly, the jth reused S-box can be expressed as Sbox_mj1_mj2.

The architecture of the reused S-box is shown in Figure 2.

In each group of masks, all four pair of input and output masks are either the same or negated. Thus, the same S-box can be reused by adding extra multiplexers, which leads to further reducing the number

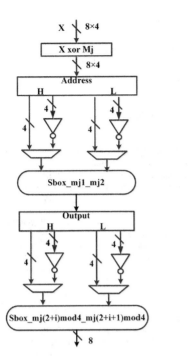

Figure 2. Architecture of the reused S-box.

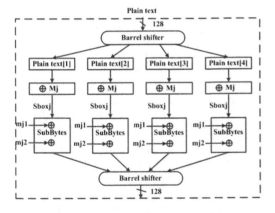

Figure 3. RSM using reused S-box for AES.

of masked S-boxes. The main principle is that when a group plain texts (32 bits) XOR one set of masks values, the address is selected according to the relationship between the high/low bits of the input mask. If the value is same, then the address is not changed, otherwise the address is inverted. Similarly, the output is selected according to the relationship between the high/low bits of the output masks. If the value is same, then the output is not changed, otherwise, the output is inverted.

3.3 RSM using reused S-box for AES

The diagram of RSM using reused S-box for AES is shown in Figure 3.

The AES encryption process is as follow: firstly, separating the plain texts into four groups, and each group has 32 bits, which expressed as Plain text[i], with $i \in [1, 4]$; secondly, each group of plain text randomly select a group of masks and XOR with these masks; finally, the results are the input of the reused S-box, and looked-up from the masked S-box four times of each group. Compared with RSM, this scheme only requires four masked S-boxes (1/4 of RSM), while RSM requires 16 masked S-boxes.

3.4 Security analysis of the proposed scheme

In this section, the security of the proposed scheme is proved by theoretical evaluation. The analysis methods have been already introduced in [8, 9].

First of all, we know that the masks group of M_j cannot be guessed by SPA (short for "Simple Power Attack"), because irrespective of $j \in \{1, 4\}$, but the M_j is selected randomly and the masks are accessed in parallel. Thus this paper considers differential SCA.

DEFINITION 1: An implementation is dth-order secure against SCA if all the tuples of d intermediate variables $I_0(X, K), \ldots, I_{d-1}(X, K)$ have the same distribution irrespective of X and K. With X is the plaintext or cipher text and the K is secret key.

DEFINITION 2: An implementation is dth-order secure against SCA if the expectation of all monomials of degree less than or equal to d in the $(I_i(X, K))_i$ knowing $(X = x, K = k)$ does depend neither x nor in k.

When $d = 1$, according the definition 2, in this paper, we know that any intermediate variable is independent on the sensitive variable, thus the masking scheme is secure against first-order SCA. In this paper the mask values is low-entropy: it does not take all possible values in F_2^8, but only a strict subset of them. The optimal correlation coefficient defined in equation (6) (Prouff 2009).

$$\rho_{opt} = \frac{\sigma[E[C \mid Z]]}{\sigma[C]} \qquad (6)$$

In order to verify the scheme against CPA and 2nd-order zero-offset CPA, this paper used a STA-solver. We know the following results:

1) Without any mask (the random variable m is deterministic), both $\rho_{opt}^{(1)}$ at first-order and $\rho_{opt}^{(2)}$ at second-order are nonzero. The mutual information leaked $I[HW(x \oplus m); x]$ is equal to 2.5442 bit for n = 8 bit;
2) The mutual information MIA = 1.8176 bit with random variable that takes two complementary values. With two complementary masks (m is uniformly distributed in a pair $\{\tilde{m}, \neg\tilde{m}\}$ for a give byte $\tilde{m}$, e.g. {0x36, 0xc9}), $\rho_{opt}^{(1)} = 0$ but $\rho_{opt}^{(2)} \neq 0$. Thus the major change is the cancellation of $\rho_{opt}^{(1)}$. Regarding the mutual information, it still remains quite large.

Figure 4. Simulation results of the proposed scheme.

Table 1. Comparisons of different scheme.

Method	XOR Operation	ROM	S-box
RSM (Nassar 2012)	31	6592	16
This work	34	3472	4

3) With the 16 masks found by the SAT-solver, $\rho_{opt}^{(1)} = \rho_{opt}^{(2)} = 0$, and can be found the mutual information $I[HW(x \oplus m); x]$ as low as 0.2168 bit.

From a leakage analysis point of view, the proposed scheme with sixteen identified masks is secure against zero-offset second-order attacks. Thus this paper scheme keeps the same security level as RSM.

3.5 Experiments and comparisons

In order to verify the functionality of the proposed scheme, the proposed masking scheme is modeled in Verilog HDL and implemented with EDA tools, the results are shown in Figure 4.

The value 'm' refers to the mask values, 'plain_text' and 'cipher_key' refer to the plaintext and key respectively. And, the value of 'cipher_text' is the corresponding cipher text. The simulation results are compared with the correct results, and proved that the proposed masking scheme is completely correct.

Table 1 lists the hardware complexity in terms of the number of XOR operation, ROM and S-box of different AES implementations: RSM and this work. The comparison results show that compared with the RSM, the proposed masking scheme reduces the hardware complexity dramatically.

4 CONCLUSION

This paper proposed an area-efficient Rotating S-boxes Masking (RSM) scheme for AES. Different from the RSM scheme, the proposed scheme reduces the masked S-boxes number from sixteen to four. And the security of this scheme is proved theoretically. This scheme can thwart first- and second-order zero-offset SCAs. The hardware implementation and simulation results are given and compared with RSM. The comparison results indicate that the proposed masking scheme has much lower hardware complexity without any loss of security.

ACKNOWLEDGMENTS

This work was sponsored by the National Natural Science Foundation of China (Grant Nos. 61604050) and (Grant Nos. 51672062).

REFERENCES

Bhasin, S., Danger, J. L., Guilley, S. & Najm, Z. (2013). A low-entropy first-degree secure provable masking scheme for resource-constrained devices.

Chari, S., Jutla, C. S., Rao, J. R. & Rohatgi, P. (1999). Towards sound approaches to counteract power-analysis attacks. *Proc. Advances in Cryptology* (CRYPTO'99). 6: 398–412.

Coron, J. S. (2014). *Higher Order Masking of Look-Up Tables. Advances in Cryptology – EUROCRYPT*. Springer Berlin Heidelberg.

Carlet, C., Goubin, L., Prouff, E., Quisquater, M. & Rivain, M. (2012). Higher-Order masking schemes for s-boxes. *International Conference on FAST Software Encryption*. Vol. 7549: 366–384.

Itoh, K., Takenaka, M. & Torii, N. (2001). DPA Countermeasure Based on the "Masking Method". *Information Security and Cryptology – Icisc 2001, International Conference Seoul, Korea.* Vol. 2288: 440–456.

Kocher. P. C. (1996). Timing Attacks on Implementations of Diffie-Hellman, RSA, DSS, and Other Systems. *International Cryptology Conference on Advances in Cryptology*. Vol. 1109: 104–113.

Mangard, S., Oswald, E. & Popp, T. (2007). Power analysis attacks: revealing the secrets of smart cards.

Nassar, M., Guilley, S. & Danger, J. L. (2011). Formal analysis of the entropy/security trade-off in first-order masking countermeasures against side-channel attacks. *International Conference on Cryptology in India*. Vol. 2011: 22–39.

Nassar, M., Souissi, Y., Guilley, S. & Danger, J. L. (2012). RSM: A small and fast countermeasure for AES, secure against 1st and 2nd-order zero-offset SCAs. *Design, Automation, Test in Europe Conference, Exhibition*. IEEE: 1173–1178.

Paul, C., Jaffe., Joshua., Jun. & Benjamin. (1999). *Differential Power Analysis. Advances in Cryptology-CRYPTO 99*. Springer Berlin Heidelberg.

Prouff, E., Rivain, M. & Bevan, R. (2009). Statistical analysis of second order differential power analysis. *IEEE Transactions on Computers*. 2010(6): 799–811.

Satoh, A., Morioka, S., Takano, K. & Munetoh, S. (2001). A Compact Rijndael Hardware Architecture with S-Box Optimization. *Advances in Cryptology-ASIACRYPT 2001*.

Yamashita, N., Minematsu, K., Okamura, T. & Tsunoo, Y. (2014). A smaller and faster variant of RSM. *Conference on Design, Automation & Test in Europe. European Design and Automation Association*, pp. 205.

Electronic Engineering – Wang (ed)
© 2018 Taylor & Francis Group, London, ISBN 978-1-138-60260-1

Lower confidence limits for complex system reliability

Q. Wang & G.Z. Zhang
Department of Mathematics, Harbin University of Science and Technology, Harbin, China

ABSTRACT: The Winterbottom–Cornish–Fisher (WCF) method is a classic method for deriving lower confidence limits of a parameter function. In this paper, we study a complex system described by the minimal path matrix. Under the condition of the lives of the subsystems obeying the logarithmic normal distribution, we derive the lower confidence limits of reliability for the system using the WCF method.

Keywords: complex system; WCF method; lower confidence limit; minimal path matrix

1 INTRODUCTION

Research into lower confidence limits for reliability of systems containing multiple subsystems has attracted much attention for a long time. Recently, there have been significant research findings with respect to some specific systems. However, the research into complex system reliability is generally based on samples of subsystems' lives, but because of the complexity of their structure, it is difficult to obtain the lower confidence limit of their reliability.

Assuming that a reliability function is given, Winterbottom (1979, 1980) offered asymptotic expansion of the lower confidence limit for the reliability of a complex system, based on the expansion method of Cornish and Fisher (Cornish & Fisher, 1937; Fisher & Cornish, 1960), which gave rise to the Winterbottom–Cornish–Fisher (WCF) method.

Yan et al. (2007) studied the confidence limit of parameters function for parameter estimation is not independent by WCF method, and this result is used to the reliability evaluation of Weibull-type components and systems containing this kind of component. Then they simulated the lower confidence limit of reliability for simple systems. Under the condition of the subsystem life obeys exponential distribution, for the sample of type-I censoring, Yu et al. (1999) studied the lower confidence limit of reliability for the system using the WCF method, and simulated series and parallel systems. The theory is applicable to particular systems. Yu and Dai (1996) and Yu et al. (2007) derived lower confidence limits by the WCF method, and this lower limit came from an assumed reliability function. Because the information for the reliability function is far from sufficient, the expression and computation for some of the parameters are complicated. Zhang (2009) defined an expression of reliability function for a complex system described by

minimal path matrix, which established the foundation for subsequent research. In 2012, for data about success and failure, Shu and Zhang (2012) gave the lower confidence limit of reliability for a general complex system described by the minimum path matrix. Liao and Zhang (2015) obtained the lower confidence limit of reliability for a complex system described by the minimum path matrix for an incomplete sample when the subsystems' lives obey an exponential distribution.

In previous studies, the subsystems' lives have obeyed an exponential distribution. Therefore, on the basis of previous research, studying the lower confidence limit of reliability for a complex system when the subsystems' lives obey a logarithmic normal distribution is a topic that has both theoretical significance and application in practice.

2 RELIABILITY FUNCTION OF COMPLEX SYSTEMS

We introduce some conclusions – Lemmas 2.1 and 2.2 – of the studies of Zhang (2009), which are further detailed therein.

Lemma 2.1: Suppose that the life distribution function of independent subsystem S_i is $F_i(t)$, and let $R_i(t) = 1 - F_i(t), i = 1, 2, \ldots, m$. The minimal path matrix of system S is $A_{m \times k}$, life distribution function is $F_{1,A}(t)$, and reliability function is $R_{1,A}(t) = 1 - F_{1,A}(t)$. Denote $C = \begin{pmatrix} 1^T \\ A \end{pmatrix} = (c_1, c_2, \ldots, c_k)$, where $1^T = (1, 1, \ldots, 1)$ and $\tilde{C}_j$ is $(m+1) \times C_k^j$ matrix, then

$$R_{1,A}(t) = -\sum_{j=1}^{k} \tilde{R}^{\tilde{C}_j}(t) \triangleq \Psi_{1,A}\left(R_1(t), R_2(t), \cdots, R_m(t)\right)$$

where $\tilde{R}(t) = (-1, R_1(t), R_2(t), \cdots, R_m(t))^T$.

Lemma 2.2: The minimal path matrix of complex system S is $A_{m \times k}$, and the system reliability function is $\Psi_{1,A}(R_1(t), R_2(t), \ldots, R_m(t))$.

Let $C = \begin{pmatrix} 1^T \\ A \end{pmatrix} = (c_1, c_2, \cdots, c_k)$, where $1^T = (1,1,\cdots,1)$, $\tilde{C}_j$ is $(m+1) \times C_k^j$ matrix, then

$$\frac{\partial R_{1,A}}{\partial R_i} = -\sum_{v=1}^{k} \tilde{R}^{d(\tilde{C}_v, j+1)}$$

$$\frac{\partial^2 R_{1,A}}{\partial R_i \partial R_j} = \begin{cases} -\sum_{v=1}^{k} \tilde{R}^{d\left(d(\tilde{C}_v, j+1), j+1\right)}, i \neq j \\ 0 \qquad\qquad , i = j \end{cases}$$

Therefore, their estimators are:

$$\frac{\partial \hat{R}_{1,A}}{\partial \hat{R}_i} = -\sum_{v=1}^{k} \hat{\tilde{R}}^{d(\tilde{C}_v, j+1)}$$

$$\frac{\partial^2 \hat{R}_{1,A}}{\partial \hat{R}_i \partial \hat{R}_j} = \begin{cases} -\sum_{v=1}^{k} \hat{\tilde{R}}^{d\left(d(\tilde{C}_v, j+1), j+1\right)}, i \neq j \\ 0 \qquad\qquad , i = j \end{cases}$$

3 WCF METHOD

Suppose that there are m distribution functions; the uth distribution contains m_u parameters. Let $\theta^{(u)} = (\theta_1^{(u)}, \ldots, \theta_{m_u}^{(u)})$ be the parameter vector of the uth distribution, and n_u be the sample size of the uth distribution, $n = \min_{1 \leq u \leq m} n_u$, $\lambda_u = \frac{n_u}{n}$, $\lambda_u \geq 1$.
$\hat{\theta}^{(u)} = (\hat{\theta}_1^{(u)}, \ldots, \hat{\theta}_{m_u}^{(u)})$ is the estimator for $\theta^{(u)}$ and supposed $\hat{\theta}_i^{(u)} - \theta_i^{(u)}$ are:

$$E\left(\hat{\theta}_i^{(u)} - \theta_i^{(u)}\right) = \frac{\mu_i^{(u)}}{n_u} + o\left(n_u^{-2}\right)$$

$$E\left(\hat{\theta}_i^{(u)} - \theta_i^{(u)}\right)\left(\hat{\theta}_j^{(u)} - \theta_j^{(u)}\right) = \frac{\mu_{ij}^{(u)}}{n_u} + o\left(n_u^{-2}\right)$$

$$E\left(\hat{\theta}_i^{(u)} - \theta_i^{(u)}\right)\left(\hat{\theta}_j^{(u)} - \theta_j^{(u)}\right)\left(\hat{\theta}_k^{(u)} - \theta_k^{(u)}\right) = \frac{\mu_{ijk}^{(u)}}{n_u^2} + o\left(n_u^{-3}\right)$$

$$E\left(\hat{\theta}_i^{(u)} - \theta_i^{(u)}\right)\left(\hat{\theta}_j^{(u)} - \theta_j^{(u)}\right)\left(\hat{\theta}_k^{(u)} - \theta_k^{(u)}\right)\left(\hat{\theta}_l^{(u)} - \theta_l^{(u)}\right)$$
$$= \frac{\mu_{ijkl}^{(u)}}{n_u^2} + o\left(n_u^{-3}\right)$$

where $u = 1, \ldots, m$; $i, j, k, l = 1, \ldots, m_u$; $\mu_i^{(u)}$ is only correlated with $\theta_i^{(u)}$; $\mu_{ij}^{(u)}$ is only correlated with $\theta_i^{(u)}$, $\theta_j^{(u)}$; $\mu_{ijk}^{(u)}$ is only correlated with $\theta_i^{(u)}$, $\theta_j^{(u)}$, $\theta_k^{(u)}$; $\mu_{ijkl}^{(u)}$ is only correlated with $\theta_i^{(u)}$, $\theta_j^{(u)}$, $\theta_k^{(u)}$, $\theta_l^{(u)}$.

Let $\theta = (\theta^{(1)}, \cdots, \theta^{(m)})$, $\hat{\theta} = (\hat{\theta}^{(1)}, \cdots, \hat{\theta}^{(m)})$, $h = H(\theta)$, $g_1 = G_1(\theta)$, $g_3 = G_3(\theta)$, $h_i^{(u)} = \frac{\partial h}{\partial \theta_i^{(u)}}$, $\Psi_i^{(u)} = \frac{\partial \Psi}{\partial \theta_i^{(u)}}$,

$$\Psi_{ij}^{(u)} = \frac{\partial^2 \Psi}{\partial \theta_i^{(u)} \partial \theta_j^{(u)}}, \quad i, j = 1, \cdots m_u, \quad u = 1, \cdots, m,$$

$$\Psi_{ij}^{(uv)} = \frac{\partial^2 \Psi}{\partial \theta_i^{(u)} \partial \theta_j^{(v)}}, \quad i = 1, \cdots m_u, \quad j = 1, \cdots m_v,$$

$u, v = 1, \ldots, m; u \neq v$. From reference [10] we have:

$$h = \left(\sum_u \sum_{i,j} \Psi_i^{(u)} \Psi_j^{(u)} \frac{\mu_{ij}}{\lambda_u}\right)^{-\frac{1}{2}}$$

$$g_1 = -\frac{3}{2} A_1 + \frac{1}{6} A_2$$

$$g_3 = h^2 \left(\frac{1}{2} A_1 - \frac{1}{6} A_2\right)$$

where

$$A_1 = h \sum_u \sum_i \Psi_i^{(u)} \frac{\mu_i^{(u)}}{\lambda_u} + \frac{h}{2} \sum_u \sum_{i,j} \Psi_{ij}^{(u)} \frac{\mu_{ij}^{(u)}}{\lambda_u}$$
$$+ \sum_u \sum_{i,j} h_i^{(u)} \Psi_i^{(u)} \frac{\mu_{ij}^{(u)}}{\lambda_u}$$

$$A_2 = h^3 \left[\sum_u \sum_{i,j,k} \Psi_i^{(u)} \Psi_j^{(u)} \Psi_k^{(u)} \frac{\mu_{ijk}^{(u)}}{\lambda_u^2} \right.$$
$$\left. + 3 \sum_{\substack{u,v \\ u \neq v}} \sum_{i,j,k} \Psi_i^{(u)} \Psi_j^{(u)} \Psi_k^{(v)} \frac{\mu_{ij}^{(u)} \mu_k^{(v)}}{\lambda_u \lambda_v} \right]$$

$$+ \frac{3}{2} h^3 \left[\sum_u \sum_{i,j,k,l} \Psi_i^{(u)} \Psi_j^{(u)} \Psi_{kl}^{(u)} \frac{\mu_{ijkl}^{(u)}}{\lambda_u^2} \right.$$
$$+ \sum_{\substack{u,v \\ u \neq v}} \sum_{i,j,k,l} \Psi_i^{(u)} \Psi_j^{(u)} \Psi_{kl}^{(v)} \frac{\mu_{ij}^{(u)} \mu_{kl}^{(v)}}{\lambda_u \lambda_v}$$
$$\left. + 4 \sum_{\substack{u,v \\ u \neq v}} \sum_{i,j,k,l} \Psi_i^{(u)} \Psi_j^{(v)} \Psi_{kl}^{(uv)} \frac{\mu_{ik}^{(u)} \mu_{jl}^{(v)}}{\lambda_u \lambda_v} \right]$$

$$+ 3h^2 \left[\sum_u \sum_{i,j,k,l} h_i^{(u)} \Psi_j^{(u)} \Psi_k^{(u)} \Psi_l^{(u)} \frac{\mu_{ijkl}^{(u)}}{\lambda_u^2} \right.$$
$$\left. + 3 \sum_{\substack{u,v \\ u \neq v}} \sum_{i,j,k,l} h_i^{(u)} \Psi_j^{(u)} \Psi_k^{(v)} \Psi_l^{(v)} \frac{\mu_{ij}^{(u)} \mu_{kl}^{(v)}}{\lambda_u \lambda_v} \right]$$

Let $\hat{\mu}_{ij}^{(u)} = \mu_{ij}^{(u)}(\hat{\theta})$, $\hat{\mu}_{ijk}^{(u)} = \mu_{ijk}^{(u)}(\hat{\theta})$, $\hat{\mu}_{ijkl}^{(u)} = \mu_{ijkl}^{(u)}(\hat{\theta})$, $\hat{\Psi}_i^{(u)} = \Psi_i^{(u)}(\hat{\theta})$, $\hat{\Psi}_{ij}^{(u)} = \Psi_{ij}^{(u)}(\hat{\theta})$, $\hat{h}_i^{(u)} = h_i^{(u)}(\hat{\theta})$, $i, j, k,$

$l = 1, \ldots, m_u; \quad u = 1, \ldots, m; \quad \hat{\Psi}_{ij}^{(uv)} = \Psi_{ij}^{(uv)}(\hat{\theta}), \quad i = 1, \ldots m_u; j = 1, \ldots m_v; u, v = 1, \ldots, m; u \neq v$, then

$$H = \left(\sum_u \sum_{i,j} \hat{\Psi}_i^{(u)} \hat{\Psi}_j^{(u)} \frac{\hat{\mu}_{ij}}{\lambda_u} \right)^{-\frac{1}{2}}$$

$$G_1 = -\frac{3}{2}\hat{A}_1 + \frac{1}{6}\hat{A}_2$$

$$G_3 = H^2 (\frac{1}{2}\hat{A}_1 - \frac{1}{6}\hat{A}_2)$$

Denote ϕ_α as α quintile of standard normal distribution, then the approximate lower confidence limit of Ψ at confidence level $1 - \alpha$ is:

$$\hat{\Psi}_L = \hat{\Psi} - n^{-\frac{1}{2}}\phi_\alpha H^{-1} + n^{-1}H^{-3}\left(G_3\phi_\alpha^2 + G_1 H^2\right)$$

4 LOWER CONFIDENCE LIMITS FOR COMPLEX SYSTEMS

Our model makes two initial assumptions:

(1) Suppose a complex system S is composed of m independent subsystems, $S_1, \ldots, S_m$,

Let $C = \begin{pmatrix} 1^T \\ A \end{pmatrix} = (c_1, c_2, \ldots, c_k)$, where $1^T = (1, 1, \ldots, 1)$, and $\tilde{C}_j$ is $(m + 1) \times C_k^j$ matrix.

(2) Subsystem S_u life obeys logarithmic normal distribution with parameter μ_u and σ_u^2, where $u = 1, \ldots, m$.

For this model, in relation to the complete sample, we study the lower confidence limit of reliability of the system with the confidence level of $1 - \alpha$.

First, we introduce Lemma 4.1: suppose there are n samples to be tested with lives obeying a logarithmic normal distribution. The sample lives are $t_1, \ldots, t_n$; $\hat{\mu}$, $\hat{\sigma}^2$ are the estimators for μ, σ^2; then:

$$E\left(\hat{\mu} - \mu\right) = 0$$

$$E\left(\hat{\mu} - \mu\right)^2 = \frac{\sigma^2}{n}$$

$$E\left(\hat{\mu} - \mu\right)^3 = 0$$

$$E\left(\hat{\mu} - \mu\right)^4 = \frac{3\sigma^4}{n^2}$$

$$E\left(\hat{\sigma}^2 - \sigma^2\right) = 0$$

$$E\left(\hat{\sigma}^2 - \sigma^2\right)^2 = \frac{2\sigma^4}{n-1}$$

$$E\left(\hat{\sigma}^2 - \sigma^2\right)^3 = \frac{8\sigma^6}{(n-1)^2}$$

$$E\left(\hat{\sigma}^2 - \sigma^2\right)^4 = \frac{12(n+3)\sigma^8}{(n-1)^3}$$

Now suppose a complex system S is constituted by m independent subsystems, $S_1, S_2, \ldots, S_m$; all subsystems' lives obey a logarithmic normal distribution (μ_u, σ_u^2 are the parameters of the uth subsystem); n_u is the sample life of the uth subsystem. Then, from Lemma 4.1, we get:

$$E\left(\hat{\mu}_u - \mu_u\right) = 0$$

$$E\left(\hat{\mu}_u - \mu_u\right)^2 = \frac{\sigma_u^2}{n}$$

$$E\left(\hat{\mu}_u - \mu_u\right)^3 = 0$$

$$E\left(\hat{\mu}_u - \mu_u\right)^4 = \frac{3\sigma_u^4}{n^2}$$

$$E\left(\hat{\sigma}_u^2 - \sigma_u^2\right) = 0$$

$$E\left(\hat{\sigma}_u^2 - \sigma_u^2\right)^2 = \frac{2\sigma_u^4}{n-1}$$

$$E\left(\hat{\sigma}_u^2 - \sigma_u^2\right)^3 = \frac{8\sigma_u^6}{(n-1)^2}$$

$$E\left(\hat{\sigma}_u^2 - \sigma_u^2\right)^4 = \frac{12(n+3)\sigma_u^8}{(n-1)^3}$$

then $E(\hat{\mu}_u - \mu_u)^2(\hat{\sigma}_u^2 - \sigma_u^2)^2 = \frac{2\sigma^6}{n(n-1)}$,

From the WCF method, we have:

$$\hat{\mu}_{11}^{(u)} = \hat{\sigma}_u^2, \quad \hat{\mu}_{22}^{(u)} = 2\hat{\sigma}_u^4, \quad \hat{\mu}_{12}^{(u)} = 0, \hat{\mu}_{111}^{(u)} = 0,$$

$$\hat{\mu}_{222}^{(u)} = 8\hat{\sigma}_u^6, \quad \hat{\mu}_{112}^{(u)} = 0, \quad \hat{\mu}_{122}^{(u)} = 0, \hat{\mu}_{1111}^{(u)} = 3\hat{\sigma}_u^4,$$

$$\hat{\mu}_{2222}^{(u)} = 12\hat{\sigma}_u^8, \quad \hat{\mu}_{1122}^{(u)} = 2\hat{\sigma}_u^6, \quad \hat{\mu}_{1222}^{(u)} = 0, \quad \hat{\mu}_{1112}^{(u)} = 0.$$

The following is an analytic expression of lower confidence limit for complex system reliability:

Theorem 4.1: for the system described in the model, each subsystem is tested and n_u is the sample life of the uth subsystem. $T_{u1}, \ldots, T_{un_u}$ are sample lives. Let:

$$n = \min\{n_u\}, \quad \lambda_u = \frac{n_u}{n}, \quad \hat{\mu}_u = \frac{1}{n_u}\sum_{i=1}^{n_u}\ln T_{ui} = \ln \overline{T}_u,$$

$$\hat{\sigma}_u^2 = \frac{1}{n_u-1}\sum_{i=1}^{n_u}\left(\ln T_{ui} - \ln \overline{T}_u\right), \quad \hat{R}_u = 1 - \Phi\left(\frac{\ln t - \hat{\mu}_u}{\hat{\sigma}_u}\right).$$

where $u = 1, \ldots, m$; then, for the given time t, the WCF lower confidence limit of reliability of this complex system at confidence level $1 - \alpha$ is:

$$\hat{\Psi}_{1,A}^L = \hat{\Psi}_{1,A} - n^{-\frac{1}{2}}\varphi_\alpha \sqrt{\sum_u \zeta_{uu} \hat{p}_u^2(\ln t) \frac{\hat{\sigma}_u^2}{\lambda_u}\left(1 + 2\left(\ln t - \hat{\mu}_u\right)^2\right)}$$

$$+ n^{-1}\left[\sum_u \zeta_{uu} \hat{p}_u^2(\ln t) \frac{\hat{\sigma}_u^2}{\lambda_u}\left(1 + 2\left(\ln t - \hat{\mu}_u\right)^2\right)\right]^{-1}$$

165

$$
\begin{bmatrix}
\frac{1}{2}(3-\varphi_a^2)\sum_{\substack{u,v\\u\neq v}}\frac{\zeta_{uv(uv)}\hat{p}_u^2(\ln t)\hat{p}_v^2(\ln t)\frac{\hat{\sigma}_u^2\hat{\sigma}_v^2}{\lambda_u\lambda_v}}{\left(1+2(\ln t-\hat{\mu}_u)^2\right)\left(1+2(\ln t-\hat{\mu}_v)^2\right)}\\
+(1-\varphi_a^2)\begin{pmatrix}\frac{4}{3}\sum_u \zeta_{uuu}\hat{p}_u^3(\ln t)\frac{\hat{\sigma}_u^3(\ln t-\hat{\mu}_u)^3}{\lambda_u^2}\\
+\sum_{\substack{u,v\\u\neq v}}\frac{\zeta_{uv(uv)}\hat{p}_u^2(\ln t)\hat{p}_v^2(\ln t)\frac{\hat{\sigma}_u^2\hat{\sigma}_v^2}{\lambda_u\lambda_v}}{\left(1+2(\ln t-\hat{\mu}_u)^2\right.}\\
\left.+4(\ln t-\hat{\mu}_u)^2(\ln t-\hat{\mu}_v)^2\right)\end{pmatrix}\\
-\frac{1}{2}\left[\sum_u\frac{\zeta_{uu}\hat{p}_u^2(\ln t)\frac{\hat{\sigma}_u^2}{\lambda_u}}{\left(1+2(\ln t-\hat{\mu}_u)^2\right)}\right]^{-1}\\
\sum_{\substack{u,v\\u\neq v}}\zeta_{uuuv(uv)}\hat{p}_u^4(\ln t)\hat{p}_v^2(\ln t)\\
\frac{\hat{\sigma}_u^4\hat{\sigma}_v^2}{\lambda_u^2\lambda_v}\left(1+2(\ln t-\hat{\mu}_v)^2\right)\\
\left(3+2(\ln t-\hat{\mu}_u)^2+12(\ln t-\hat{\mu}_u)^4\right)\\
\zeta_{uvvv(uv)}\hat{p}_u^2(\ln t)\hat{p}_v^4(\ln t)\\
+3\sum_{\substack{u,v\\u\neq v}}\frac{\hat{\sigma}_u^2\hat{\sigma}_v^4}{\lambda_u\lambda_v^2}\left(1+2(\ln t-\hat{\mu}_v)^2\right)\\
\left(1+2(\ln t-\hat{\mu}_u)^2\\
+4(\ln t-\hat{\mu}_u)^2(\ln t-\hat{\mu}_v)^2\right)
\end{bmatrix}
$$

where $\hat{\Psi}_{1,A} = -\sum_{j=1}^{k}\tilde{R}^{\tilde{C}_j}(t)$, $\hat{\tilde{R}} = (-1,\hat{R}_1,\cdots,\hat{R}_m)^T$,

$$\zeta_{uuu} = \sum_{v_1=1}^{k}\sum_{v_2=1}^{k}\sum_{v_3=1}^{k}\hat{\tilde{R}}^{d(\tilde{C}_{v_1},u+1)\mp d(\tilde{C}_{v_2},u+1)\mp d(\tilde{C}_{v_3},u+1)},$$

$$\zeta_{uu} = \sum_{v_1=1}^{k}\sum_{v_2=1}^{k}\hat{\tilde{R}}^{d(\tilde{C}_{v_1},u+1)\mp d(\tilde{C}_{v_2},u+1)},$$

$$\zeta_{uv(uv)} = \sum_{v_1=1}^{k}\sum_{v_2=1}^{k}\sum_{v_3=1}^{k}\hat{\tilde{R}}^{d(\tilde{C}_{v_1},u+1)\mp d(\tilde{C}_{v_3},v+1)\mp d(d(\tilde{C}_{v_2},u+1),v+1)},$$

$$\zeta_{uuuv(uv)} = \sum_{v_1=1}^{k}\sum_{v_2=1}^{k}\sum_{v_3=1}^{k}\sum_{v_4=1}^{k}\sum_{v_5=1}^{k}\hat{\tilde{R}}^{d(\tilde{C}_{v_1},u+1)\mp d(\tilde{C}_{v_2},u+1)\mp d(\tilde{C}_{v_3},u+1)\mp d(\tilde{C}_{v_4},v+1)\mp d(d(\tilde{C}_{v_2},u+1),v+1)},$$

ϕ_α is α quantile of standard normal distribution.

5 NUMERICAL EXAMPLE: BRIDGE SYSTEM

In this section, we will simulate lower confidence limits for the reliability of a bridge system, as shown in Figure 1 and described by a minimal path matrix based on Theorem 4.1. The result can be shown in three aspects as follows:

(1) Coverage Rate (CR): the ratio of the simulated lower confidence limit of reliability at confidence level $1-\alpha$ is less than the true value of reliability to the true value of reliability, and the one that is closer to $1-\alpha$ is better;
(2) Mean: the average of lower confidence limit of reliability simulated by simulation, the one is closer to true value is better with the same coverage rate;
(3) Standard Deviation (SD): the standard deviation of lower confidence limit of reliability simulated is smaller is better.

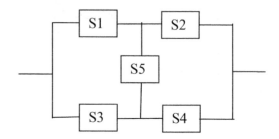

Figure 1. A bridge system, S, constituted by five independent subsystems.

The minimal path matrix is:

$$A = \begin{pmatrix} 1 & 0 & 1 & 0 \\ 1 & 0 & 0 & 1 \\ 0 & 1 & 0 & 1 \\ 0 & 1 & 1 & 0 \\ 0 & 0 & 1 & 1 \end{pmatrix}$$

The steps in the simulation are as follows:

(1) After input of the minimal path matrix A, the analytic expression of the lower confidence limit can be derived from Theorem 4.1;
(2) Supposing that the life of every subsystem has identical exponential distribution taking t, μ, σ; let $x = \ln t$, we can calculate the true value of reliability for each subsystem at time t; furthermore, the true value of bridge system reliability R_{ture} can be calculated from Lemma 2.1;
(3) For the uth subsystem, take samples from exponential distribution with size n_u, calculate $\hat{\Psi}_{1,A}^L$ at confidence level $1-\alpha$;
(4) Repeat step 3 1,000 times, then obtain 1,000 $\hat{\Psi}_{1,A}^L$;
(5) Calculate standard deviation ε, mean μ and coverage rate θ of $\hat{\Psi}_{1,A}^L$;
(6) Select different t, μ, σ; calculate the true value of bridge system reliability R_{ture}; repeat steps 3, 4 and 5.

Table 1 shows the results of this simulation.
Analysis of the result of this simulation shows that as the sample size increases, the result improves. When the sample size is 15, coverage rate is almost equal to confidence level and standard deviation is relatively small; when the sample size is 50, mean is close to the true value of reliability, standard deviation is relatively

Table 1. Result of WCF method.

$n1=n2=$ $n3=n4$ $=n5$	$\alpha = 0.05$		
	$R_{ture} = 0.7905853$		
	$x = 9.5, \mu = 10, \sigma = 1$		
	ε	μ	θ
5	0.1541	0.6034	0.893
15	0.0753	0.6787	0.936
30	0.0501	0.7141	0.944
50	0.0361	0.7321	0.953
$n1=n2=$ $n3=n4$ $=n5$	$\alpha = 0.05$		
	$R_{ture} = 0.7055574$		
	$x = 9.5, \mu = 10, \sigma = 1.5$		
	ε	μ	θ
5	0.1721	0.4721	0.918
15	0.0794	0.5763	0.952
30	0.0502	0.6125	0.968
50	0.0391	0.6338	0.964
$n1=n2=$ $n3=n4$ $=n5$	$\alpha = 0.05$		
	$R_{ture} = 0.8585704$		
	$x = 14, \mu = 15, \sigma = 1.5$		
	ε	μ	θ
5	0.1445	0.6799	0.894
15	0.0650	0.7471	0.952
30	0.0422	0.7792	0.956
50	0.0308	0.7948	0.960

Table 2. Simplified results of WCF method.

$n1=n2=$ $n3=n4$ $=n5$	$\alpha = 0.05$		
	$R_{ture} = 0.7905853$		
	$x = 9.5, \mu = 10, \sigma = 1$		
	ε	μ	θ
5	0.1529	0.5952	0.906
15	0.0747	0.6743	0.943
30	0.0498	0.7116	0.951
50	0.0359	0.7305	0.957

matrix; assuming that the subsystems' lives obey the logarithmic normal distribution, we derived the reliability confidence limit of the system by the WCF method.

It is very convenient for theoretical research and application values with the expression of reliability lower confidence limit for complex system. With the aid of computer programming, by inputting the minimal path matrix of system and sample lives of every subsystem, we can derive the reliability confidence limit for the system.

The result of the reliability confidence limit estimation of complex systems is satisfactory. In particular, when the sample size is very small, the WCF method can produce good results. In many situations, the simplified expression can also achieve a satisfactory result.

small, and coverage rate is almost equal to confidence level.

However, because the expression in Theorem 4.1 is complex, we try to simplify the expression and the simulation, as shown in Table 2.

In comparison, we find that the results in Table 2 are similar to those in Table 1, so we can use the following expression when the result we need is not very accurate:

$$\hat{\Psi}_{1,A}^{L} = \hat{\Psi}_{1,A} - n^{-\frac{1}{2}}\varphi_{\alpha}\sqrt{\sum_{u}\zeta_{uu}\hat{p}_u^2(\ln t)\frac{\hat{\sigma}_u^2}{\lambda_u}\left(1+2\left(\ln t - \hat{\mu}_u\right)^2\right)}$$

$$+n^{-1}\left[\sum_{u}\zeta_{uu}\hat{p}_u^2(\ln t)\frac{\hat{\sigma}_u^2}{\lambda_u}\left(1+2\left(\ln t - \hat{\mu}_u\right)^2\right)\right]^{-1}$$

$$\cdot\frac{1}{2}\left(3-\varphi_{\alpha}^2\right)\sum_{\substack{u,v\\u\neq v}}\frac{\zeta_{uv(uv)}\hat{p}_u^2(\ln t)\hat{p}_v^2(\ln t)\frac{\hat{\sigma}_u^2\hat{\sigma}_v^2}{\lambda_u\lambda_v}}{\left(1+2\left(\ln t - \hat{\mu}_u\right)^2\right)\left(1+2\left(\ln t - \hat{\mu}_v\right)^2\right)}$$

6 CONCLUSION

A complex system is usually described by the minimum path matrix or the minimum cut-set matrix. The system we studied is described by the minimum path

REFERENCES

Cornish, E.A. & Fisher, R.A. (1937). Moments and cumulants in the specification of distributions. *Review of the International Statistical Institute*, 5, 307–322.

Fisher, R.A. & Cornish, E.A. (1960). The percentile points of distributions having known cumulants. *Technometrics*, 2, 209–225.

Liao, C.F. & Zhang, G.Z. (2015). Study of reliability lower confidence limit for complex system in incomplete data. Harbin: Harbin University of Science and Technology, 1–33.

Shu, Y. & Zhang, G.Z. (2012). Study of reliability lower confidence limit for complex system described by the minimum path matrix. Harbin: Harbin University of Science and Technology, 12–31.

Winterbottom, A. (1979). Cornish–Fisher expansions for confidence limits. *Journal of the Royal Statistical Society B*, 41, 69–75.

Winterbottom, A. (1980). Asymptotic expansions to improve large sample confidence intervals for system reliability. *Biometrika*, 67, 351–357.

Yan, X., Yu, D. & Li, G. (2007). The generalized WCF method for calculating confidence bounds of parametric functions and its applications. *Acta Mathematica Scientia. Series A. (Chinese Edition)*, 27(2), 229–239.

Yu, D. & Dai, S.S. (1996). Study on comprehensive evaluation method of complex system reliability. Chinese academy of sciences institute of systems science.

Yu, D. et al. (1999). Reliability lower confidence limit under life obeys exponential distribution for the sample of type-I censoring. *Systems Science and Mathematics*, 19(2), 240–245.

Yu, D., Li, X.J. & Jiang, N.N. (2007). Some statistical inference problems and research progress in the reliability analysis of complex systems. *Systems Science and Mathematics*, 27(1), 68–81.

Zhang, G.Z. (2009). Study of reliability for complex system. Beijing: Beijing University of Technology, 19–60.

APPENDIX

Proof of Theorem 4.1:

Because $R_u(\ln t) = 1 - \Phi\left(\frac{\ln t - \mu_u}{\sigma_u}\right)$, from WCF method we have:

$$\Psi_1^{(u)} = \frac{\partial \Psi_{1,A}}{\partial \mu_u} = \frac{\partial \Psi_{1,A}}{\partial R_u} \cdot \frac{\partial R_u}{\partial \mu_u} = -\sum_{v=1}^{k} \tilde{R}^{d(\tilde{C}_v, u+1)} \cdot p_u(\ln t) \ ,$$

$$\Psi_2^{(u)} = \frac{\partial \Psi_{1,A}}{\partial \sigma_u^2} = \frac{\partial \Psi_{1,A}}{\partial R_u} \cdot \frac{\partial R_u}{\partial \sigma_u^2}$$

$$= -\sum_{v=1}^{k} \tilde{R}^{d(\tilde{C}_v, u+1)} \cdot p_u(\ln t) \cdot \frac{\ln t - \mu_u}{\sigma_u},$$

$$\Psi_{ij}^{(u)} = 0 \ , i,j = 1,2 \ , u = 1,\cdots,m \ ,$$

$$\Psi_{11}^{(uv)} = -\sum_{v=1}^{k} \tilde{R}^{d(d(\tilde{C}_v, u+1), v+1)} \cdot p_u(\ln t) \cdot p_v(\ln t),$$

$$\Psi_{12}^{(uv)} = -\sum_{v=1}^{k} \tilde{R}^{d(d(\tilde{C}_v, u+1), v+1)} \cdot p_u(\ln t) \cdot p_v(\ln t) \cdot \frac{\ln t - \mu_v}{\sigma_v},$$

$$\Psi_{22}^{(uv)} = -\sum_{v=1}^{k} \tilde{R}^{d(d(\tilde{C}_v, u+1), v+1)} \cdot p_u(\ln t) \cdot$$

$$p_v(\ln t) \cdot \frac{\ln t - \mu_u}{\sigma_u} \cdot \frac{\ln t - \mu_v}{\sigma_v} \ ,$$

$$u, v = 1, \cdots, m , u \neq v \ . \text{then,}$$

$$h = \left(\sum_u \sum_{i,j} \Psi_i^{(u)} \Psi_j^{(u)} \frac{\mu_{ij}^{(u)}}{\lambda_u} \right)^{-\frac{1}{2}}$$

$$= \left[\sum_u \left(\Psi_1^{2(u)} \frac{\mu_{11}^{(u)}}{\lambda_u} + \Psi_2^{2(u)} \frac{\mu_{22}^{(u)}}{\lambda_u} \right) \right]^{-\frac{1}{2}}$$

$$= \left[\sum_u \frac{\sum\limits_{v_1=1}^{k} \sum\limits_{v_2=1}^{k} \tilde{R}^{d(\tilde{C}_{v_1}, u+1) \mp d(d(\tilde{C}_{v_2}, u+1)}}{p_u^2(\ln t) \frac{\sigma_u^2}{\lambda_u}(1 + 2(\ln t - \mu_u)^2)} \right]^{-\frac{1}{2}}$$

$$h = \left(\sum_u \sum_{i,j} \Psi_i^{(u)} \Psi_j^{(u)} \frac{\mu_{ij}^{(u)}}{\lambda_u} \right)^{-\frac{1}{2}}$$

$$= \left[\sum_u \left(\Psi_1^{2(u)} \frac{\mu_{11}^{(u)}}{\lambda_u} + \Psi_2^{2(u)} \frac{\mu_{22}^{(u)}}{\lambda_u} \right) \right]^{-\frac{1}{2}}$$

$$= \left[\sum_u \frac{\sum\limits_{v_1=1}^{k} \sum\limits_{v_2=1}^{k} \tilde{R}^{d(\tilde{C}_{v_1}, u+1) \mp d(\tilde{C}_{v_2}, u+1)}}{p_u^2(\ln t) \frac{\sigma_u^2}{\lambda_u}(1 + 2(\ln t - \mu_u)^2)} \right]^{-\frac{1}{2}}$$

$$h_1^{(u)} = -h^3 p_u(\ln t) \sum_v \frac{\sum\limits_{v_1=1}^{k} \sum\limits_{v_2=1}^{k} \tilde{R}^{d(\tilde{C}_{v_1}, v+1) \mp d(d(\tilde{C}_{v_2}, u+1), v+1)}}{p_v^2(\ln t) \frac{\sigma_v^2}{\lambda_v}(1 + 2(\ln t - \mu_v)^2)}$$

$$h_2^{(u)} = -h^3 p_u(\ln t) \frac{\ln t - \mu_u}{\sigma_u}$$

$$\sum_v \frac{\sum\limits_{v_1=1}^{k} \sum\limits_{v_2=1}^{k} \tilde{R}^{d(\tilde{C}_{v_1}, v+1) \mp d(d(\tilde{C}_{v_2}, u+1), v+1)}}{p_v^2(\ln t) \frac{\sigma_v^2}{\lambda_v}(1 + 2(\ln t - \mu_v)^2)}$$

$u, v = 1, \ldots, m$, then from the WCF method and Theorem 4.1 we have:

$$\hat{A}_1 = \sum_u \sum_{i,j} \hat{h}_i^{(u)} \hat{\Psi}_i^{(u)} \frac{\hat{\mu}_{ij}^{(u)}}{\lambda_u}$$

$$= \sum_u \left(\hat{h}_1^{(u)} \hat{\Psi}_1^{(u)} \frac{\hat{\mu}_{11}^{(u)}}{\lambda_u} + \hat{h}_2^{(u)} \hat{\Psi}_2^{(u)} \frac{\hat{\mu}_{22}^{(u)}}{\lambda_u} \right)$$

$$= -H^3 \sum_{\substack{u,v \\ u \neq v}} \frac{\zeta_{uv(uv)} \hat{p}_u^2(\ln t) \hat{p}_v^2(\ln t) \frac{\hat{\sigma}_u^2 \hat{\sigma}_v^2}{\lambda_u \lambda_v}}{\left(1 + 2\left(\ln t - \hat{\mu}_u\right)^2\right)\left(1 + 2\left(\ln t - \hat{\mu}_v\right)^2\right)}$$

$$\hat{A}_2 = H^3 \left[\begin{array}{l} \sum\limits_u \sum\limits_{i,j,k} \hat{\Psi}_i^{(u)} \hat{\Psi}_j^{(u)} \hat{\Psi}_k^{(u)} \frac{\hat{\mu}_{ijk}^{(u)}}{\lambda_u^2} \\ +6 \sum\limits_{\substack{u,v \\ u \neq v}} \sum\limits_{i,j,k,l} \hat{\Psi}_i^{(u)} \hat{\Psi}_j^{(u)} \hat{\Psi}_{kl}^{(uv)} \frac{\hat{\mu}_{ik}^{(u)} \hat{\mu}_{jl}^{(v)}}{\lambda_u \lambda_v} \end{array} \right]$$

$$+3H^2 \left[\begin{array}{l} \sum\limits_u \sum\limits_{i,j,k,l} \hat{h}_i^{(u)} \hat{\Psi}_j^{(u)} \hat{\Psi}_k^{(u)} \hat{\Psi}_l^{(u)} \frac{\hat{\mu}_{ijkl}^{(u)}}{\lambda_u^2} \\ +3 \sum\limits_{\substack{u,v \\ u \neq v}} \sum\limits_{i,j,k,l} \hat{h}_i^{(u)} \hat{\Psi}_j^{(u)} \hat{\Psi}_k^{(v)} \hat{\Psi}_l^{(v)} \frac{\hat{\mu}_{ij}^{(u)} \hat{\mu}_{kl}^{(v)}}{\lambda_u \lambda_v} \end{array} \right]$$

and,

$$\sum_u \sum_{i,j,k} \hat{\Psi}_i^{(u)} \hat{\Psi}_j^{(u)} \hat{\Psi}_k^{(u)} \frac{\hat{\mu}_{ijk}^{(u)}}{\lambda_u^2}$$

$$= \sum_u \zeta_{uuu} \hat{p}_u^3(\ln t) \frac{\hat{\sigma}_u^3 (\ln t - \hat{\mu}_u)^3}{\lambda_u^2}$$

$$\sum_{\substack{u,v \\ u \neq v}} \sum_{i,j,k,l} \hat{\Psi}_i^{(u)} \hat{\Psi}_j^{(v)} \hat{\Psi}_{kl}^{(uv)} \frac{\hat{\mu}_{ik}^{(u)} \hat{\mu}_{jl}^{(v)}}{\lambda_u \lambda_v}$$

$$= \sum_{\substack{u,v \\ u \neq v}} \frac{\zeta_{uv(uv)} \hat{p}_u^2(\ln t) \hat{p}_v^2(\ln t) \frac{\hat{\sigma}_u^2 \hat{\sigma}_v^2}{\lambda_u \lambda_v}}{\left(1 + 2\left(\ln t - \hat{\mu}_u\right)^2 + 4\left(\ln t - \hat{\mu}_u\right)^2 \left(\ln t - \hat{\mu}_v\right)^2\right)}$$

168

$$\sum_u \sum_{i,j,k,l} \hat{h}_i^{(u)} \hat{\Psi}_j^{(u)} \hat{\Psi}_k^{(u)} \hat{\Psi}_l^{(u)} \frac{\hat{\mu}_{ijkl}^{(u)}}{\lambda_u^2}$$

$$\zeta_{uuuv(uv)} \hat{p}_u^4 (\ln t) \, \hat{p}_v^2 (\ln t) \frac{\hat{\sigma}_u^4 \hat{\sigma}_v^2}{\lambda_u^2 \lambda_v}$$

$$= -H^3 \sum_{\substack{u,v \\ u \neq v}} \left(1 + 2 \left(\ln t - \hat{\mu}_v \right)^2 \right)$$

$$\left(3 + 2 \left(\ln t - \hat{\mu}_u \right)^2 + 12 \left(\ln t - \hat{\mu}_u \right)^4 \right)$$

$$\sum_{\substack{u,v \\ u \neq v}} \sum_{i,j,k,l} \hat{h}_i^{(u)} \hat{\Psi}_j^{(u)} \hat{\Psi}_k^{(v)} \hat{\Psi}_l^{(v)} \frac{\hat{\mu}_{ij}^{(u)} \hat{\mu}_{kl}^{(v)}}{\lambda_u \lambda_v}$$

$$\zeta_{uvvvv(uv)} \hat{p}_u^2 (\ln t) \, \hat{p}_v^4 (\ln t)$$

$$= -H^3 \sum_{\substack{u,v \\ u \neq v}} \frac{\hat{\sigma}_u^2 \hat{\sigma}_v^4}{\lambda_u \lambda_v^2} \left(1 + 2 \left(\ln t - \hat{\mu}_v \right)^2 \right)$$

$$\left(\begin{array}{l} 1 + 2 \left(\ln t - \hat{\mu}_u \right)^2 \\ + 4 \left(\ln t - \hat{\mu}_u \right)^2 \left(\ln t - \hat{\mu}_v \right)^2 \end{array} \right)$$

From the above result, we can obtain H, G_1, G_3; we can then obtain Theorem 4.1 via the following:

$$\hat{\Psi}_{1,A}^{\ L} = \hat{\Psi}_{1,A} - n^{-\frac{1}{2}} \phi_\alpha H^{-1} + n^{-1} H^{-3} \left(G_3 \phi_\alpha^2 + G_1 H^2 \right)$$

Electronic Engineering – Wang (ed)
© 2018 Taylor & Francis Group, London, ISBN 978-1-138-60260-1

Stiffness analysis of heavy spindle tops

B.W. Gao
College of Mechanical and Power Engineering, Harbin University of Science and Technology, Harbin, China

W.L. Han
Power Station Equipment Filiale Harbin Fenghua Co. Ltd., China Aerospace Science & Industry Corporation, Harbin, China

ABSTRACT: Spindle tops are an important component used to reinforce the shaft workpiece on machine tools, so that the spindle can meet the highest efficiency precision requirements. In actual production, the height-to-diameter ratios of spindle tops in the plant have 1:4 and 1:7 types; however, in use, there is no stipulation of the principle of selection in a variety of different load cases. A model of the spindle top will be created in Siemens Unigraphics NX ("UG") software. Then, through the data interface between UG and ANSYS software, the model in UG is changed into a finite element model in ANSYS, and finite element analysis is carried out on the regular deformation of spindle tops with height-to-diameter ratios of 1:4 and 1:7 in a variety of different load cases. Based on the regular deformation, the selection principle for the height-to-diameter ratio of the spindle tops can be established.

Keywords: Spindle top; finite element analysis; height-to-diameter ratio; regular deformation

1 INTRODUCTION

Heavy spindle tops are one of the important components in the direct bearing of the weight and the reaction force of associated workpieces. The top and the workpiece rely on the cone contact of the centre hole to ensure that the workpiece centre and the centre of head and tail frame are consistent (Chen, 2013). At present, the research on the process and stress of the spindle is not adequate. In the design and calculation of the top structure of such spindles, the method used is simple and rarely even calculated. The design work has blind spots, which seriously affect spindle performance (Xu & Sheng, 2008; Li, 2007). For workpieces of different weights, the use of the angle and the tightening force of the top are not aligned (Wang & Yu, 2002). In actual factory production, the ratio of the height and diameter of the spindle top is either 1:4 or 1:7, and the principle of selection of a variety of different load conditions in the process of use is not stipulated, but is solved in accordance with empirical data that has been in the field for a long time. With the progress of modern science and technology, a variety of advanced machines continue to emerge; thus, a variety of parts of complex shape and high precision can be produced. Among them, the processing of the rotary body accounts for a large proportion, which creates a higher demand of the process performance and carrying capacity of the spindle top (Cui & Wang, 2006).

Using ANSYS finite element analysis software to analyse spindle tops with height-to-diameter ratios of 1:4 and 1:7 in a variety of different load cases, the regular deformation of various types of spindle top in a variety of different load cases can be achieved. Based on the regular deformation, the selection principle for the appropriate height-to-diameter ratio of the spindle top can be established.

2 TYPE OF CONSTRUCTION OF THE SPINDLE TOP

2.1 *Type of top*

The top is the essential tool for mechanical centre positioning and processing centre, divided into two categories of fixed top and rotary top, and used in general lathes, crankshafts, other cylindrical grinding machines, some gear processing machines, and in some measuring instruments and equipment (Cai, 2012). In addition, in the actual work, the height-to-diameter ratio of spindle tops have values of 1:4, 1:7 or 1:9, and the 1:4 and 1:7 types are often used in the plant. This paper therefore focuses on the tops of these two types, shown in Figures 1 and 2, and analyses them under different loads.

Figure 1 shows the short-taper shank flange fixed form, and this structure ensures that the fixed surface and the spindle face of the top have a 0.2 mm gap by processing, so the top of this form has a better contact stiffness. However, the tailstock mandrel cone hole cannot be equipped with drills and other tools, as it is not easy to hit the centre hole.

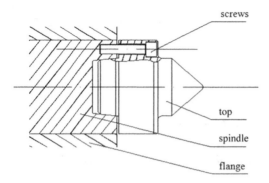

Figure 1. Top of 1:4 type.

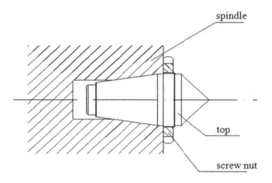

Figure 2. Top of 1:7 type.

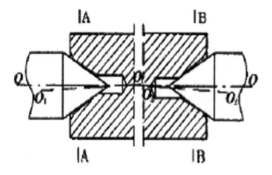

Figure 3. Schematic of centre location.

Figure 2 is the long-taper shank top, usually with 1:7 metric taper fit, such that the top stiffness of this suspension structure is poor. However, the tailstock mandrel cone hole can be equipped with drills and other tools, as it is easy to hit the centre hole.

2.2 *Positioning of top*

The positioning method of an ordinary top is as follows: the two tops are pre-tightened at both ends of the chamfered hole of the workpiece, and the front fixed tip connected with the spindle inner cone, and rear live tops are in the tailstock sleeve (Zhang, 2010), as shown in Figure 3.

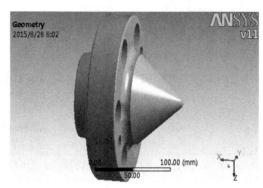

Figure 4. 1:4 top model imported from UG to ANSYS.

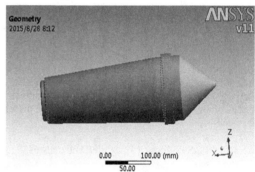

Figure 5. 1:7 top model imported from UG to ANSYS.

3 RIGIDITY ANALYSIS OF SPINDLE TOP

In practical work, the acting force of the workpiece is not entirely sustained by the top. The spindle and the end of the rack also bear part of the force, but these latter bear a smaller force, which can be negligible (Kiswanto & Zariatin, 2015). Therefore, only analysis of the top is necessary to facilitate calculation and analysis.

3.1 *Establish model of top*

In this paper, Siemens Unigraphics NX ("UG") software is used to model the top, and this is imported into ANSYS software for finite element analysis. Because the UG Computer-Aided Design (CAD) system has a very strong modelling function, the operation is convenient and quick; first establish the model in the CAD system, then induct to the ANSYS software to continue the analysis. Because of space limitations, only the specification models of 1:4 and 1:7 spindle tops are analysed here, and these are shown in Figures 4 and 5.

3.2 *Finite element analysis of top*

After the finite element modelling, it is necessary to set the cell properties, material parameters, and then divide the grid of the model, and generate nodes and units. The process of meshing involves setting the cell

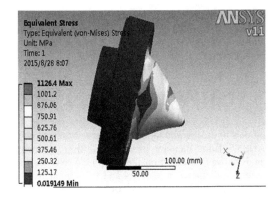

Figure 6. Equivalent stress result for the 1:4 top.

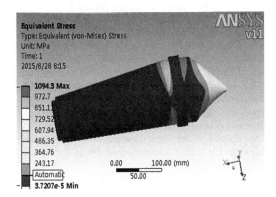

Figure 7. Equivalent stress result for the 1:7 top.

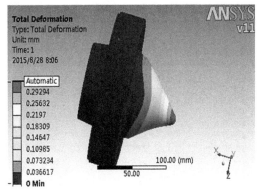

Figure 8. Total deformation result of the 1:4 top.

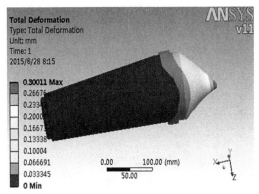

Figure 9. Total deformation result of the 1:7 top.

properties, setting the mesh control, and generating the mesh (Sheng & Xu, 2007).

Because the top is fixed in the taper hole of the spindle during actual production, this constraint should be added to the taper of the top tails in the analysis of the top.

Finite element model loading is used in this paper. In practice, the top not only bears the radial pressure, but also bears the axial force and the effect of torque, although the latter two forces have a smaller effect on the deformation of the top, so these are deemed negligible for the sake of simplicity. In addition, the actual contact surface of the top and the workpiece is not the entire cone, so the cone of the top will be divided into two equal parts, with the approximation that the first half of the cone makes contact with the workpiece: the load is applied to this half.

The stress and strain results of the two 1:4 and 1:7 models are obtained as shown in Figures 6, 7, 8 and 9.

The materials of the tops used in the company are 9Cr2Mo and GCr15SiMo, and the yield limits are 1180 MPa and 1230 MPa, respectively. The stiffness analysis was carried out for the two different 1:4 and 1:7 top specifications for the two different materials but, because of space limitations, the detailed results of the stress and strain analyses are not listed here one by one. The maximum stress in the case where the yield limit is not exceeded, and the maximum strain at maximum stress can be obtained.

In practical work, the acting force of the workpiece is not entirely sustained by the top; the spindle and the end of the rack also bear part of the force. At the same time, the workpiece will be subject to rotation friction in machining. Although the tightening force and torque experienced by the top have little deformation effect on the top, there are still some deviations. Therefore, this paper gives a safety factor for the above analysis results, taking into account the actual work of the top; the safety factor is 3.5. The weight-bearing situation for the spindle top for two different materials is summarised in Table 1.

3.3 Results analysis

The following conclusions can be drawn from the analysis results for the top under a variety of different types and specifications:

(1) The load capacities of the spindle tops of the two materials 9Cr2Mo and GCr15SiMo are essentially the same. This result is consistent with the fact that the chemical composition and mechanical properties of 9Cr2Mo and GCr15SiMo are not very different from one another.

Table 1. Maximum bearing weight of top.

Specification	Maximum bearing weight (t) 9Cr2Mo	GCr15SiMo
1:4-80	25.7	19.4
1:4-100	47.1	44.3
1:4-125	60.0	74.3
1:4-140	74.3	80.0
1:4-160	97.1	102.9
1:4-200	145	160
1:4-250	228.6	271.4
1:4-400	571.4	600
1:7-120	28.6	31.4
1:7-140	37.1	40

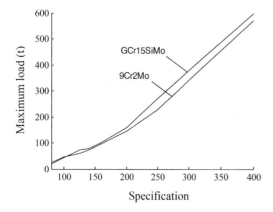

Figure 10. Maximum load curves of the different tops.

(2) For each material, with an increase in top size, the load capacity increases, and the deformation also increases.
(3) Comparing the 1:4-type top with the 1:7-type top, the load capacity of the 1:4 flange-type top is greater. The ANSYS analysis results show that the load capacity of the 1:4-type top is about twice that of the 1:7-type top for the same specification. Moreover, for the same specification, the size of the 1:7-type top is larger than the 1:4-type top, thus increasing the waste of raw materials. In addition, the spindle associated with a 1:7-type top is a hollow shaft, but as the weight of the spindle reduces, the difficulty of processing and the man-hours required increase. Therefore, for similar situations, the 1:4 flange-type top should be used.

4 FORMULATION OF SELECTION PRINCIPLES FOR THE TOP

Based on the above results, the load-bearing capacity of different materials, types and specifications is established, and the function curves for the maximum load bearing of the 1:4-type top with different materials are shown in Figure 10.

These curves clearly describe the maximum load that can be tolerated by tops of different specification, and these curves can be referred to in order to select the appropriate top for future production.

5 CONCLUSION

By using UG and ANSYS software data interfaces, models of the spindle tops in UG were converted into ANSYS finite element models, and finite element analysis was executed on them. Spindle tops with a height-to-diameter ratio of 1:4 and 1:7 were analysed in a variety of different load cases using the ANSYS finite element analysis software, obtaining the regular deformation of spindle top of various types in a variety of different load cases. Based on the regular deformation, the selection principle for the height-to-diameter ratio of the spindle tops was established. Based on the ANSYS analysis results for stress and deformation of tops of different materials, different models and different specifications, we summarised some of the associated principles and conclusions. Finally, the maximum load capacity of tops of two kinds of material with different specifications were calculated, and function curves of the maximum load capacities varying according to top specification were drawn, which provides the basis for selection of tops in the future.

REFERENCES

Cai, G.Y. (2012). Finite element analysis of heavy roll grinder head spindle top component. *Precise Manufacturing & Automation*, *1*, 23–24.
Chen, W.J. (2013). Analyze the top displacement and tailstock spindle back phenomenon of top grinding machine tool. *Precise Manufacturing & Automation*, *1*, 38–39.
Cui, H.B. & Wang, Y.C. (2006). Methods of improving positioning accuracy of centers during machining accurate spindles. *Science and Technology Information*, *4*, 17.
Kiswanto, G. & Zariatin, D.L. (2015). The effect of spindle speed, feed-rate and machining time to the surface roughness and burr formation of aluminum alloy 1100 in micro-milling operation. *Journal of Manufacturing Processes*, *16*(4), 435–450.
Li, P. (2007). Refinement method of positioning accuracy of top when machining precision spindle. *Science and Technology Consulting Herald*, *14*, 1.
Sheng, Z.C. & Xu, N.F. (2007). The spindle's top eccentric adjusting mechanism in the application of CNC hobbing machine. *Mechanical Engineer*, *10*, 145–146.
Wang, W.K. & Yu, X.R. (2002). Methods of improving positioning accuracy of centers during machining accurate spindles. *Tool Engineering*, *36*(12), 45–46.
Xu, N.F. & Sheng, Z.C. (2008). A new spindle top eccentric adjusting mechanism. *China's High-Tech Enterprises*, *1*, 84.
Zhang, J. (2010). Skillfully take the broken top in the spindle taper. *Metal Processing (Cold)*, *12*, 48.

Electronic Engineering – Wang (ed)
© 2018 Taylor & Francis Group, London, ISBN 978-1-138-60260-1

Experimental research into fuzzy control strategy of hydraulic robot actuator

G.H. Han, C.J. Zhang & Y.N. Liu
Department of Mechanical Power Engineering, Harbin University of Science and Technology, Harbin, China

Y.C. Shi
Harbin Dongan Automotive Engine Manufacturing Co. Ltd., Harbin, China

ABSTRACT: The joint movement of a hydraulic quadruped robot is controlled precisely by the actuator positioning system. In order to reduce the positioning deviation of the actuator, a Fuzzy Weight Function and Fuzzy Integral (FWFFI) algorithm was introduced to the fuzzy controller to increase the weight factor in the fuzzy rules of the deviation e when e is large, and to increase the weight factor in the fuzzy rules of the variation of deviation Δe when the deviation is small. Only when $e \cdot \Delta e > 0$ or $\Delta e = 0$ and $e \neq 0$ is the deviation integral action carried out. A comparison experiment with variable inertial and variable elastic loads was conducted on the hardware-in-loop simulation workbench according to the walking environments and motion poses of the robot. The experimental results show that a FWFFI controller can adapt well to a change of robot operating conditions.

Keywords: Actuator; positioning servo; fuzzy integral; fuzzy weight function

1 INTRODUCTION

The motion of each joint of a hydraulic quadruped robot is realised by controlling the extension and the retraction of each actuator piston rod. Therefore, the robot actuator position control is key to the robot joint movement and the entire efficiency. Li et al. (2011), Kong et al. (2013a, 2013b, 2015) and Sun et al. (2015) have launched their own hydraulic quadruped robot actuator. Robots are designed to have independent movement in a complex environment, and the motion control system must have strong robustness and contingency in order to adapt to different terrain conditions. There are the parameters of time variation and external force interference for the robot joints when the external environment changes. Fuzzy control is especially suitable for a control system that is nonlinear, time-variant, complex, and where the model is difficult to obtain. Liang and Zhou (2014) studied fuzzy control mainly in terms of the correction factor and the variable universe. Moreover, fuzzy control has been studied in combination with other algorithms; for example, Cerman and Hušek (2012) applied fuzzy rules to the sliding mode controller, Yang et al. (2015) studied fuzzy neural networks, Zhu et al. (2014) designed fuzzy adaptive controls, Sinthipsomboon et al. (2011) used fuzzy control to adjust the Proportional-Integral-Derivative (PID) parameters, and Shao et al. (2008) applied fuzzy control and PID control in parallel selection. In summary,

research into the fuzzy control of electro-hydraulic servo actuators has achieved much, but research with the parameters of time variation and external disturbance force is relatively limited. Changes in these parameters will greatly affect the performance of a robot actuator positioning servo system. For this reason, combined with the actual motion of the robot and a hardware-in-loop experimental simulation workbench, a fuzzy control strategy based on Fuzzy Weight Function and Fuzzy Integral (FWFFI) is proposed in this paper, which is used in the positioning servo system of a hydraulic robot actuator. The tracking performance is studied under the effect of the parameters of time variation and external disturbance force.

2 CONTROL STRATEGY DESIGN

2.1 *Fuzzy weight function algorithm*

The positioning servo system of the robot actuator is a closed-loop system, involving an integrated servo valve and a hydraulic cylinder equipped with a displacement sensor, as shown in Figure 1. By controlling the opening of the servo valve to control the direction and flow of the hydraulic cylinder we can, ultimately, control the displacement of the hydraulic cylinder.

In a two-dimensional fuzzy control system, in order to enable the output control level to achieve the goal of eliminating deviation as soon as possible, we should

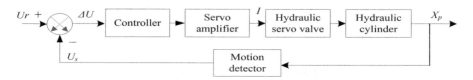

Figure 1. Principle diagram of the actuator positioning servo system.

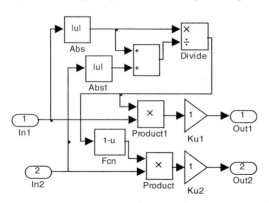

Figure 2. Fuzzy weight function simulation model.

increase the weight factor of the deviation E in the fuzzy control rules when the deviation E is large. In order to guarantee the stability of the system and reduce overshoot, we should increase the weight factor of the deviation in the fuzzy control rules to increase the effect of the variation of deviation change EC when the deviation is small.

The analytic expression of the fuzzy control rules of fuzzy weight function is:

$$U = \langle ae \times E + aec \times EC \rangle \quad (1)$$

The correction factors ae and aec are calculated by the fuzzy calculation method of the weight function, $aec = 1-ae$; then ae and aec are input to the fuzzy controller to adjust the input deviation e and deviation change ec, the fuzzy control rules are adjusted in real time by adjusting the correction factor online automatically. The model is built in MathWorks MATLAB/Simulink software, as shown in Figure 2.

2.2 Fuzzy integration controller

The integral control process in PID control is shown in Figure 3, and the initial overshoot of the system appears in intervals (a–b) and (b–c). The integral action at (0–a) is difficult to offset and change the symbol. In the (c–d) interval, the positive control is increased by the integral action, which is favourable for the reduction of the recoil. In the (d–e) interval, because the role of the integral continues to increase, it is easy to cause an overshoot to appear again (in the e–g interval). Therefore, a fuzzy integral is proposed in this paper. As shown in Figure 3(4), when the deviation is increased, the integral action is carried out at (a–b), (c–d), (e–f) and (g–h). However, when the deviation is

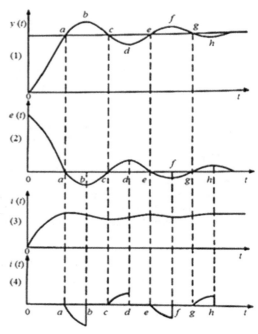

Figure 3. Curves of dynamic response error and integral error.

Table 1. Symbol changes of e and Δe.

	0–a	a	a–b	b	b–c	c	c–d	d	d–e
e	>0	=0	<0	<0	<0	=0	>0	>0	>0
Δe	<0	<0	<0	=0	>0	>0	>0	=0	<0
$e \cdot \Delta e$	<0	=0	>0	=0	<0	=0	>0	=0	<0

reduced, the integral action is not carried out, so that the system can achieve steady state transition with the aid of inertia in (0–a), (b–c), (d–e) and (f–g) intervals.

Based on Figure 3(2), we can obtain Table 1; only when $e \cdot \Delta e > 0$ or $\Delta e = 0$ and $e \neq 0$ is the deviation integral applied.

2.3 Control strategy

The control strategy is shown in Figure 4, and the rule is shown in Equation 2. When $EC \& E < 0$, the system is developing and changing in the direction of decreasing the deviation. Therefore, the integral control is not

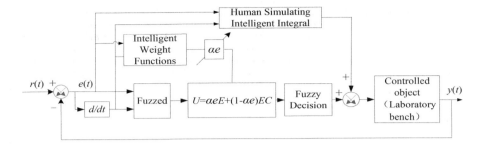

Figure 4. Principle of control strategy.

Figure 5. Experiment workbench.

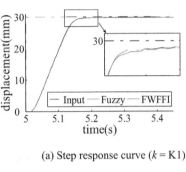

(a) Step response curve ($k = K1$)

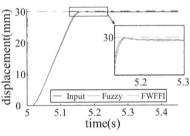

(b) Step response curve ($k = K2$)

Figure 6. Step response of variable elastic load experiment.

applied, and only the fuzzy weight function controller controls the system; when $EC \& E > 0$, the system is developing in the direction of deviation, so the fuzzy weight function and fuzzy integral (FWFFI) controller are used to control the object.

$$U = \begin{cases} \langle \alpha eE + (1-\alpha e)EC \rangle & E \cdot EC < 0 或 E = 0 \\ \langle \alpha eE + (1-\alpha e)EC + \beta \sum_{i=1}^{k} E_i \rangle & E \cdot EC > 0 或 EC = 0 且 E \neq 0 \end{cases} \quad (2)$$

3 EXPERIMENTAL STUDIES

3.1 Variable elastic load experiment

The electro-hydraulic servo hardware-in-loop experimental simulation workbench is mainly composed of an oil source, hydraulic power mechanism and measurement control system, as shown in Figure 5.

When walking on the ground in different environments, such as snow, sand or cement, the equivalent load stiffness of the robot is also different. Therefore, the elastic load stiffness is changed by adjusting the connecting blocks at both ends of the plate spring to simulate the situation when the robot is in contact with different surfaces. When the connecting piece is on both ends of the plate spring, the spring stiffness is K1, and when the connecting piece is in the middle of the plate spring, the spring stiffness is K2, as shown in Figure 5. After measurement, K1 = 2727 N/mm, and K2 = 3750 N/mm.

There is no load mass in the experiment, and the input signal is kept at 1.5 V. When the input signal is positive, one end of the hydraulic cylinder piston rod with a plate spring is exposed to a fixed plate, which causes the piston rod to be subjected to elastic load force. The step response curves of the variable elastic load are shown in Figure 6.

The fuzzy controller and the FWFFI controller can effectively suppress the overshoot under the condition of variable elastic load stiffness, and the steady state phase is entered more quickly, while the FWFFI controller is better in reducing the steady state error and suppressing overshoot. After calculation, the phase lag angle is 9.5°, which meets the requirements of the robot dual 10 indicators.

Table 2. Performance parameters of variable elastic load experiment.

Controller	Mass load (kg)	Rise time (s)	Overshoot (%)	Steady state time (s)	Steady state error (mm)	Amplitude error (%)
Fuzzy controller	m1 = 30	0.105	1.0521	0.11	−0.10142	−1
	m2 = 60	0.105	1.5569	0.111	−0.13787	−1.5
	m3 = 120	0.109	1.9165	0.113	−0.14329	−1.5
	m4 = 180	0.109	1.373	0.114	−0.11212	−1
FWFFI controller	m1 = 30	0.104	0.6308	0.109	−0.09281	0.5
	m2 = 60	0.105	1.3242	0.112	−0.13024	0.4
	m3 = 120	0.108	1.784	0.113	−0.13185	0.2
	m4 = 180	0.109	1.354	0.115	−0.10941	0.5

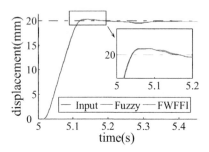

(a) Step response curve ($m1$ = 30 kg)

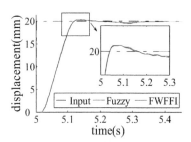

(b) Step response curve ($m3$ = 120 kg)

Figure 7. Step response of variable inertial load experiment.

3.2 Variable inertial load experiment

The motion conditions of the quadruped robot will change when the external environment changes, so the dynamic change of the equivalent load is also dynamic. For example, when the robot climbs, the equivalent load on the actuator is different from when the robot walks on the ground.

According to the actual conditions of the experiment, load mass is added to the experimental workbench to simulate the change. Experiments are carried out on working conditions with two mass blocks (30 kg), four (60 kg), eight (120 kg) or 12 (180 kg). In the experiment, the input signal is 1 V, and the elastic load is not added. The step response curve of the variable inertial load is shown in Figure 7. The performance parameters of the response curves are shown in Table 2. From Figure 7 and Table 2, the fuzzy controller and FWFFI controller have significantly improved overshoot, steady state adjustment time, steady state error, and amplitude tracking aspects when the inertial load quality changes.

However, the FWFFI controller is better at suppressing the overshoot and reducing the steady state error and amplitude tracking, and its amplitude error is smaller: the maximum amplitude error is 0.5%, which is closer to the input curve. After calculation, the phase lag angle is 8.3°, which satisfies the requirement of the robot positioning control.

4 CONCLUSIONS

The FWFFI controller is designed to adapt to robot working conditions, in which the online self-tuning factor of the fuzzy weight function is introduced to suit the occasion of the changing working conditions, and the fuzzy integral is combined to correct the steady state output error of the fuzzy controller.

The hardware-in-loop simulation experiment is completed with variable inertial load and elastic load, according to the actual robot working conditions. By comparison with the fuzzy controller, the FWFFI controller can adapt to the change of parameters and external disturbance, and the effect is significant in reducing the steady state error.

ACKNOWLEDGEMENT

This work is supported by the national Natural Science Foundation (51405113).

REFERENCES

Cerman, O. & Hušek, P. (2012). Adaptive fuzzy sliding mode control for electro-hydraulic servo mechanism. *Expert Systems with Applications*, 39(11), 10269–10277.

Chiang, M.H., Lee, L.W. & Liu, H.H. (2014). Adaptive fuzzy controller with self-tuning fuzzy sliding-mode compensation for position control of an electro-hydraulic displacement-controlled system. *Journal of Intelligent & Fuzzy Systems*, 26(2), 815–830.

Kong, X.D., Yu, B. & Quan, L.X. (2013a). Trajectory sensitivity analysis of hydraulic drive unit of quadruped bionic robot. *Journal of Mechanical Engineering*, 49(14), 170–175.

Kong, X.D., Yu, B. & Quan, L.X. (2013b). Effect of parameters perturbation on displacement control characteristics of hydraulic drive unit of quadruped robot. *Journal of Mechanical & Electrical Engineering, 30*(10), 1169–1171.

Kong, X.D., Yu, B. & Quan, L.X. (2015). Characteristic parameters sensitivity of position servo control for hydraulic drive unit of a quadruped robot in trotting gait. *ROBOT, 37*(1), 63–73.

Li, Y.B., Li, B. & Rong, X.W. (2011). Mechanical design and gait planning of a hydraulically actuated quadruped bionic robot. *Journal of Shandong University (Engineering Science), 5*, 32–36.

Liang, S.C. (2014). Research and application on VFPID control algorithm for electro-hydraulic servo system (in Chinese). *Central South University.*

Shao, J.P., Han, G.H. & Dong Y.H. (2008). Model identification and control method of electro-hydraulic position servo system. *Central South University (Science and Technology), 2*, 22–24.

Sinthipsomboon, K., Hunsacharoonroj, I. & Khedari, J. (2011). A hybrid of fuzzy and fuzzy self-tuning PID controller for servo electro-hydraulic system. *Industrial Electronics and Applications (ICIEA), 49*(20), 220–225.

Sun, G., Shao, J., Zhao, X. & Liu, X. (2015). Hydraulic robot actuator modeling and joint angle tracking control. *Chinese Journal of Scientific Instrument, 3*, 584–591.

Yang, T., Wang, S.L. & Dai, J.B. (2015). Active control of spatial structure based on GMM actuator and T-S type fuzzy neural network. *Journal of Vibration and Shock, 34*(24), 1–6.

Zhou, D. & Hazima, E.M. (2014). Variable universe double fuzzy control for huge elastic loaded servo system. *Journal of Mechanical Engineering, 50*(13), 165–169.

Zhu, Y., Guang, J.B. & Li, W. (2014). Leg compliance control of hexapod robot based on adaptive fuzzy control. *Journal of Zhejiang University (Engineering Science), 48*(8), 1419–1426.

Electronic Engineering – Wang (ed)
© 2018 Taylor & Francis Group, London, ISBN 978-1-138-60260-1

Research into DDoS attack detection based on a decision tree constructed in parallel on the basis of mutual information

C. Zhao, H.Q. Wang, F.F. Guo & G.N. Hao
Harbin Engineering University, Harbin, China

ABSTRACT: A decision tree construction algorithm was proposed based on measuring attribute importance in terms of average mutual information, and the algorithm was optimised using a parallel architecture based on MapReduce to improve efficiency. In addition, a parallel architecture was used to detect Distributed Denial of Service (DDoS) attacks. Experiments show that our methods effectively improve the efficiency of decision tree construction and DDoS attack detection, and decrease the false positive and false negative rates associated with the latter.

Keywords: DDoS; attack detection; decision tree; MapReduce

1 INTRODUCTION

Distributed Denial of Service (DDoS) attacks (Jaber et al., 2015) use a large number of computers distributed in different locations to simultaneously send a large number of packets in a coordinated fashion to a target host, exceeding the capacity of the target host and constantly consuming network bandwidth and/or system resources of the target host to make it incapable of providing normal service, even causing its paralysis. Owing to their power, difficulty in tracking, and low cost, DDoS attacks have become frequent in recent years. On 21 October 2016, the most important Domain Name System (DNS) service provider in the United States, Dyn, suffered a large-scale DDoS attack for several hours. During that time, hundreds of sites, such as Twitter, Spotify, Netflix, GitHub, Airbnb, Visa and CNN, were unable to be accessed or logged into, and people's lives were seriously affected.

DDoS attacks have been paid more and more attention because of their increasing frequency. Many researchers and institutions have studied DDoS detection and defence. A DDoS attack detection algorithm was proposed based on traffic fluctuation (Zhang & Huang, 2013), which assumed that normal traffic presents random fluctuation, and abnormal traffic, namely the traffic with DDoS attacks, shows a continuous and gradual trend. A method of clustering detection of DDoS attacks was proposed using ant colony algorithms (Zhang et al., 2011). This method was used to detect whether a particular session was a DDoS attack by calculating the similarity of user sessions from web logs, and comparing the differences between normal and abnormal browsing modes.

Information entropy can effectively measure the changes of traffic characteristics. Based on this, a variety of methods are proposed to distinguish normal traffic and attack traffic as the basis of traffic analysis. Two new information indicators have been proposed, namely, the generalised entropy and the information distance, to effectively monitor DDoS attacks at lower rates and to reduce error rates (Kang et al., 2015). A method was proposed to decompose the flow chain entropy time series into two parts, namely, the change trend and the randomness (Dzurenda et al., 2015). DDoS attacks with increasing attack intensity were detected through the prediction of long-term network traffic changes.

With the development of cloud computing technology, more and more researchers have used cloud computing to handle DDoS attack detection. A schedule-based DDoS attack detection mechanism was proposed for traditional DDoS attacks (Xu et al., 2015). The time spent per byte on requesting a URL was calculated with MapReduce and compared with the normal situation to detect attacks.

Current DDoS detection methods have limitations in data processing capability, scalability and accuracy, which can be difficult to apply in a cloud environment. Therefore, a DDoS attack detection method is proposed based on decision trees, measuring the importance of attributes with average mutual information, and the use of parallel computing of the MapReduce architecture, to improve efficiency.

2 A DDOS ATTACK DETECTION SYSTEM ARCHITECTURE BASED ON DECISION TREES

To ensure comprehensive analysis, data from as many sources as possible needs to be obtained. Multi-source

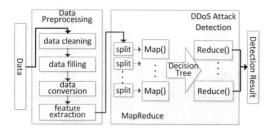

Figure 1. System architecture of DDoS attack detection based on decision tree.

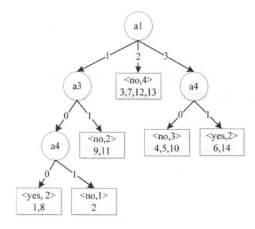

Figure 2. Decision tree.

data are frequently incomplete and redundant, therefore data preprocessing is necessary. Data preprocessing includes data cleaning, data filing, data transformation, and feature extraction. The main purpose of data preprocessing is to improve the integrity and standardisation of data and to prepare for the follow-up training and detection.

DDoS attack detection is the focus of this paper. First, rapidly construct a decision tree using parallel computing with the architecture of MapReduce. Then, capture possible attacks through anomaly detection methods. Finally, detect DDoS attack by extracting the characteristics according to the generated decision tree, which improves the accuracy of detection. The core of the above process is to build an efficient and accurate decision tree.

3 CONSTRUCTION METHOD OF DECISION TREE BASED ON MAPREDUCE

Constructing a decision tree is a process of generating a decision tree through an example set. There are two key issues in this process: one is to choose which attribute to divide, and the other is to determine the condition for generating a leaf node. Attack detection is the process of distinguishing new instances via the generated decision tree, and also the ultimate purpose of constructing the decision tree.

In a decision tree, root and branch nodes correspond to attributes, and leaf nodes correspond to decision classifications. To construct a decision tree, we first select an attribute as the root node, and the value of this attribute as branches, according to the specified rules. Then, for each branch, we choose one of the remaining attributes as a child node. If the node does not meet the conditions for a leaf node, we continue this process until the decision tree is complete. For a new instance, attack detection is a process that starts from the root node, with the appropriate branch then selected according to the value of the attribute, until arrival at a leaf node.

3.1 Decision tree construction based on average mutual information

The purpose of the decision tree algorithm is to reduce unnecessary attributes and improve efficiency without affecting the decision-making ability. Therefore, the key to a decision tree construction algorithm is an appropriate choice of the attributes.

In the construction of a decision tree, the tree that can generate the minimum rule base is the optimal decision tree, but it has been proven that constructing an optimal decision tree is an NP-hard problem and therefore cannot be solved exactly. The concept of entropy from statistical thermodynamics expresses the degree of confusion in a system. Information entropy represents the uncertainty of a random variable. In this paper, average mutual information is used to measure the importance of attributes for decision-making, and a greedy strategy is used to gradually construct a decision tree.

S denotes the sets of training samples with which to construct a decision tree. X stands for the classification of a piece of data containing n different classifications, namely $x_1, x_2, x_3, \ldots, x_n$; Y stands for an attribute, $y_1, y_2, y_3, \ldots, y_m$, respectively. Then the uncertainty of X can be shown as the information entropy $H(X)$:

$$H(X) = -\sum_{i=1}^{n} p(x_i) \log_2 p(x_i) \qquad (1)$$

where $p(x_i)$ represents the probability that X equals x_i.

Conditional entropy refers to the uncertainty degree of another variable in the case of a known variable. Conditional entropy $H_c(X|Y)$ is defined as:

$$H_c(X|Y) = -\sum_{i=1}^{n}\sum_{j=1}^{m} p(x_i, y_j) \log p(x_i|y_j) \qquad (2)$$

Average mutual information represents the information content of two variables jointly owned. Average mutual information $I(X; Y)$ is defined as:

$$I(X;Y) = H(X) - H_c(X|Y) \qquad (3)$$

The uncertainty of decision classification attribute X minus the uncertainty of X in the situation of known attribute Y indicates how much information Y can provide for decision-making, so mutual information is able to measure the importance of attributes for decision-making.

The process of constructing the decision tree is to build the tree from top to bottom, starting from the empty tree T. First, the mutual information $I(X;Y)$ of each attribute and decision classification is calculated. Then, the attribute with the largest average mutual information is selected as the splitting attribute on the basis of a greedy strategy. Each value of this attribute is taken as a branch. If the child node is pure or the average mutual information of all remaining attributes and decision classes is less than the threshold, the child node is a leaf node; otherwise the above process is repeated until all leaf nodes are obtained.

3.2 Parallel optimisation of the algorithm based on MapReduce

The process of constructing a decision tree is iterative. It is very time-consuming to construct a decision tree serially and it is difficult to implement when the data volume is large. Therefore, a parallel method of constructing a decision tree is designed, based on MapReduce.

The MapReduce-based parallel method of constructing a decision tree principally follows the approach of "divide and conquer":

Step 1: according to the performance of a virtual machine, the training data is divided into n data blocks of fixed size according to the capacity of the memory;
Step 2: using the Map() function, various numerical of calculating average mutual information are counted;
Step 3: the Reduce() function is used to integrate data, calculate average mutual information, and select splitting attributes;
Step 4: treating each branch as a new task, the decision tree is constructed in parallel;
Step 5: the preceding steps are repeated until a complete decision tree is generated.

3.3 DDoS attack detection

Based on the algorithm proposed in this paper, a system of DDoS attack detection based on MapReduce is implemented using the parallel design scheme of Section 3.2. DDoS attack detection systems take parallel detections, build a Hadoop platform, and experiment by simulating a large number of DDoS attacks in the system to be detected, to ensure the diversity and comprehensiveness of rules mining.

As a record management tool for network users and system behaviour, log files record information about the system kernel, related applications, network conditions, and so on. Therefore, log files are selected as analysis data. Linux system logs and firewall logs from the system to be detected are collected, and the preprocessed data is transmitted to the DDoS attack detection system for analysis.

Table 1. Parameters of experimental environment.

Name	Parameter
Server	Intel[R] server
CPU	Core[TM] i5-3470
Frequency	3.20 GHz
Memory	2.00 GB
Operating system	Linux 2.6.18
Kernel version	Centos 5.5 x86-64
Hadoop version	2.2.4

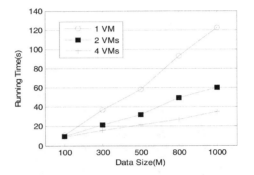

Figure 3. Comparison of time required to build a decision tree with different quantities of virtual machines (VMs).

4 EXPERIMENT AND PERFORMANCE ANALYSIS

4.1 Efficiency analysis of the decision tree parallel construction algorithm

First, the efficiency of the decision tree parallel construction algorithm is analysed. The parameters of the experimental environment are shown in Table 1.

Nursery data sets were selected from the University of California, Irvine (UCI) Machine Learning Repository, and data blocks of 100 MB, 300 MB, 500 MB, 800 MB and 1 GB were obtained by duplication, and processed by one, two and four Virtual Machines (VMs), respectively. The experimental results are shown in Figure 3.

Figure 3 shows that as the number of virtual machines increases, the time required to construct the decision tree decreases, and as the amount of data increases, this decrease in time becomes more and more obvious. The experiments indicate that parallel algorithms based on MapReduce effectively improve the efficiency of decision tree construction.

4.2 DDoS attack detection accuracy analysis

To check the accuracy of the DDoS attack detection system, 1,000 MB of data were collected from the injected system, of which 400 MB were firewall logs, 450 MB were message logs, and 150 MB were wtmp/utmp logs, and a comparison was made with the Fuzzy Pattern Decision Tree (FPDT) method of Zhang et al. (2014). Four VMs analyse in parallel, including

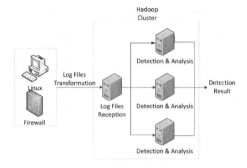

Figure 4. Experimental environment.

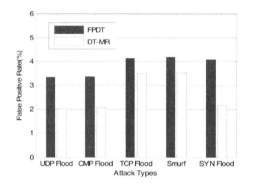

Figure 5. False positive rates of different detection methods.

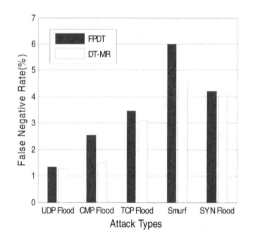

Figure 6. False negative rate of different detection methods.

a VM running the main program for data division and merging, and the other three VMs calculating. The experimental system uses the Java language and the experimental environment is shown in Figure 4.

The data is first preprocessed and is then input to the detection system. The experimental system analyses the algorithm from two perspectives: false positive rate and false negative rate.

The false positive rates of the two algorithms, FPDT and our method (DT-MR), are compared in Figure 5. The experimental results show that for a variety of types of DDoS attack, DT-MR has lower false positive rates, by 1.321%, 1.300%, 0.635%, 0.669%, and 1.916%, respectively. Proportionately, false positive rates are greatly decreased, especially for DDoS attacks with large traffic changes, and also for the most common types of DDoS attack: UDP flood, ICMP flood, and SYN flood attacks.

The false negative rates of the two algorithms are compared in Figure 6. These results show that for a variety of types of DDoS attacks, the false negative rates are reduced by the DT-MR algorithm by 0.051%, 1.036%, 0.369%, 1.465%, and 0.200%, respectively.

The DT-MR algorithm reduces detection time by 67.01% due to the parallel detection afforded by MapReduce, which significantly improves detection efficiency.

5 CONCLUSIONS

In response to the problem of low efficiency in constructing a decision tree and low accuracy of attack detection, an algorithm for decision tree construction was proposed, which measures the importance of attributes based on average mutual information. A parallel architecture optimisation algorithm based on MapReduce improved efficiency. In addition, a parallel architecture was used to detect DDoS attacks. Experimental results show that our methods effectively improve the efficiency of decision tree construction, reducing the false positive rate by 1.55% and the false negative rate by 0.62%, and potentially improving detection efficiency.

REFERENCES

Dzurenda, P., Martinasek, Z. & Malina, L. (2015). Network protection against DDoS attacks. *International Journal of Advances in Telecommunications, Electrotechnics, Signals and Systems*, 4(1), 8–14.

Jaber, A.N., Zolkipli, M.F. & Majid, M.A. (2015). Security everywhere cloud: An intensive review of DoS and DDoS attacks in cloud computing. *Journal of Advanced & Applied Sciences (JAAS)*, 3(5), 152–158.

Kang, J., Yang, M. & Zhang, J. (2015). Accurately identifying new QoS violation driven by high-distributed low-rate denial of service attacks based on multiple observed features. *Journal of Sensors*, 2015(1), 1–11.

Xu, G., Yu, W. & Chen, Z. (2015). A cloud computing based system for cyber security management. *International Journal of Parallel, Emergent and Distributed Systems*, 30(1), 29–45.

Zhang, J. & Huang, M.L. (2013). Visual analytics model for intrusion detection in flood attack. In *Proceedings of the 12th IEEE International Conference on Trust, Security and Privacy in Computing and Communications, Melbourne* (pp. 277–284). New York, NY: IEEE.

Zhang, T., Zhang, H. & Fu, L. (2014). Script virus detection algorithm based on fusion of fuzzy pattern and decision tree. *Journal of Electronics and Information Technology*, 36(1), 108–113.

Zhang, W., Jia, Z. & Li, X. (2011). One method for application layer DDoS detection using ant clustering. *Computer Engineering and Applications*, 47(14), 99–102.

Random addition-chains based DPA countermeasure for AES

H. Huang
School of Software, Harbin University of Science and Technology, Harbin, China

X.X. Feng
School of Computer Sciences and Technology, Harbin University of Science and Technology, Harbin, China

J. Hou, Y.Y. Zhao & J.X. Jiang
School of Applied Sciences, Harbin University of Science and Technology, Harbin, China

ABSTRACT: To counteract differential power attacks (DPA), various higher-order masking schemes were proposed in recent years. The main issue while applying masking scheme to a block cipher implementation is to develop an effective scheme for evaluating the S-box. This paper proposes a new masking scheme based on random addition-chains for the AES to thwart DPA. The proposed scheme adopts sixteen different addition chains with the same numbers of field square and multiplication, and randomly selects one of them to evaluate the objective S-box. Compared to existing addition-chain based masking schemes, the proposed scheme offers higher degree security, but does not bring any additional area-overhead.

Keywords: Differential Power Attacks (DPA); Random Addition-Chains; S-box reusing

1 INTRODUCTION

The security of the block ciphers is thwarted by the Side-channel analysis attacks (SCA). Among them, the differential power attacks (DPA) are the most widely and effective attack method (Paul 1999). To thwart DPA, many important efforts have been carried out. Most of existing schemes against DPA are in the basis of secret sharing (Shamir 1979), which is also called masking. The higher-order masking schemes had been proposed to counteract higher-order DPA (HODPA) (Schramm 2006). The principle of this masking scheme is to split every intermediate value x into $d+1$ shares $x_0, x_1, \ldots, x_d$, which satisfied the following operation,

$$x_0 \oplus x_1 \oplus \cdots \oplus x_d = x \qquad (1)$$

where $\oplus$ denotes exclusive-OR (XOR) operation; the d of shares $x_1, x_2 \ldots, x_d$ are called the masks, whose values are randomly picked up; and the last one x_0 is known as the masked variable. When every sensitive variable is masked by d random masks, the masking order is d. Therefore, only attackers exploiting some leakages which are related to at least $d+1$ intermediate variables may succeed in obtaining useful information. However, the difficulty and the cost of a successful attack against the masked algorithm increases exponentially with the increase of masking order. The most difficult part lies in masking the S-boxes when developing the higher-order masking

for the cryptographic algorithms. Because the S-boxes are only non-linear operations in block ciphers, e.g., advanced encryption standard (AES). As a result, the design of efficient masking schemes for S-boxes of different algorithms have become a hot topic.

Most of the existing masking schemes for S-boxes were based on Look-Up-Table (LUT) (Coron 2014) or composite field (Satoh 2001). However, in recently years, a new type higher-order masking scheme for the S-Boxes was proposed. The main idea of this scheme is to transform the S-box operation into power function $x \mapsto x^{254}$ over the Galois Field GF(2^n), then masked the power function using The ISW scheme (Ishai 2003). These schemes are based on different addition-chains, which composed of the field squaring operations and the field multiplications, such as the masking schemes in Rivain (2010); Carlet (2012) and Roy (2013). However, the computing complexity exponentially increases with regard to the order d because each share is required to separately operate, which significantly increase the masking complexity.

Different from existing addition-chain schemes, a new masking scheme for S-box is proposed in this paper by combing the addition-chain with shuffling. In the proposed scheme, sixteen different addition chains with the same number of field squaring as well as field multiplication are adopted to construct sixteen S-boxes. For every S-box, one of them is randomly adopted, and sixteen S-boxes are operated simultaneously. Moreover, in every round the addition chain can also be randomly selected. Compared to previous addition-chain schemes, this random masking scheme provides higher order security without bringing any area overhead, and the running time is almost the same as the other ones.

The remainder of this paper is organized as follow. Section 2 describes related works. Section 3 presents the proposed random addition-chain based scheme for the AES. Section 4 concludes the paper.

Figure 1. Normal addition-chain.

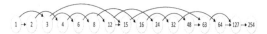

Figure 2. Cyclotomic-class addition-chain.

Figure 3. Polynomial addition chain.

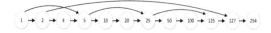

Figure 4. Extended addition chain.

2 PREVIOUS WORKS

The ISW scheme was proposed by Ishai et al. (2003). to secure a multiplication $c = a \times b$ over $GF(2^n)$ for any n greater than '1'. The detail algorithm is recalled hereafter, which is shown in Algorithm 1.

Algorithm 1. SecMult – d^{th}-order secure multiplication over GF_2^n

INPUT: shares a_i satisfying $\oplus_i a_i = a$, shares b_i satisfying $\oplus_i b_i = b$
OUTPUT: shares c_i satisfying $\oplus_i c_i = c = ab$

1. for $i = 0$ to d do
2.   for $j = i + 1$ to d do
3. $r_{i,j} \leftarrow \text{rand}(n)$
4. $r_{j,i} \leftarrow (r_{i,j} \oplus a_i b_j) \oplus a_j b_i$
5. for $i = 0$ to d do
6. $c_i \leftarrow a_i b_i$
7.   for $j = 0$ to d, $j \neq i$ do $c_i \leftarrow c_i \oplus r_{i,j}$

As stated in Rivain (2010), to ensure the security of the Algorithm 1 at any order d, the inputs masks $(a_i)_{i\geq 1}$ and $(b_i)_{i\geq 1}$ must be independent. If this condition is not satisfied, Rivain and Prouff (2010) proposed a mask refreshing scheme to secure the multiplication processing. Thus, it is effective and secure to evaluate the he power-of-254 over $GF(2^8)$ through the whole operation. The addition-chain proposed in Rivain (2010) is shown in Figure 1.

In Figure 1, there are four field multiplications and seven field squares required to construct the normal addition chain.

In 2012, Carlet and Goubin proposed a new masking scheme which could effectively protect any S-box at any order (Carlet & Goubin 2012). Most important of all, they adopted an optimal scheme for the set of power functions. The addition-chain, named cyclotomic-class addition-chain, proposed in this scheme is shown in Figure 2.

In Figure 2, there were also has four field multiplications in the cyclotomic-classes addition-chain.

In 2013, Roy and Vivek studied a variant of addition chain (Roy & Vivek 2013) to explore optimal schemes for exponentiation over $GF(2^8)$. They made an analysis and improvement of the generic higher-order masking scheme based on FSE 2012 (Carlet 2012). In addition, they defined the notion of F_2^n-polynomial chain, and reduced the number of non-linear field multiplications by taking advantage of it. The F_2^n-polynomial addition-chain is shown in Figure 3.

In Figure 3, there are four field multiplications and seven field squares. Thus, it has the same computing-complexity with the scheme in Rivain (2010). Grosso and Prouff (2014) paid more attention on the scheme proposed by Carlet, et al. at FSE 2012 (Carlet 2012), and latter improved by Roy at CHES 2013 (Roy 2013). They proposed a new type of addition-chain named extended addition chains. The extended addition-chain is shown in Figure 4.

In Figure 4, it showed that the extended addition chain adopted four field multiplications and seven squares. In conclusion, the addition-chain presented above all provided different methods to improve performance. For the AES S-box, in terms of the number of the non-linear multiplications, the existing addition-chains obtained the same results. The reason for this was that they all followed a similar approach. Based on this observation, if the masking-complexity is to be further reduced, then new scheme should be developed.

3 ADDITION-CHAIN BASED AES AND ITS HIGHER-ORDER MASKING SCHEME

3.1 Addition-chain based masking scheme for S-box

To address above issue, in this section we propose an efficient DPA countermeasure by coming the addition-chain scheme with shuffling operation. For the AES, there are four different transformations, which can be split into linear and nonlinear operations. As mentioned above, to mask the non-linear operation, i.e., S-box, is the much more difficult than the linear one. Thus, we focus on the masking scheme for the S-box. The AES S-box operation is consist of multiplicative inverse $x^{-1} = x^{254}$ over $GF(2^8)$ and the affine transformation. The multiplicative inverse is further split into

Table 1. Sixteen different addition-chains for AES S-box.

$x \rightarrow x^2 \rightarrow x^3 \rightarrow x^6 \rightarrow x^{12} \rightarrow x^{15} \rightarrow x^{30} \rightarrow x^{60} \rightarrow x^{120} \rightarrow x^{240} \rightarrow x^{252} \rightarrow x^{254}$
$x \rightarrow x^2 \rightarrow x^4 \rightarrow x^8 \rightarrow x^{16} \rightarrow x^{17} \rightarrow x^{34} \rightarrow x^{68} \rightarrow x^{136} \rightarrow x^{204} \rightarrow x^{238} \rightarrow x^{254}$
$x \rightarrow x^2 \rightarrow x^4 \rightarrow x^5 \rightarrow x^{10} \rightarrow x^{20} \rightarrow x^{25} \rightarrow x^{50} \rightarrow x^{100} \rightarrow x^{125} \rightarrow x^{127} \rightarrow x^{254}$
$x \rightarrow x^2 \rightarrow x^4 \rightarrow x^6 \rightarrow x^{12} \rightarrow x^{24} \rightarrow x^{48} \rightarrow x^{96} \rightarrow x^{192} \rightarrow x^{240} \rightarrow x^{252} \rightarrow x^{254}$
$x \rightarrow x^2 \rightarrow x^4 \rightarrow x^5 \rightarrow x^{10} \rightarrow x^{15} \rightarrow x^{30} \rightarrow x^{60} \rightarrow x^{120} \rightarrow x^{240} \rightarrow x^{250} \rightarrow x^{254}$
$x \rightarrow x^2 \rightarrow x^3 \rightarrow x^4 \rightarrow x^7 \rightarrow x^{14} \rightarrow x^{28} \rightarrow x^{56} \rightarrow x^{112} \rightarrow x^{224} \rightarrow x^{252} \rightarrow x^{254}$
$x \rightarrow x^2 \rightarrow x^3 \rightarrow x^4 \rightarrow x^7 \rightarrow x^{14} \rightarrow x^{15} \rightarrow x^{30} \rightarrow x^{60} \rightarrow x^{120} \rightarrow x^{127} \rightarrow x^{254}$
$x \rightarrow x^2 \rightarrow x^4 \rightarrow x^5 \rightarrow x^{10} \rightarrow x^{20} \rightarrow x^{25} \rightarrow x^{50} \rightarrow x^{100} \rightarrow x^{125} \rightarrow x^{250} \rightarrow x^{254}$
$x \rightarrow x^2 \rightarrow x^4 \rightarrow x^5 \rightarrow x^{10} \rightarrow x^{20} \rightarrow x^{40} \rightarrow x^{50} \rightarrow x^{100} \rightarrow x^{200} \rightarrow x^{250} \rightarrow x^{254}$
$x \rightarrow x^2 \rightarrow x^4 \rightarrow x^5 \rightarrow x^{10} \rightarrow x^{20} \rightarrow x^{30} \rightarrow x^{60} \rightarrow x^{120} \rightarrow x^{240} \rightarrow x^{250} \rightarrow x^{254}$
$x \rightarrow x^2 \rightarrow x^4 \rightarrow x^8 \rightarrow x^{16} \rightarrow x^{32} \rightarrow x^{36} \rightarrow x^{72} \rightarrow x^{144} \rightarrow x^{216} \rightarrow x^{252} \rightarrow x^{254}$
$x \rightarrow x^2 \rightarrow x^4 \rightarrow x^8 \rightarrow x^{16} \rightarrow x^{17} \rightarrow x^{34} \rightarrow x^{68} \rightarrow x^{102} \rightarrow x^{204} \rightarrow x^{238} \rightarrow x^{254}$
$x \rightarrow x^2 \rightarrow x^4 \rightarrow x^5 \rightarrow x^7 \rightarrow x^{14} \rightarrow x^{28} \rightarrow x^{56} \rightarrow x^{112} \rightarrow x^{224} \rightarrow x^{252} \rightarrow x^{254}$
$x \rightarrow x^2 \rightarrow x^4 \rightarrow x^8 \rightarrow x^{12} \rightarrow x^{24} \rightarrow x^{48} \rightarrow x^{96} \rightarrow x^{192} \rightarrow x^{240} \rightarrow x^{252} \rightarrow x^{254}$
$x \rightarrow x^2 \rightarrow x^3 \rightarrow x^6 \rightarrow x^{12} \rightarrow x^{24} \rightarrow x^{48} \rightarrow x^{96} \rightarrow x^{192} \rightarrow x^{240} \rightarrow x^{252} \rightarrow x^{254}$
$x \rightarrow x^2 \rightarrow x^3 \rightarrow x^4 \rightarrow x^7 \rightarrow x^{14} \rightarrow x^{28} \rightarrow x^{30} \rightarrow x^{60} \rightarrow x^{120} \rightarrow x^{127} \rightarrow x^{254}$

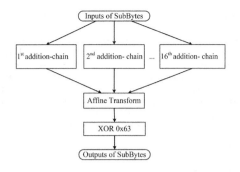

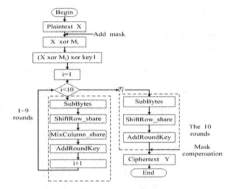

Figure 5. Random addition-chains based S-box evaluation algorithm.

some lower power functions and masked respectively. For linear function, e.g., power-of-two, the Boolean masking can be directly adopted, since

$$x_0^2 \oplus x_1^2 \oplus \cdots \oplus x_d^2 = x^2 \qquad (2)$$

While, for the non-linear operation, e.g., field multiplication, the ISW scheme is adopted, which has been well studied in published literatures.

3.2 Proposed random addition-chain for S-box

As mentioned in the Section 2, some addition-chain-based masking schemes have been proposed to against side-channel attacks. However, these schemes only have one addition-chain, and each byte for every round adopt the same addition-chain, whose security order is limited to the masking order. To thwart higher-order DPA, their masking complexity will be dramatically increased. To solve this problem, a new masking scheme combing addition-chain with shuffling operation is proposed. Firstly, 16 different addition-chains which have the same number of field multiplication and square are constructed; Secondly, sixteen S-boxes based on different addition-chains are developed; Finally, a random generator is developed to control the shuffling operation. In Carlet (2012), we know that the masking complexity in terms of nonlinear multiplication for AES is four. The different addition-chains used in this paper are shown in Table 1.

In Table 1, each addition-chain has four field multiplications and seven field squares, while the permutation order is different. Therefore, the power trace of each addition-chain is different even if the input is same. The random addition-chains based S-box evaluation algorithm is shown in Figure 5.

In the proposed scheme, sixteen inputs bytes are processed simultaneously. The inputs are randomly shuffled according to the number generated from random generator before processed. Then, each byte is evaluated by one of the sixteen addition-chain. After that, the output are rearranged to the original order.

Figure 6. Procedure of mask scheme.

Since all of the chains has same functionality, the output is always right. While, the power trace of the chain is absolutely different from each other, which enhance the security by dramatically increasing the number of traces required by the DPA.

The masking procedure of the whole AES is shown in Figure 6.

In Figure 6, it shows that the masks are added in the initial 'AddRoundKey' and compensated at the end of the last round in this proposed scheme. The correctness of the proposed masking schemes is proved by comparing with the classic LUT-based scheme; the comparison results indicate that this masking scheme is completely correct. This random chain-additions based scheme dramatically increases the difficulty of the DPA attacks, which can further improve the security of the AES. The scheme protects all immediate variables and we prove its dth-order security in the following section.

3.3 Security analysis of proposed masking scheme

The security of our scheme can be easily proved by the proofs in Rivain (2010). It is easy to prove the security for each operation, regardless of linear or nonlinear operation. The notion of dth-order SCA security can be formally defined as that a random encryption algorithm is said to achieve d^{th}-order SCA security, if every d-tuple of its intermediate variables is independent of any sensitive variable.

Table 2. Comparisons of running time for different schemes.

Method	$d=1$	$d=2$	$d=3$
Normal addition-chain	4.12 s	6.36 s	9.87 s
Polynomial addition-chain	4.14 s	6.63 s	10.00 s
Extended addition-chain	4.40 s	6.77 s	10.18 s
This paper	4.29 s	6.76 s	9.97 s

Table 3. Comparisons of various addition chain-based S-boxes.

Method	Number of Multiplication
Normal addition-chain	4
Cyclotomic addition-chain	4
Parity-Split addition-chain	6
Polynomial addition-chain	4
Extended addition-chain	4
This paper	4

The most sensitive part of this proposed method is the masked multiplication algorithm which is based on the generalized ISW scheme (Algorithm 1). To prove this scheme, let a and b be two sensitive variables; let $(a_i)_{0 \leq i \leq d}$ and $(b_i)_{0 \leq i \leq d}$ be two sets of intermediate variables of the input of the Algorithm 1, which satisfy $a = \oplus_{0 \leq i \leq d} a_i$ and $b = \oplus_{0 \leq i \leq d} b_i$ where $(a_i)_{i \geq 1}$ and $(b_i)_{i \geq 1}$ are randomly generated. Then, the distribution of every tuple of d or less intermediate variables in Algorithm 1 is independent of (a, b). Each sub-routine of the evaluation procedures is provably secure and the security of the proposed masking scheme for the whole AES can be proven. The detail proof can be found in the full version of the literature (Rivain 2010).

3.4 Experiments and efficiency comparisons

In this section, in order to verify the proposed scheme, different order masking schemes for AES are modeled in C language. Table 2 shows the simulation results of running time for AES implementations (the Rivain (2010), Roy (2013), Grosso (2014) and this paper) with different order schemes ($d = 1, 2, 3$).

In the Table 2, the running time results show that the running time is almost same with other masking schemes. While, the security of the proposed masking scheme is significantly improved. Compared to the extended addition chain, the running time of the proposed scheme is reduced by about 2.5%, 0.14%, 2.1%. Moreover, compared to the other two schemes, the maximum running time is about 6.2%.

Table 3 lists the number of the multiplication of various addition-chain implementations for S-box: the normal, cyclotomic, parity-split, polynomial, extended addition chain and this paper. For various addition chain masking schemes, the number of multiplication is all four in addition to parity-split scheme. As already pointed out, the number of multiplication is at least four. Therefore, the masking scheme offers higher degree security without bringing

any additional area-overhead in terms of the number of multiplication.

4 CONCLUSION

In this paper, we propose a new higher-order masking method based on addition-chain algorithm. Different from the existing masking schemes, the proposed masking scheme adopts multiple addition chains to protect the S-box. Moreover, this new masking scheme is applied to AES which offers higher security than other similar schemes. The detailed performance of the implementation results are given. Our future work will focus on the research on hardware implementation of the proposed scheme and extend this scheme for other block ciphers.

ACKNOWLEDGEMENT

This work was sponsored by the National Natural Science Foundation of China (Grant Nos. 61604050) and (Grant Nos. 51672062).

REFERENCES

Carlet, C., Goubin, L., Prouff, E., Quisquater, M. & Rivain, M. (2012). Higher-Order masking schemes for s-boxes. *International Conference on FAST Software Encryption* Vol. 7549: 366–384.

Coron, J. S. (2014). *Higher Order Masking of Look-Up Tables. Advances in Cryptology – EUROCRYPT 2014.* Springer Berlin Heidelberg.

Grosso, V., Prouff, E. & Standaert, F. X. (2014). *Efficient Masked S-Boxes Processing – A Step Forward –. Progress in Cryptology – AFRICACRYPT 2014.* Springer International Publishing.

Ishai, Y., Sahai, A. & Wagner, D. (2003). Private circuits: securing hardware against probing attacks. *Lecture Notes in Computer Science, 2729*: 463–481.

Paul, C., Jaffe, Joshua, Jun, & Benjamin. (1999). *Differential Power Analysis. Advances in Cryptology — CRYPTO'99.* Springer Berlin Heidelberg.

Rivain, M. & Prouff, E. (2010). Provably Secure Higher-Order Masking of AES. *International Workshop on Cryptographic Hardware and Embedded Systems* Vol. 6225: 413–427.

Roy, A. & Vivek, S. (2013). *Analysis and Improvement of the Generic Higher-Order Masking Scheme of FSE 2012. Cryptographic Hardware and Embedded Systems – CHES 2013.* Springer Berlin Heidelberg.

Satoh, A., Morioka, S., Takano, K. & Munetoh, S. (2001). *A Compact Rijndael Hardware Architecture with S-Box Optimization. Advances in Cryptology – ASIACRYPT 2001.* Springer Berlin Heidelberg.

Schramm, K. & Paar, C. (2006). Higher Order Masking of the AES. *Topics in Cryptology – CT-RSA 2006, The Cryptographers' Track at the RSA Conference 2006, San Jose, CA, USA, February 13–17, 2006, Proceedings*, Vol. 3860: 208–225.

Shamir, A. (1979). How to share a secret. *Communications of the Acm.* 22(11): 612–613.

Electronic Engineering – Wang (ed)
© 2018 Taylor & Francis Group, London, ISBN 978-1-138-60260-1

Diagonal gait planning and simulation of hydraulic quadruped robot based on ADAMS

J.P. Shao & S.K. Wang

College of Mechanical and Power Engineering, Harbin University of Science and Technology, Harbin, China

ABSTRACT: In order to realize the high adaptability movement requirement of the high load and uneven ground of the robot, a diagonal gait of a hydraulically driven quadruped robot is designed. The kinematics of the quadruped robot is analyzed and simulated based on the robot kinematics theory. Firstly, the D-H parameters are analyzed by combining with the structure of the real robot, and the forward and inverse kinematics solutions are deduced. Then, a new compound cycloidal trajectory is proposed by combining sinusoidal and cycloidal trajectories. Finally, the simulation results of ADAMS are used to analyze the position of the center of mass, the centroid velocity of the body and the position of the foot end, which are combined with kinematics analysis and diagonal gait trajectory planning. Simulation results verify the rationality of robot gait planning.

Keywords: quadruped robot; hydraulic drive; gait planning; ADAMS

1 INTRODUCTION

The quadruped robot can walk in unstructured environment and walk in a dynamic way by choosing the appropriate landing point, which reflects the good adaptability of the quadruped robot to the ground (Gao et al. 2015).

For the quadruped robot gait planning, domestic and foreign scholars have done some research. Raibert, who first developed a class of inverted pendulum single-foot bounce robot, and the success of dynamic analysis and feedback control applied to a single-foot bounce robot, so that a single-legged robot to achieve a stable dynamic jump, forward speed of up to 1 m/s (Raibert & Chepponis 1986). University of Tokyo, Japan, introduced a quadruped bio-leg structure of the robot Kenken, by simulating biological rhythm control, the robot can do the shape of quadruped creatures running action. Boston Dynamics company successfully developed a quadruped robot Little Dog, the control algorithm, this robot can learn the state of each movement, and optimize the control parameters and logical structure, so it can stably move in unstructured and complex roads with strong capacity of passing over the obstacles.

In domestic, Rong of Shandong University has designed a method of the three foot polynomial and linear combination of the foot trajectory planning, which is successfully applied to the gait planning of the quadruped robot (He & Ma 2005). Wang of Beijing Institute of Technology has improved the cycloid equation and applied it to the foot-end trajectory planning of the quadruped robot, reducing the impact force of the foot-end touchdown. Cai of National University of Defense Technology uses Time-Pose control method to realize the diagonal gait walking of quadruped robot under flat topography and slope topography (Yu et al. 2013). Liu and Zhang of Tong ji University used gait planning method of neural network based on Central Pattern Generator (CPG) to complete gait walking of AIBO quadruped robot in unknown environment (Kurazume et al. 2001).

2 KINEMATICS ANALYSIS OF QUADRUPED ROBOT

In order to achieve the diagonal gait planning of the hydraulic quadruped robot, it is necessary to carry on the kinematics analysis of the robot. First of all, the method of modified Denavit-Hartenberg was used to establish the single leg MDH model of quadruped robot. As shown in Figure 1, it gives the robot the right front (FR) schematic diagram of the robot joint Figure 1, the robot has three degrees of freedom, one out swing pendulum degrees, two roll degrees of freedom.

The parameters of each joint in the right foreleg are shown in Table 1. Which a_i represents the length of the rod joints in the robot α_i is the corresponding rod twist angle, d_i is the joint distance and θ_i is joint angle.

189

homogeneous transformation matrix formula is used, the robot tip position matrix is

$$^BT_4 = {}^BT_1 {}^1T_2 {}^2T_3 {}^3T_4 =$$

$$\begin{bmatrix} s_{23} & c_{23} & 0 & L_2s_2+L_3s_{23}+a \\ s_1c_{23} & -s_1s_{23} & -c_1 & L_1s_1+L_2s_1c_2+L_3s_1c_{23}+b \\ -c_1c_{23} & c_1s_{23} & -s_1 & -L_1c_1-L_2c_1c_2-L_3c_1c_{23}+c \\ 0 & 0 & 0 & 1 \end{bmatrix} \quad (1)$$

s_1 represent $\sin\theta_1$, c_1 represents $\cos\theta_1$, s_{23} represents $\sin(\theta_2+\theta_3)$, c_{23} represents $\cos(\theta_2+\theta_3)$, Thus, the position of the toe of the right front leg in the body coordinate system (the positive kinematics equation of the quadruped robot) is

$$^BP_{TOE} = \begin{bmatrix} L_2s_2+L_3s_{23}+a \\ L_1s_1+L_2s_1c_2+L_3s_1c_{23}+b \\ -L_1c_1-L_2c_1c_2-L_3c_1c_{23}+c \end{bmatrix} \quad (2)$$

The inverse kinematics equation of quadruped robot can be obtained by solving the inverse kinematics of quadruped robot. Thus we can get the value of each joint angle of the robot

$$\theta_1 = \arctan\left(-\frac{y-b}{z-c}\right) \quad (3)$$

$$\theta_2 = \arctan\frac{x-a}{\sqrt{(y-b)^2+(z-c)^2-L_1}} - \varphi \quad (4)$$

$$\theta_3 = \frac{L_2+L_3}{L_3}\varphi \quad (5)$$

Among them, the articulation angle θ_1 and θ_2 in the middle of the variable φ representation

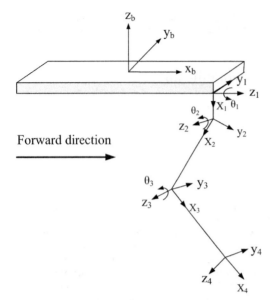

Figure 1. MDH model of a leg of quadruped robot.

Table 1. MDH parameters of right front leg.

Joint i	a_i	α_i	d_i	θ_i
1	0	0	0	θ_1
2	L_1	$\frac{\pi}{2}$	0	θ_2
3	L_2	0	0	θ_3
4	L_3	0	0	0

Assuming that the coordinates of the right front hip joint in the body coordinate system is, and the

$$\varphi = \arccos\frac{L_2^2+(x-a)^2+(y-b-L_1\sin\theta_1)^2+(z-c+L_1\cos\theta_1)^2-L_3^2}{2L_2\sqrt{(x-a)^2+(y-b-L_1\sin\theta_1)^2+(z-c+L_1\cos\theta_1)^2}} \quad (6)$$

Similarly, the other three legs of the kinematics positive and negative solutions use the same method.

3 GAIT TRAJECTORY PLANNING OF QUADRUPED ROBOT

When the quadruped robot is walking on the ideal road surface or in steady state, the robot can be basically stable by using the hip joint 2 and the knee joint, combined with the appropriate gait planning. In order to make the model easy to analyze, the motion of the hip joint 1 is neglected, so the analysis of one-legged model of the quadruped robot is simplified as a two-dimensional plane problem, and the plane of the leg is always perpendicular to the ground.

In order to meet the stability of the foot when the foot end touches the ground and increase the walking stability, the speed and acceleration of the swing leg should be as small as possible when touching and lifting the leg. Combined with the sinusoidal trajectory and on the basis of the existing complex cycloid correction, the new cycloidal trajectory is defined as follows:

$0 \leq t \leq T_y$

$$p_x = S\left(\frac{t}{T_y}-\frac{1}{2\pi}\sin\frac{2\pi t}{T_y}\right)-\frac{S}{2} \quad (7)$$

$$\begin{cases} p_z = 2H\left(\frac{t}{T_y}-\frac{1}{4\pi}\sin\frac{4\pi t}{T_y}\right) & 0 \leq t \leq \frac{T_y}{2} \\ p_z = 2H\left[\frac{T_y-t}{T_y}-\frac{1}{4\pi}\sin\frac{4\pi(T_y-t)}{T_y}\right] & \frac{T_y}{2} \leq t \leq T_y \end{cases} \quad (8)$$

$T_y \leq t \leq T$

$$p_x = \frac{S}{2}-S\left[\frac{t-T_y}{T-T_y}-\frac{1}{2\pi}\sin\frac{2\pi(t-T_y)}{T-T_y}\right] \quad (9)$$

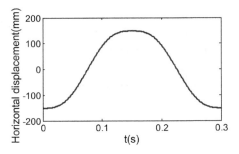

Figure 2. Motion trajectory of the toe in X axis direction.

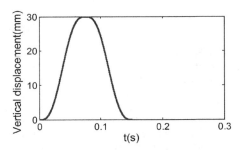

Figure 3. Motion trajectory of the toe in Z axis direction.

$$p_z = 0 \qquad (10)$$

P_x and P_z denote the foot-end advancement direction and the vertical direction position, respectively, S and H denote the step length and the leg lift height of the quadruped robot, t represents the sampling time of the gait trajectory, T represents the gait cycle, T_y represents the time of the swing phase.

Based on the characteristics of the diagonal gait, $T_y = T/2$; According to the initial position of the hydraulic quadruped robot, the step width $S = 300$ mm, $H = 30$ mm, $T = 0.3$ s, without considering the ground contact of the two-dimensional foot trajectory shown in Figures 2 to 4.

4 MOTION SIMULATION AND RESULT ANALYSIS

Set the robot material for the aluminum alloy, the ground for the wooden plane, the robot load of 90 kg; According to the actual movement of the quadruped robot, the joint constraint is set, the leg of the robot and the joint of the body and the leg are bounded by the rotating pair, the hydraulic cylinder is controlled by the moving pair, the passive degree of the leg is restrained by the moving pair. Set the simulation time to 6 s and the simulation step to 600. At the end of simulation, we can see the result of kinematics analysis in ADAMS/Processor, and verify the gait of the designed robot.

It can analyze the kinematics of the quadruped robot in the diagonal gait and judge the stability performance of the robot by measuring the displacement, velocity

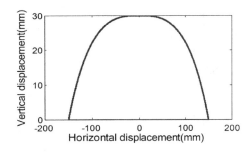

Figure 4. Motion trajectory of the toe in sagital plane.

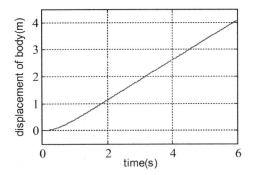

Figure 5. The centroid displacement of the body in the forward direction.

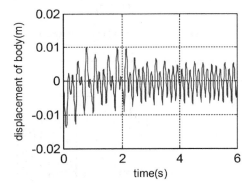

Figure 6. The centroid displacement of the body in the vertical direction.

and foot end position of the body centroid in the space direction.

The positions of robots in forward direction (axis_x), vertical direction (axis_z), and lateral direction (axis_y) displacements of the robot body centroid, as shown in the figure.

In Figure 5, the body center of mass in the overall movement of the process of moving more smoothly, in Figure 6, the body center of mass in the vertical direction to 10mm amplitude fluctuations, Although the initially set single-leg trajectory is to ensure the support foot phase when the foot-end height constant, but in order to reduce the impact of ground impact on the stability of the fuselage in the calf added to the passive constraint, will cause the body center of

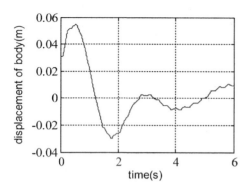

Figure 7. The centroid displacement of the body in the lateral direction.

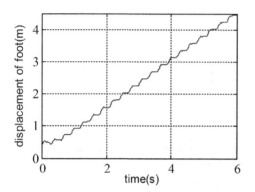

Figure 8. Position of the foot end in the forward direction.

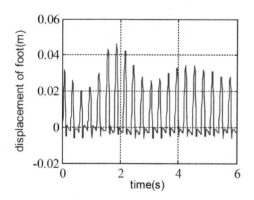

Figure 9. Position of the foot end in the vertical direction.

mass. In the vertical direction of the small amplitude fluctuations, This is also permissible for the stability of the robot. In Figure 7, the body center of mass will move relative to the direction of lateral movement, but each time there will be a short distance in the process of moving the opposite direction of the phenomenon of movement. This is due to the quadrilateral robot diagonal gait movement due to the fuselage along the support diagonal tilt leading to the diagonal legs of the front legs first ground, resulting in relatively large ground reaction force to move the robot body center of mass direction.

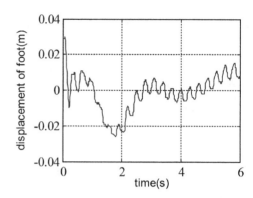

Figure 10. Position of the foot end in the lateral direction.

The robot moves in the forward direction (axis_x), the vertical direction (axis_z), and the foot end position on the lateral direction (axis_y), as shown in the figure.

In Figure 8, the foot end position in the forward direction of a regular and relatively smooths forward movement. In Figure 9, the amplitude of the foot end position in the vertical direction is substantially equal to 30 mm. In Figure 10, the foot end position in the transverse direction has a relatively large amplitude changes, gradually becoming a small amplitude changes. The rationality of diagonal gait is verified by simulation and results analysis.

5 CONCLUSION

Through the kinematics analysis and the trajectory of the food end, this paper plans the trot gait of hydraulic quadruped robot. And then, based on the simulations of ADMAS toward trot gait, the simulations verify the gait rationality, and analyze the simulation results, including the centroid position of the body, the centroid velocity of the body and the position of the foot end. The simulation results validate the rationality of gait planning and provide references for further research on gait of hydraulic quadruped robot.

REFERENCES

Gao, B.W., Shao, J.P. & S, G.T. (2015). Structural Design and Gait Planning of Hydraulically Actuated Quadruped Bionic Robot [J]. *Journal of the Chinese Society of Mechanical Engineers.* 36(5): 421–430.

He, D.Q. & Ma, P.S. (2005). Dynamic Walking Simulation and Walking Stability Analysis of Quadruped Robot [J]. *Computer Simulation.* 22(2): 146–149.

Kurazume, R., Hirose, S. & Yoneda, K. (2001). Feedforward and Feedback Dynamic Trot Gait Control for a Quadruped Walking Vehicle. *IEEE Int. Conf on Robotics and Automation, Korea.* 3172–3180.

Raibert, M. & Chepponis, M. (1986). Running on Four Legs as though They Were One [J]. *IEEE Journal of Robotics and Automation.* 2(2):7.

Yu, H.T., Li, M.T. & Cai, H.G. (2013). Analysis on the Performance of the SLIP Runner with Nonlinear Spring Leg [J]. *Chinese Journal of Mechanical Engineering.* 26(5):892–899.

Electronic Engineering – Wang (ed)
© 2018 Taylor & Francis Group, London, ISBN 978-1-138-60260-1

Design and implementation of a mobile intelligent programming learning platform based on Android

C.Z. Ji & B. Yu
School of Software, Harbin University of Science and Technology, Harbin, China

Y. Wei
Foreign Language Institute, Harbin University of Science and Technology, Harbin, China

Z.L. Song
School of Software, Harbin University of Science and Technology, Harbin, China

ABSTRACT: In order to solve the problem in the programming learning field of a lack of mobile intelligent programming learning platforms both at home and abroad, which help effective study with the 4A method (Anyone, Anytime, Anywhere, Any Style) of mobile learning, we designed and implemented an application software system based on Android mobile intelligent programming – Code Pass. This application software, incorporating high-quality programming learning resources and learning experiences, can support the efficient learning, practice and testing of programming language knowledge. Its most distinctive characteristic is to allow the quick and convenient query of the meaning of programming symbols just by scanning or typing. Meanwhile, it enables developers to enter program code by scanning in order to edit and compile it. By means of a running instance, this new and multinational mobile intelligent programming learning system proved itself effective and practical.

Keywords: Mobile intelligent learning platform; OCR; scanning; enquiring; scanning code

1 INTRODUCTION

With the "Internet Plus" era approaching and mobile Internet technology becoming increasingly mature, the software industry is becoming the key pillar of the global information industry. In the 12th Five-Year Plan, it was proposed that the aim for the current domestic Chinese software industry was to triple production. Therefore, education in software programming has become an essential part of higher education. In 2015 alone, seven million high school graduates entered universities. According to conservative estimates, at least one million of them chose to learn programming. Like all domestic and foreign programming learners, they will encounter two particular difficulties as follows: first, due to the abundance of programming symbols, learners will have difficulty understanding them. Second, it is inconvenient to carry learning resources and programming tools and there is a lack of intelligent, simplified and humanised mobile programming learning platforms. It is these problems that limit the learning of programming and cannot satisfy users' high-efficiency requirements in the Internet Plus era (Peng et al., 2013). This project is carried out to mitigate the above disadvantages of traditional programming learning by designing and implementing an intelligent programming learning platform. Incorporating the innovative idea of Internet Plus, it adopts new-generation technology updating learning system, such as Optical Character Recognition (OCR), a scanning and identifying system, and proposes revolutionary functions to improve the learning platform, thus maximising the efficiency and possibility of learning programming. Using the OCR scanning and typing functions provided by the platform, entry time is dramatically shortened, and it is possible to not only study online and type in, scan, and search keywords, but to practice programming also. This, therefore, considerably increases the efficiency of programming, and facilitates the learners to a large extent.

2 RELEVANT WORKS

Currently, the primary software of mobile programming learning based on Android is C4droid. This software uses the Tiny C Compiler (TCC) as its default compiler, but the GCC compiler can be installed as a plug-in to highlight the syntax. The time taken to compile will be decided by the basic frequency of the CPU. The higher this frequency, the faster the compilation. The 4.7.2 plug-in version of GCC supplies

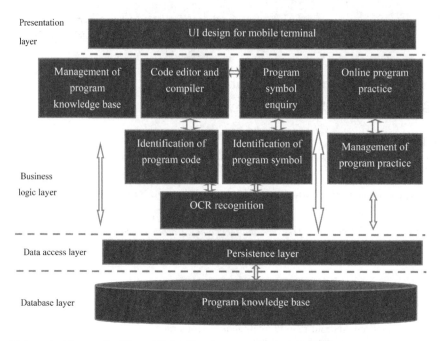

Figure 1. Main content framework of the mobile intelligent programming learning platform.

a sample program, including SDL (Liu et al., 2014), native Android, and Qt libraries, as well as source code. This software made great progress in mobile programming while still having some drawbacks like single function, complex operation and high error rate. For these drawbacks, a mobile intelligent programming system, Code Pass, which is founded on C4droid, embeds the scanning function of OCR and collects high-quality programming learning resources and learning experiences, which efficiently support learning, practice, and tests of learning. It can help students enquire about the meaning of programming symbols, such as function, class, keyword and term, by scanning or typing and they learn useful information from sharing experiences of others' programming symbols learning. Meanwhile, it enables developers to enter program code by scanning in order to edit and compile it. The software increases the efficiency of the program significantly as well as saving the time required for programming, making it possible for users to learn programming anytime and anywhere.

3 OUTLINE DESIGN

The mobile intelligent programming learning platform is based upon a client/server framework, which includes the four layers of presentation, business, data access, and database, respectively. The main content of the framework is shown in Figure 1.

3.1 Mobile user interface design

For the user, the mobile application interface design takes a high priority, and some users believe that the User Interface (UI) is the user's application. The desktop application has a very complex application interface. As the interface screen of the mobile terminal is much smaller than the desktop screen, so the design of the system desktop display controls to do a strict limit, did not take PC input operation.

Due to the instability of the wireless network, the design of the system with a local charge or data cache is better than wireless access to data. In one screen update, the application interface can get more information that will not involve a long waiting time. For different types of mobile terminal, the screen size and resolution are different; the system has been tested for all sizes of mobile phones to be compatible with most phone screen types.

3.2 Programming knowledge base design

Programming knowledge base management: access to the programming database is through the persistence layer module. Learners can use the module to make queries about related programming symbols, such as functions, classes, keywords, terms and other methods of use, and other experienced programmers can share their use of methods. At the same time, learners can upload their own understanding of some programming symbols and experience to the database server. Through others' feedback they can not only improve their own programming level, but can also help other learners solve problems.

For unsatisfied programming knowledge sharing, learners can modify or delete their own programming knowledge. At the same time, the module will help all kinds of documents – for example, e-books – become mobile and much easier for learners to carry

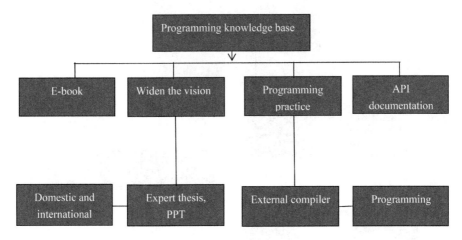

Figure 2. Knowledge base structure.

and learn from. In addition, in the background of the server the administrator can use the module to manage the programming knowledge base. The knowledge base interface includes access to e-book materials and programming exercises, broadening horizons, and access to Application Programming Interface (API) documentation, which as the main video format is well-known at home and abroad, gives access to a large amount of learning material. The structure of the knowledge base is shown in Figure 2.

3.3 Code editor compilation design

This module includes programming symbol recognition and query, and programming code recognition queries. The programming recognition module can be used to resolve programming symbols or code, and then through the persistence layer and access to the programming knowledge base, users can get other learners' answers and share experiences.

Compilation is divided into front-end processing, back-end processing and compiler analysis.

Front-end processing is mainly responsible for parsing the input source code, with the parser and semantic parser working together. The parser is responsible for the source code of 'words' (tokens); the semantic analyser classifies the words according to the predefined grammar and then assembles them into meaningful expressions, statements, functions, and so on.

For example, "a = b + c"; the front-end parser gives "+, a, b, c", and the semantic parser analyses this according to grammatical definitions and first assembles it into the expression "b + c", and then into the full "a = b + c" statement (Zhang et al., 2015) The front-end is also responsible for the inspection of the semantics, such as testing whether to participate in the operation of variable is of the same type and simple error handling.

The final result normally is an abstract syntax tree, so the back-end can carry out further optimisation and handle it on this basis. The back-end compiler is responsible for analysing, optimising and generating the machine code. Generally speaking, all the compiler analysis and optimisation variants can be divided into two categories: either *within* one function or *between* one function and another. Obviously, the analysis between functions, optimisation, is more accurate but takes more time to complete.

Various analyses and optimisations take place at the most appropriate layer of intermediate code.

Common compiler analyses include variable definition and application, definition use chain, variable alias analysis, pointer analysis, and data dependency analysis based on the function call tree and control flow chart.

The results of program analysis are preconditions for compiler optimisation and program transformation.

Common optimisations and transformations cover functions in linking, useless code deletion, normalised loop structures, loop body unwinding, loop merging, splitting, array stuffing, and so on. The purpose of optimisation and transformation is to decrease the length of the code, improve the utilisation of the memory and cache, and reduce the frequency of access to the disk and network. The more advanced optimisation can even convert serialised code into multicoloured code, which supports parallel computation. The generation of machine code is the process of transforming optimised intermediate code into machine instructions. Modern compilers mainly adopt the method of generating assembler code, instead of generating binary object code directly.

Even in the code generation stage, high-level compilers still have to analyse, optimise and transform heavily. For instance, they need to allocate registers, choose the appropriate machine instructions and merge code, among other things. This application is available for programming and compiling in mobile technology.

At the same time, the program code can be input through the character recognition system. If the code is resolved inaccurately, it can be edited manually and code added.

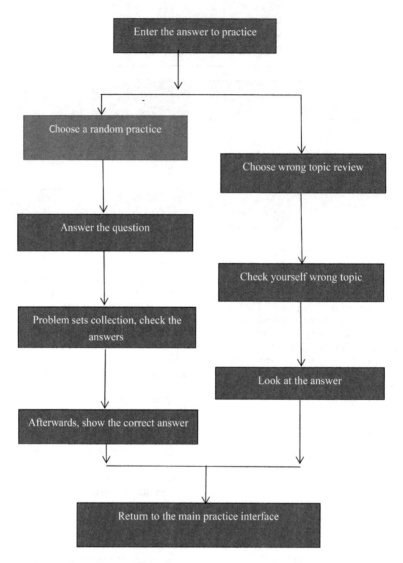

Figure 3. Programming practice management flow chart.

For possible compilation errors, effective solutions can be obtained via the search module for programmed symbols. There are teaching videos at the bottom of the menu on the home page.

As the era of Big Data arrives, video display is not only confined to PCs, and online video in mobile technology is becoming a big trend. Our users can learn to program by watching teaching videos from famous professors, anytime and anywhere. The videos are stored in the network cloud, which take up no local storage space.

3.4 *Online programming design practice and management*

This module includes two parts: online programming practice and programming practice management.

The online programming practice module is used to add, delete and manage programming exercises through the persistence layer. Individual learners can add a title and share good topics in the programming knowledge base. The topic sources mainly include classic programming practice subjects, computer grade examinations, every big company pen test questions, and so on.

The programming practice management module provides the programming practice and can track individual user progress, including such things as studying the wrong topic. This process is shown in Figure 3.

3.5 *Programming query and OCR scanning identification symbols*

The programming symbol or the program code is mainly composed of English letters or numbers; this

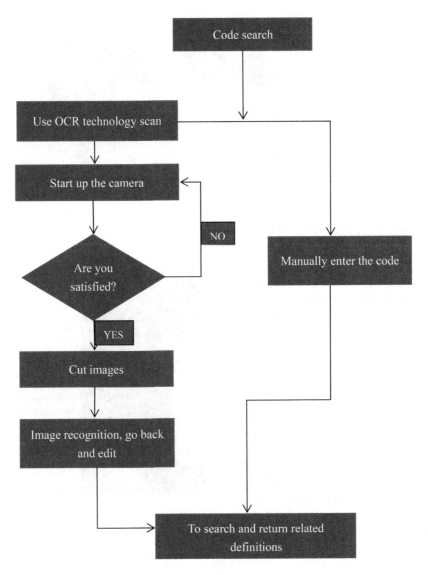

Figure 4. Flow chart of compiled code editor.

project uses OCR technology to complete the input of programming symbols or program code. The use of OCR Software Development Kits (SDKs) for various open source comparison tests, and to acquire programming languages for identification, can produce higher recognition rates. The flowchart is displayed in Figure 4.

3.6 Data access and database management

The third layer is the data access layer, which can also be referred to as the persistence layer and is mainly responsible for adding, deleting, modifying, and querying the programming knowledge information.

The last layer is the database layer, mainly responsible for the information storage of the programming knowledge, using a relational database.

4 IMPLEMENTATION OF THE SYSTEM AND RUNNING RESULTS

The system architecture is based on a client/server framework. The technology architecture used is shown in Figure 2. The functional module is called the business layer and the main functional components are shown in Figure 5. This system uses the MySQL relational database as a background data store. Because MySQL is stateless, once users set up a request, it

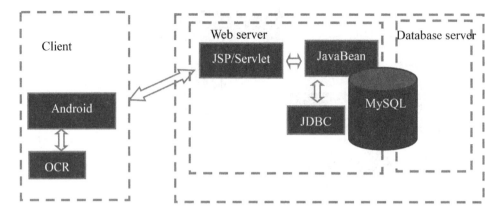

Figure 5. Project technical framework.

will remain active until users opt to exit and the thread will stay live for a long time. Thus, a MySQL server possesses the same hardware resource, which prevents the system from disconnecting when in the operation process.

4.1 Implementation of top-down UI

The top-down design of the mobile programming learning system UI makes it friendly and easy to use, and enables students to learn to program quickly and conveniently. At present, the main characteristic of the graphical UI is a mode employing mended, icons, buttons and windows. The main colour of the system interface is blue. Putting the menu bar at the top of the screen avoids users becoming tired of the bottom menu bar of a traditional mobile phone app. Matching the blue colour with red and yellow avoids visual discomfort because a variety of colours are used and boredom due to single colour usage is avoided.

4.2 Implementation of enquiry in knowledge base

The knowledge base addresses queries about programming codes and symbols. A lot of program language keywords are written into the system, which can give an explanation for the keywords together with the parsing program. Users can enter programming symbols or keywords directly and execute an enquiry.

4.3 Implementation of OCR

OCR technology is mainly divided into scanning, analysing, character-matching, and displaying information. Because the core programming knowledge base contains numerous keywords, users can get program language information through shooting or mobile terminal itself image recognition. First, having placed the client information into a document edit box, the internal processing mechanism will convert it into

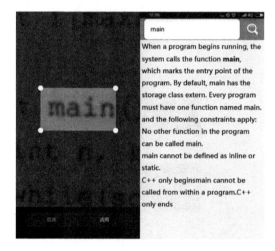

Figure 6. OCR identification code.

a string object. Then the server system will call the API for document matching, and return the string information through HTML. Finally, it will be displayed in the mobile phone client.

4.4 Database access and management

The system is designed to integrate into the C4droid system installation package. The first time users log in, manual installation is required. At the second time of logging into the system, they can program directly. The compiler supports GCC, GNU, G++ GNU and several other compilers. Figure 6 shows the code identification process using OCR.

5 CONCLUSION

Throughout implementation and testing of the various modules, the system realises the original design purpose. This system fulfils the demand for mobile programming learning and supports some functions such as studying online, code scanning, and keyword queries. It overcomes some drawbacks of conventional mobile program software such as single functions and complex operation, giving users a fresher experience. However, this system also has some faults, such as low agility when scanning keywords and distinguishing words from the background, low speed when scanning codes, and low accuracy when identifying. So, on account of these problems, a possible next direction for research is improving the speed and accuracy of scanning to further promote the system's intended function.

REFERENCES

Liu, M., Yang, Z.H., Xie, Y.L., Xie, D.Q. & Tang, C.M. (2014). Android recognition system design and implementation of the graphic synchronization. *Computer Engineering and Design*, 36(6), 2207–2207.

Peng, F.L., Du, X.G., Wang, H.H. & Pu, J.H. (2013). Android mobile phones and network PC communication system design and reality. *Computer Engineering and Design*, 34(7), 2333–05.

Zhang, R., Chen R.X. & Li, T. (2015) Analysis of research and implementation of OCR beyond Android platform. *Industrial Design*, 4, 94–95.

Electronic Engineering – Wang (ed)
© 2018 Taylor & Francis Group, London, ISBN 978-1-138-60260-1

Design and implementation of a UDP/IP hardware module based on a field-programmable gate array

Q. Li, G. Luo & G.M. Song
Chengdu Technological University, Chengdu, China

S.Y. Jiang, F. Yuan & C. Wu
School of Automation Engineering, University of Electronic Science and Technology, Chengdu, China

ABSTRACT: In order to solve the problem of delay and delay jitter that a large volume of real-time data has when the data is transmitted simultaneously, and to reduce the burden for the CPU or microprocessor of servers in handling UDP/IP protocols, this paper proposes a design and implementation scheme for a UDP/IP hardware module based on a Field-Programmable Gate Array (FPGA). The Verilog Hardware Description Language (HDL) supports the implementation of the UDP/IP protocol in an independent hardware module, which relocates the processing of UDP/IP protocols from the software of the server CPU to a hardware chip, reducing the consumption of CPU resources, and enhancing the network rate. Experimental results show that this UDP/IP protocol hardware module can accomplish UDP/IP communication independently and provide a UDP/IP offload engine.

Keywords: Real-time data; UDP/IP protocol; HDL; UDP/IP offload engine

1 INTRODUCTION

The User Datagram Protocol (UDP) is an interface of the Internet Protocol (IP) network and provides connectionless, unreliable but efficient transport services to an upper-level protocol, having the characteristics of real-time data transmission (Li et al., 2013; Qu, 2014). Thus, it plays an important role in mainstream applications. With the rapid development of Internet technology, we need higher quality requirements and real-time data transmission rates for applications such as Voice over Internet Protocol (VoIP) telephony, online games, and video conferencing. The network performance requirements of these applications are high bandwidth, high definition, high quality, high network throughput, low power consumption, low latency, low CPU overhead, and so on (Ge et al., 2011; Du et al., 2015). At the same time as the network bandwidth continues to expand (from the original 10 M/100 M to the present 1 G/10 G), people have a higher requirement for speed. The processing of the traditional Ethernet UDP/IP protocol is completed by software running on the server's CPU or microprocessor. When the speed of the Internet is more than 1 G/10 G, the server will expend huge CPU resources to handle heavy UDP/IP protocols. This not only affects the speed of the server, but also makes the CPU overhead soar, and results in a reduction in the overall performance level of the server (Li & Chen, 2016; Zhou et al., 2013; Cui et al., 2014).

Implementing the UDP/IP protocol in hardware can eliminate the server CPU software processing overhead. It causes UDP/IP protocol processing tasks from the server CPU to be loaded into a separate UDP/IP protocol chip in the hardware. Through this kind of realisation of UDP/IP hardware technology, not only is the burden on server CPU resources reduced, but the rate of network performance is increased as well (Zhang et al., 2014; Chen & Zeng, 2009; Liu et al., 2008; Ye et al., 2009).

2 OVERALL SYSTEM DESIGN

In this paper, the design of a UDP/IP protocol module is principally composed of a UDP sending module (UDP_TX_BLOCK), a UDP receiving module (UDP_RX_BLOCK) and an IP module (IP_BLOCK). The UDP/IP protocol hardware design block diagram is shown in Figure 1.

In Figure 1, the UDP sending module (UDP_TX_BLOCK) is responsible for receiving the user data packet in the application layer and encapsulating, processing and sending the processed datagram to the IP layer. The UDP receiver module's task is to check if there is a datagram in the IP module of the network layer. If the answer is yes, it begins to receive data and calculates the checksum. If the checksum result is correct, the UDP receiver module reverses the process of

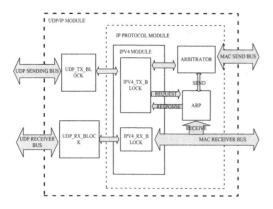

Figure 1. UDP/IP protocol module hardware design.

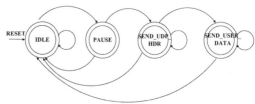

Figure 2. UDP sending state machine.

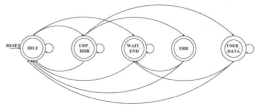

Figure 3. UDP receiving state machine.

encapsulation for the received IP packets to remove the header. Thus, UDP gets a UDP packet that is the same as the sender, and the received UDP packet is passed to the upper application layer according to the data reported in the destination port. If the checksum result is not correct, the UDP receiver discards the datagram and awaits the arrival of the next packet. The IP module in this paper mainly uses IP version 4 (IPV4) protocol, which is widely used in today's Internet (Chen, 2011). This paper realises the IP datagram processing function module through an arbitration mechanism module (ARBITRATOR) and an Address Resolution Protocol (ARP) module.

Thus, our design realises the function of the UDP/IP protocol through the mutual coordination of the three major modules, and establishes a good data communication channel for the application layer and data link layer.

3 DESIGN OF MAIN MODULES

3.1 UDP sending module

The UDP sending module (UDP_TX_BLOCK) is responsible for the entire user data packet reception, encapsulating and sending to the IP layer control network. The UDP sending state machine which realises the function of the UDP sending module is shown in Figure 2. When the application layer needs data to be sent, it first inspects the status of the UDP sending module. If the UDP module is free, it can send data. When the UDP sending module detects the data to be sent, it will put the data into the data buffer first, then add the UDP header in front of the data, that is, UDP encapsulation. Once this data encapsulation is complete, the sending state machine tracks the status of the IP delivery module, when IP send module is idle, after encapsulating data read from the send buffer, transmits to the IP module. It returns to an idle state, awaiting the next arrival of data following the completion of the sending state machine operating cycle. The UDP transmission process has no guarantee of reliability and doesn't need to get any response after the IP layer has received the data message, other than for encapsulation of data, so the whole process of data packet transmission is straightforward.

3.2 UDP receiving module

The UDP receiving state machine is shown in Figure 3. The UDP receiving module (UDP_RX_BLOCK) checks whether there is data in the network layer. If the answer is yes, it begins to receive the data through the unlocked IP layer to remove the IP header of the UDP datagram; the UDP checksum is calculated after receiving the data message, and if the checksum is correct, the UDP receiving module sends the datagram to the corresponding application layer program according to the target port in the datagram. If the checksum is not correct, it discards the datagram accordingly and awaits the arrival of the next datagram.

3.3 IP protocol module

The Internet is the most fundamental, core layer in the Internet protocol. The IP layer is also a connectionless and unreliable packet transmission system (Zhang & Yang, 2006). The design of the IP protocol module consists of four main parts: the IPV4 receiving module (IPV4_RX_BLOCK), the IPV4 sending module (IPV4_TX_BLOCK), the arbitration mechanism module (ARBITRATOR) and the ARP module, as shown in Figure 1.

3.3.1 IPV4 receiving module

When the Media Access Control (MAC) layer has data, the IPV4 receiving module receives the data from the MAC data link layer, unseals it and reads the data. The state of the IPV4 receiving module can be divided into six: IDLE, ETH_HDR, IP_HDR, USER_DATA, ERR, WAIT_END. The specific definitions of each state machine value are shown in

Table 1. IPV4 receiver module state machine definitions.

State machine name	State definition
IDLE	Ready to receive new data
ETH_HDR	Receive Ethernet data check whether type fields of 0x0800 (shows that IP datagram), if they are entering a state IDLE, otherwise enter WAIT_END state.
IP_HDR	Receiving IP header, analysis the version number, whether it is 4, if it is 4 began to receive, if header part information matched, entering a state of USER_DATA
USER_DATA	Receive the data part of the IP datagram.
ERR	When received data is wrong, if there is still has data in the MAC to IP data transmission module, then into the WAIT_END state, otherwise into IDLE state.
WAIT_END	Receives datagram is completed, enter the initial state IDLE is ready to receive new data.

Table 2. IPV4 sender module state machine definitions.

State machine name	State definition
IDLE	Prepare for new IP data. If the length of IP data to be sent is more than 1480 bytes, the IP sending module will divide the datagram into slices; the length of the last data frame will be between 1 and 1480 bytes. Once the previous frame has been sent, the next will be sent.
WAIT_MAC	Check if MAC is ready; if it is ready to enter the SEND_ETH_HDR state, invoke this.
WAIT_CHN	If the MAC gives the received signal, it is ready to send data.
SEND_ETH_HDR	Send Ethernet frame.
SEND_IP_HDR	Send IP header data.
SEND_USER_DATA	Send the user's data; when finished, return to the IDLE state.

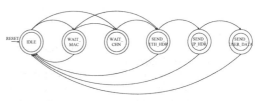

Figure 5. IPV4 sending state machine.

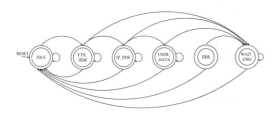

Figure 4. IPV4 receiving state machine.

Table 1. The corresponding state machine diagram is shown in Figure 4.

3.3.2 *IPV4 sending module*

The IPV4 sending module adds an IP header in front of the UDP data which is received from the UDP layer; the UDP datagram forms the data area of the IP datagram. The ARP module maps the IP address of the destination in the IP header to the corresponding physical address. When the data link layer is free, the IP data is transmitted to the MAC with the corresponding physical address (Zhang & Yang, 2006). The IPV4 sending module state machine can be divided into IDLE, WAIT_MAC, WAIT_CHN, SEND_ETH_HDR, SEND_IP_HDR, and SEND_USER_DATA states, as defined in Table 2. The corresponding state machine is represented in Figure 5.

3.3.3 *Arbitration mechanism module*

The arbitration mechanism module is mainly responsible for the arbitration of the handshake signals sent to the bus, and for processing the multiplexed signal. It is the bridge by which the MAC data and IPV4 data transmission block are transferred correctly. Its primary code is as follows:

```
case (state)
  s1: if (req1= = 1) // If signal 1 ask
    allow <= Msg1; // Transmit the
    information of the first channel;
    else if (req2 == 1) // Otherwise, if
    the 2-way signal sends a request
    allow <= Msg2; // Transmit the
    information of the second channel;
  s2: if (req2 == 1) // if signal 2 sends
a request
    allow <= Msg2; // Transmit the
    information of the second channel;
  else
    allow <= Msg1; // Otherwise,
    transmit the information of the first
    channel;
end case;
```

3.3.4 *Address resolution protocol module*

The ARP is located below the IP protocol; it is through the ARP that the IP protocol carries out the conversion between the IP address and the physical address. There are two kinds of addresses in the Internet, namely, the physical address and the IP address. The IP address is encapsulated in the IP message and the physical

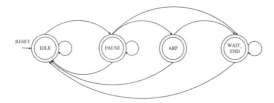

Figure 6. ARP receiving state machine.

address is encapsulated in the MAC frame, according to the protocol. In each forwarding node, the IP layer determines the next forwarding node according to the routing table, obtaining the associated IP address. The IP layer sends the information to the MAC of the data link layer and this process analyses the IP address of the next forwarding node (Zhang & Yang, 2006). The ARP parses the IP address to obtain the corresponding physical address, and the IP datagram provides the data to the MAC of the corresponding data link layer according to the physical address derived in the sending process (Ming et al., 2012; Wang et al., 2013). In the design, the ARP module is much more functional and such modules are more complex to implement, so we only provide a brief introduction to the main ARP receiving module here.

The ARP receiving module is mainly responsible for processing ARP query requests and the ARP message response. At the same time, it receives ARP cache list queries and it is also responsible for the task of updating the ARP cache list. The ARP receiving state machine that realises the ARP receiving function is shown in Figure 6. Each time an ARP reply is received, the host maps the destination IP address and physical address into the ARP cache. If the IP address of the target is already in the ARP list, an ARP valid entry signal is issued; at this point, the corresponding MAC address will be output (Lu, 2012). When new IP information is added to the ARP list, it first checks whether the list has the information already. If the IP information already exists, it will be stored in the latest location of the ARP list, and the address will be updated. If there is no information present in the previous ARP list, the latest IP information will be added and any older information will be deleted.

4 MODULE SIMULATION ANALYSIS

Our design was implemented in the ISE Design Suite 13.3 platform (Xilinx Inc., San Jose, CA). On the basis of the UDP/IP protocol standard, it uses the Verilog Hardware Description Language (HDL) to complete the design, synthesis, simulation, mapping and layout of the whole UDP/IP protocol module, after which it is used to carry out the verification of the UDP/IP protocol modules. Because there are several such modules, we will not introduce them one by one but focus on an analysis of the simulation results as a whole. The design simulation uses a system clock at 125 MHz. When the simulation is initiated, we set the target IP

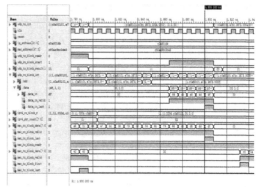

Figure 7. UDP/IP protocol functional simulation.

address to 192.168.1.107. The physical address of the corresponding 48-bit MAC is C8-0A-A9-9A-3D-E8, the target port is "3873", and the source IP address is C0A80101, namely, 192.168.1.1, with a source port of E7BC. Through the course of the simulation, data "d3, d8, dd, df" is sent by the user through port E7BC. After handling by the UDP module, they are sent to the IP module for packaging. Then, based on the ARP, the IP address is parsed to the corresponding physical address, and the corresponding port of the MAC module is received accurately. At the same time, the data sent by the MAC to the UDP/IP module – "40, 41, 42, 43" – can also be properly received by the corresponding port of the UDP module. The simulation results show that this design fulfils the function of the UDP/IP protocol stack. The UDP/IP protocol functional simulation is shown in Figure 7.

5 CONCLUSION

This paper proposes the design and implementation of a UDP/IP protocol module, and the function of each module is simulated and verified. The results show that the design can achieve end-to-end real-time data communication in UDP/IP. The design method is simple, low cost and easy to operate; the hardware performance is stable, has a high transmission efficiency and is easy to use in a network server. If this protocol module function is further improved, such as by the incorporation of the TCP/IP protocol, MAC core, peripheral interface, and so on, it could be made into an independent Integrated Circuit (IC) chip. This would play a positive role in promoting the development of the Internet in China.

ACKNOWLEDGEMENTS

This work was supported by the Program of National Nature Science Foundation of China under Grant Nos. 61471407, 60971036 and 60934002, and the Application Fundamental Research Funds of the Department of Science and Technology of Sichuan Province under Grant No. 2013JY0192.

REFERENCES

Chen, D.N. (2011). Analysis of the IP network of the rapid fault recovery. *Information and Computer (Theoretical Version)*, 5, 98–100.

Chen, X.X. & Zeng, W.H. (2009). Design and implementation of a TCP/IP unloading and lead structure. *Application of Computer Systems*, 1, 132–134.

Cui, H., Liu, Y.Q. & Sheng, J.J. (2014). Research and implementation of IP/UDP protocol stack based on FPGA. *Journal of Changchun University of Science and Technology (Natural Science Edition)*, 2, 133–137.

Du, Q.Y., Qian, S.S. & Li, T. (2015). Data transmission scheme for millimeter wave detection based on UDP/IP protocol. *Microwave Journal*, S2, 307–310.

Ge, L.Y., Wu, Q.L. & Tian, Y.M. (2011). IEEE 802.11e capacity analysis and improvement of VoIP in wireless local area network. *Computer Engineering and Applications*, 18, 73–75.

Li, J. & Chen, S.H. (2016). TCP data receiving and unloading based on multi core. *Computer Engineering and Science*, 38(7).

Li, S.Q., Wang, C.A. & Zhu, W.G. (2013). Reliability assurance of remote data transmission based on UDP. *Automation & Instrumentation*, 1, 142–145.

Liu, W., Bai, R.L. & Yao, Y. (2008). The design of high speed embedded Ethernet interface based on single chip. *Computer Engineering and Design*, 12, 3082–3084.

Lu, W. (2012). 802.11 standard MAC frame parsing. *Communications Management and Technology*, 6, 58–59.

Ming, J.T., Shen, Z. & Xia, D.H. (2012). United MAC/IP header compression design and implementation of the point-to-multipoint communication. *Study on Optical Communications*, 21(3), 14–17.

Qu, X. (2014). Analysis of video conference system based on IP network. *Electronic Production*, 18, 106–106.

Wang, L.Y., Wang, J.F. & Cao, K. (2013). Implementation of VLAN and gigabit Ethernet in MAC. *Radio Engineering*, 1, 1–3.

Ye, H., Zhang, H.S. & Wang, Q. (2009). Embedded TCP communication method for traffic state acquisition. *Chinese Journal of Scientific Instrument*, 4, 685–688.

Zhang, Z.H., Wu, Q.B. & Shao, L.S. (2014). Design and implementation of intelligent platform based on TOE protocol stack. *Computer Technology and Development*, 7, 22–23.

Zhang, Z.K. & Yang, X.H. (2006). *Computer networks* (pp. 212–291). Beijing, China: Tsinghua University Press.

Zhou, J.L., Wang, Z.H. & Jiang, M.H. (2013). Acceleration technology of TCP/IP protocol stack based on multi core processor. *Network New Media Technology*, 1, 58–64.

Electronic Engineering – Wang (ed)
© 2018 Taylor & Francis Group, London, ISBN 978-1-138-60260-1

A method of building CAD models based on algebraic optimisation of geometric constraints

X.Y. Gao & Y.T. Liu
School of Computer Science and Technology, Harbin University of Science and Technology, Harbin, China

C.X. Zhang
School of Software, Harbin University of Science and Technology, Harbin, China

ABSTRACT: Solving geometric constraints automatically is a new trend in the field of Computer-Aided Design (CAD). It can improve the efficiency of building CAD models and decrease costs. In this paper, a feature modelling method is proposed in which geometric constraints in models are transformed into a set of algebraic equations and optimisation algorithms are applied to solve them. At the same time, an architecture for building CAD models based on constraint optimisation solving is described, together with a method of describing two-dimensional geometric constraints. Experimental results show the proposed method is feasible.

Keywords: CAD models; feature modelling; algebraic equations; optimisation algorithms

1 INTRODUCTION

The technology of feature modelling is important in Computer-Aided Design (CAD), Computer-Aided Industrial Design (CAID) and Computer-Aided Manufacturing (CAM). It can improve the efficiency of building CAD models and decrease design costs. Now, many researchers are focusing on this problem both within China and outside it, giving rise to much relevant research.

Thus, for example, a new method of integrating feature modelling technology into complex development environments has been described. The purpose was to extract requirements to which feature modelling methods can be applied in complex systems (Reiser & Kolagari, 2007). CAD systems can also be used to retain the design information of past mechanisms and to reconstruct the engineering technology of previous epochs. This approach can promote knowledge and saves on design costs (Timofeev et al., 2017). Reverse engineering technology has been applied to reconstruct 3D CAD models. For example, Zhang and Wen (2017) scanned the cartoon character Mickey Mouse and obtained a cloud of data points. Reverse engineering software was used to segment the data, and high-level surface building and quality evaluation functions were adopted to reconstruct a 3D model. Similarly, a new method has been described for the model-driven design of deep drawing tools, which represents a graphical modelling language. This method

makes modelling of parametric relations and dependencies easy (Scheffler et al., 2016). Tang et al. (2015) designed a new method to transform CAD data into a virtual reality model. They parsed CAD data in order to obtain mesh data, and then used an improved quadric error metrics algorithm to simplify it. Finally, the mesh data was transformed into CAD data.

Geometric models are often used for simulations. Parametric CAD modelling methods have been utilised to build geometric models in which large areas of deviation are substituted with 3D surface-scanned data (Katona et al., 2015). Similarly, Salmi et al. (2014) proposed a new method to obtain a 3D CAD model from its physical castings. This method consisted of pre-digitising, digitising of equipment parts, surface reconstruction, and 3D CAD modelling.

Geometric constraint solving is an important step in feature modelling. Its quality has a direct influence on the performance of CAD systems. A variable step-size revisable optimisation algorithm has been defined to deal with complex geometric constraints in parametric CAD design. Experimental results show that this method is insensitive to initial variables and has a high convergence (Zhou & Zhang, 2011). Similarly, a framework has been designed to use distance geometry and decomposition methods in CAD systems with the purpose of exploiting dimension requirements and invariance in rigid body motions (Mathis & Schreck, 2014). Geometric knowledge can be applied to solve geometric constraints. This method cannot identify all of the solutions for a given geometric constraint problem but a real solution can be obtained (Imbach et al., 2014).

In this paper, a framework for building CAD models is described, in which a geometric constraint problem is transformed into a problem of solving algebraic equations. Then, optimisation algorithms are utilised to search for the best solutions of these algebraic equations.

2 FRAMEWORK FOR BUILDING CAD MODELS

The framework for building CAD models is shown in Figure 1. In the feature modelling interface, the designer can use commands to create basic geometric entities, including ellipses, rectangles, triangles, lines, echelons, circles, and so on. After the parameters of a geometric entity are input, the shape will be created and drawn automatically. These geometric entities are basic and frequently used. A complex geometric model is composed of these basic entities.

In a CAD model, there are a lot of geometric constraints among the entities. The designer can input these geometric constraints through the interface. A geometric constraint is expressed as a template. The user fills in this template with information on the entities. For a distance constraint, he need only give two points. In order to simplify the process of solving geometric constraints, a complex constraint is transformed into several simple ones. In a simple constraint, there are only two geometric entities.

There are two kinds of geometric constraints: structure and dimension. There are six structure constraints, described as follows:

(1) Parallel constraint, which is expressed as (PAR, L1, L2, A), and describes the relationship in which line L1 is parallel with line L2. When A is +1, line L1 and L2 have the same direction. When A is −1, they have the opposite direction.
(2) Perpendicular constraint, which is denoted as (PER, L1, L2, A), and describes the relationship in which line L1 is perpendicular with line L2. When A is +1, the angle between L1 and L2 is 90°. When A is −1, the angle between L1 and L2 is 270°.
(3) Tangent constraint of two circles, which is expressed as (TAN_CC, C1, C2, A), and describes the relationship in which circle C1 is tangential with circle C2. When A is +1, they are tangential with each other internally. When A is −1, they are tangential with each other externally.
(4) Tangent constraint between line and circle, which is denoted as (TAN_LC, L, C), and describes the relationship in which line L is tangent to circle C.
(5) The constraint that point P is located on line L is expressed as (ON_L, P, L).
(6) The constraint that point P is located in circle C is denoted as (ON_C, P, C).

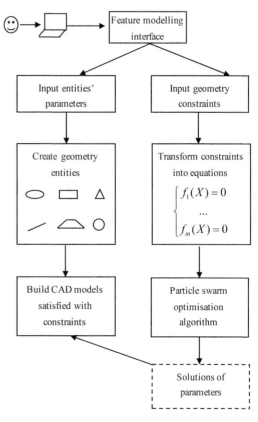

Figure 1. Framework for building CAD models based on geometric constraint solving.

There are five dimension constraints, described as follows:

(1) Distance constraint of two points, which is expressed as (D_PP, P1, P2, d), and describes the distance between point P1 and point P2 as d.
(2) Distance constraint between point and line, which is denoted as (D_PL, P, L, d), and describes the distance between point P and line L as d.
(3) Distance constraint of two lines, which is expressed as (D_LL, L1, L2, d), and describes the distance between line L1 and line L2 as d.
(4) Angle constraint of two lines, which is denoted as (ANG, L1, L2, φ), and describes the angle between line L1 and line L2 as φ.
(5) Radius constraint of a circle, which is expressed as (RAD, C, r), and describes the radius of circle C as r.

Our system uses an algebraic equation to express this geometric constraint. A group of algebraic equations can be obtained. In order to satisfy all geometric constraints, the parameters of some entities need to be determined.

The Particle Swarm Optimisation (PSO) algorithm is applied to find optimal solutions for the geometric entities' parameters. Then, the system draws geometric entities according to their parameters. When the designer modifies this CAD model, they can edit the relevant geometric constraints, delete them and add more. Although there are many tasks in modifying geometric constraints, the ease of building CAD models will be increased automatically.

3 SOLVING GEOMETRIC CONSTRAINTS BASED ON PSO ALGORITHM

The PSO algorithm is a swarm intelligence one involving a population of particles in n-dimensional space. Every particle has its position at current time and moves at some velocity. Its movement at the next time period is influenced by its local best known position and velocity. Its purpose is to move towards its best position in the search space. The best position will be updated iteratively. This can improve a candidate solution with regard to a given measure of quality. This given measure is always defined as an objective function. The purpose of PSO is to find the best solution which optimises the objective function. The best solution is the best particle's position after several iterations.

After every geometric constraint is denoted as an algebraic equation, a group of equations can be obtained. This is shown in Equation 1:

$$\begin{cases} f_1(x_1, x_2, \ldots, x_n) = 0 \\ \vdots \\ f_m(x_1, x_2, \ldots, x_n) = 0 \end{cases} \quad (1)$$

where x_i is the parameter of the geometric element. There are m geometric constraints. The objective function $F(X)$ is constructed as shown in Equation 2:

$$F(X) = \sum_{i=1}^{m} f_i^2(x_1, x_2, \ldots, x_n) \quad (2)$$

The particle swarm optimisation algorithm is applied to find X^*, which gives a minimum objective function. Here, X^* is the solution of the geometric constraint problem.

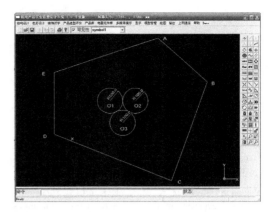

Figure 2. CAD model.

100. There are three geometric constraints, which are shown as follows:

(1) There is an external tangency constraint between circles O1 and O2. This is denoted as (TAN_CC, O1, O2, −1).
(2) Circles O1 and O3 are tangential with each other externally. This is expressed as (TAN_CC, O1, O3, −1).
(3) There is an external tangency constraint between circles O2 and O3. This is denoted as (TAN_CC, O2, O3, −1).

If we build a CAD model, we need to determine centre (x_3, y_3) of circle O3. The associated constraint equations are shown in Equation 3:

$$\begin{cases} (x_3 - 300)^2 + (y_3 - 300)^2 = (100 + 100)^2 \\ (x_3 - 500)^2 + (y_3 - 300)^2 = (100 + 100)^2 \\ (500 - 300)^2 + (300 - 300)^2 = (100 + 100)^2 \end{cases} \quad (3)$$

The objective function $F(X)$ is constructed according to Equation 3. The proposed method is applied to search for the solution (x_3^*, y_3^*) which gives a minimum objective function $F(X)$.

Circle O3 is drawn according to its centre and radius. The design model is shown in Figure 2.

4 EXPERIMENTS

The HUST-CAD system was developed by the Computer Application Institute at Harbin University of Science and Technology. It was developed with Visual C++ 6.0. In this system, users can design two-dimensional (2D) models. In the design interface, the user inputs the parameters of a geometric entity. The system creates an entity automatically according to its parameters. Then, constraint conditions between geometric entities are input by the user.

For example, there are three circles: O1, O2 and O3. The centre of circle O1 has the coordinates (300, 300) and its radius is 100. The centre of circle O2 is (500, 300) and its radius is 100. The radius of circle O3 is

5 CONCLUSIONS

A geometric constraint problem is researched in this paper. An architecture for building CAD models is described, in which geometric constraints are solved. First, constraints between geometric elements are denoted as algebraic equations. Second, the objective function is constructed according to this group of algebraic equations. Third, the particle swarm optimisation algorithm is applied to search for the best solution for the objective function. At the same time, geometric constraints are categorised and expressed. Experiments show that the proposed method is feasible.

ACKNOWLEDGEMENTS

This research was supported by the National Natural Science Foundation of China (61502124) and the Natural Science Foundation of Heilongjiang Province, China (F201420).

REFERENCES

Imbach, R., Schreck, P. & Mathis, P. (2014). Leading a continuation method by geometry for solving geometric constraints. *CAD Computer-Aided Design, 46*(1), 138–147.

Katona, S., Koch, M. & Wartzack, S. (2015). Generating hybrid geometry models for more precise simulations by combining parametric CAD-models with 3D surface scanned geometry inserts. In *Proceedings of the 20th International Conference on Engineering Design* (pp. 11–20).

Mathis, P. & Schreck, P. (2014). Coordinate-free geometry and decomposition in geometrical constraint solving. *CAD Computer-Aided Design, 50*, 51–60.

Reiser, M.O. & Kolagari, R.T. (2007). Unified feature modeling as a basis for managing complex system families. In *Proceedings of the 1st International Workshop on Variability Modelling of Software-Intensive Systems* (pp. 79–86).

Salmi, A., Atzeni, E., Calignano, F., Minetola, P. & Iuliano, L. (2014). Combined reverse engineering and CAD approach for mould modelling in casting simulation. *International Journal of Cast Metals Research, 27*(4), 213–220.

Scheffler, R., Koch, S., Wrobel, G., Pleßow, M., Buse, C. & Behrens, B-A. (2016). Modelling CAD models: Method for the model driven design of CAD models for deep drawing tools. In *The 4th International Conference on Model-Driven Engineering and Software Development* (pp. 377–383).

Tang, Y., Xu, Y. & Yuan, L-L. (2015). Geometry modeling for virtual reality based on CAD data. *Open Cybernetics & Systemics Journal*, 9(1), 2339–2343.

Timofeev, G., Egorova, O. & Grigorev, I. (2017). Applying modern CAD systems to reconstruction of old design. In P. Wenger & P. Flores (Eds.), *New Trends in Mechanism and Machine Science* (pp. 323–331). Cham, Switzerland: Springer.

Zhang, M. & Wen, J.H. (2017). Mickey Mouse 3D CAD model reconstruction based on reverse engineering. In *Proceedings of the International Conference on Intelligent and Interactive Systems and Applications* (pp. 282–289).

Zhou, Y.H. & Zhang, J.X. (2011). A variable step-size revisable optimization algorithm to solve geometry constraint. *Journal of Central South University (Science and Technology)*, 42(5), 1326–1331.

Electronic Engineering – Wang (ed)
© 2018 Taylor & Francis Group, London, ISBN 978-1-138-60260-1

Use of face similarity to compare the shapes of two CAD models

X.Y. Gao & Y.N. Chen
School of Computer Science and Technology, Harbin University of Science and Technology, Harbin, China

C.X. Zhang
School of Software, Harbin University of Science and Technology, Harbin, China

ABSTRACT: Computing model similarity in the field of Computer-Aided Design (CAD) is a new research topic. Model similarity computation improves the performance of CAD model retrieval and decreases human design cost. In this paper, a new method of computing the similarity of CAD models is proposed, in which face similarity is utilised. A CAD model is traversed and the number of edges in every face is obtained. The edge number is used to construct a face similarity matrix. A greedy algorithm is applied to scan the face similarity matrix and an optimisation sequence of face pairs between the source model and target model is extracted. Then, the equation for CAD model similarity is described, and a framework for computing model similarity based on the face similarity matrix is designed. Experiments show that the proposed method can compute the similarity of CAD models efficiently.

Keywords: Model similarity; CAD models; face similarity; greedy algorithm; face pairs

1 INTRODUCTION

The technology of retrieving models from a model library is important in Computer-Aided Design (CAD), Computer-Aided Industrial Design (CAID) and Computer-Aided Manufacturing (CAM). Existing models in the library may be used as a reference for design reuse. The user can utilise old designs in the process of designing a new CAD model, which will decrease the designer's time and costs. Model similarity calculation is very important in CAD model retrieval. With the development of CAD technology, many scholars, both inside and outside China, have conducted a large amount of research on model similarity calculation.

Region property codes have been adopted to represent face regions in a CAD model. The similarity between the source model and the target model is computed by comparing their region property codes. Experiments show that this method can retrieve 3D CAD models efficiently (Tao et al., 2017). Cuillière et al. (2011) described another method of automatically computing two models' similarity, in which shape descriptors based on vector representation are utilised. Their experiments show that the proposed method was efficient. Similarly, Furuya and Ohbuchi (2016) proposed a method to extract a set of local features from a 3D CAD model. These features are then aggregated and pooled, and a local feature is encoded into a highly sparse binary code. These sparse binary codes are summed into a compact feature vector for retrieving

the 3D model. Likewise, Sun et al. (2009) proposed a topology approximation method to retrieve CAD models with partial structures. A uniform measurement was adopted to compute the similarity between boundary faces, and a breadth-first search algorithm was used to search for initial boundaries in a model, which is satisfied by the retrieval condition.

Semantic processing methods and rule processing ones have been adopted to retrieve CAD models with design documents. At the same time, domain ontology and shallow natural language processing technology are used to deal with texts in CAD models. The purpose is to extract hidden design information from CAD models and design documents (Jeon et al., 2016). Hu et al. (2007) applied a global optimal feedback-based method to retrieve CAD models satisfied by the given condition. At the same time, the Particle Swarm Optimisation (PSO) algorithm was adopted in order to improve the performance of retrieval. Similarly, Biasotti et al. (2016) conducted comparative experiments on six methods of 3D model retrieval and classification. In order to compare these six methods, 572 synthetic textured mesh models were collected. The purpose was to deal with specific classes of geometric and texture deformations correctly. Liu et al. (2016) built a multi-modal clique-graph in order to retrieve 3D models. There are a lot of cliques in such a graph. At the same time, hyper-edges are used to link pairwise cliques. They described a new method of computing edgewise similarity, in which an image set-based clique is adopted.

Wang et al. (2014) used global features and local ones to retrieve 3D models. These are extracted from 2D views of 3D models. The global features are adopted to describe shape information of the gross exterior boundary, and local features are used to describe all interior details. Their experiments showed that the proposed method was feasible. Similarly, Wang et al. (2015) designed a sketch-based query interface in order to retrieve 3D CAD models efficiently. They also described a similarity assessment algorithm to calculate shape similarity between user sketches and those of models already contained in the library.

In this paper, a framework for computing the similarity of CAD models is described. The number of edges in a face is extracted from a CAD model. A face similarity matrix between the source CAD model and the target one is constructed. Then, the face similarity matrix is applied to compute the degree of similarity between the source model and the target one.

2 FRAMEWORK FOR CALCULATING MODEL SIMILARITY

The architecture for calculating the similarity between a source CAD model and the target one is shown in Figure 1.

There are a large number of source models in the CAD model library. The CAD model is analysed to get its face number. At the same time, the number of edges in every source face is extracted. The designer inputs the face number of the target CAD model, and also inputs the number of edges in every target face. After the edge numbers between the source face and the target one are compared, a face similarity matrix is constructed. When a greedy algorithm is used to traverse the face similarity matrix, an optimal sequence of face pairs is obtained. A face pair consists of a source face and a target one that are similar to each other in shape. The model similarity is calculated on the basis of this optimal sequence of face pairs. Source models are shown in the user interface in ascending order according to model similarity. The designer selects a source model which best satisfies their design intent.

3 CALCULATING MODEL SIMILARITY OF TWO CAD MODELS

The difference between two faces' edge numbers reflects their shape diversity. The larger the edge number difference, the greater their shape diversity. If there is a large divergence of edge number between source model and that of the target, the probability that they are similar to each other is small. Thus, edge number can be used to evaluate the similarity between source face and target one. The similarity between face f_i and face f_j is calculated as shown in Equation 1:

$$Sim_F(f_i, f_j) = 1 - \frac{|Num(f_i) - Num(f_j)|}{\max(Num(f_i), Num(f_j))} \quad (1)$$

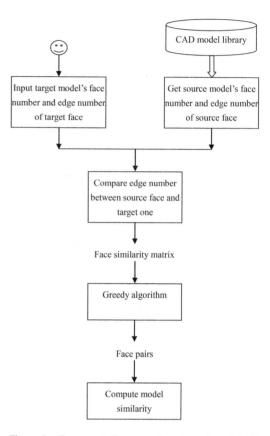

Figure 1. Framework for computing similarity of CAD models.

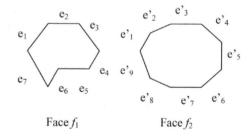

Figure 2. Faces f_1 and f_2.

where $\max(x, y)$ is an operator that selects a maximum value from x and y; $Num(X)$ is the edge number of face X. If $Sim_F(f_i, f_j)$ is larger, face f_i is more similar to face f_j.

For example, f_1 is a face that has seven edges, and f_2 is a face that has nine edges, as shown in Figure 2. The similarity $Sim_F(f_1, f_2)$ between face f_1 and face f_2 is calculated as follows:

$$Sim_F(f_1, f_2) = 1 - \frac{|Num(f_1) - Num(f_2)|}{\max(Num(f_1), Num(f_2))}$$
$$= 1 - \frac{|7-9|}{\max(7,9)}$$
$$= 0.778$$

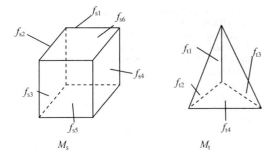

Figure 3. Source model M_s and target model M_t.

The source CAD model is traversed and each source face's edge number is obtained. At the same time, the target CAD model is also traversed and each target face's edge number is obtained. Source faces are viewed as rows and target faces are viewed as columns. Then, a face similarity matrix A is constructed. If the ith source face is completely similar with the jth target one, the value of A[i, j] is 1. If there is a large shape difference between them, the value of A[i, j] is very small. Source model M_s and target model M_t are shown in Figure 3. There are six faces in the source model M_s, including $f_{s1}, f_{s2}, f_{s3}, f_{s4}, f_{s5}$ and f_{s6}. There are four faces in the target model M_t, including f_{t1}, f_{t2}, f_{t3} and f_{t4}.

Based on Equation 1, the face similarity between the source face f_{si} ($i = 1, 2, \ldots, 6$) and target face f_{tj} ($j = 1, 2, \ldots, 4$) is calculated. Then, face similarity matrix A is constructed, which is as follows:

$$A = \begin{array}{c} f_{s1} \\ f_{s2} \\ f_{s3} \\ f_{s4} \\ f_{s5} \\ f_{s6} \end{array} \begin{pmatrix} f_{t1} & f_{t2} & f_{t3} & f_{t4} \\ 0.25 & 0.25 & 0.25 & 0.25 \\ 0.25 & 0.25 & 0.25 & 0.25 \\ 0.25 & 0.25 & 0.25 & 0.25 \\ 0.25 & 0.25 & 0.25 & 0.25 \\ 0.25 & 0.25 & 0.25 & 0.25 \\ 0.25 & 0.25 & 0.25 & 0.25 \end{pmatrix}$$

The greedy algorithm adopts a locally optimal strategy to search for solutions at each stage. The purpose is to find a global optimum. The greedy algorithm can derive locally optimal solutions that approximate a global optimal one in a reasonable time. In this paper, a greedy algorithm is adopted to scan the face similarity matrix to search for a sequence of face pairs that match with each other in shape. Then, an optimal sequence of face pairs is obtained, which is denoted as (f_{s1}, f_{t1}), $(f_{s2}, f_{t2}), \ldots, (f_{sm}, f_{tm})$. The model similarity between M_s and M_t is calculated, as shown in Equation 2:

$$Sim_M(M_s, M_t) = \frac{1}{n} \sum_{k=1}^{m} Sim_F(f_{sk}, f_{tk}) \quad (2)$$

where n is the number of faces in M_t.

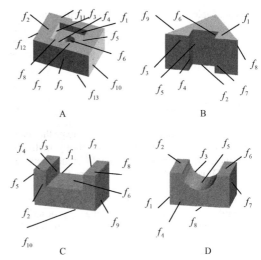

Figure 4. CAD test models.

Table 1. Similarity between source model and model A.

Source model	Face number	Model similarity
A	13	1
B	9	0.633
C	10	0.712
D	8	0.556

4 EXPERIMENTS

HUST-CAD is a feature modelling system. The designer has used it to design 3D CAD models. After many years, a large number of CAD models have been accumulated in its model library. In order to test the proposed method's performance, four models are selected, which are shown in Figure 4.

Model A is the target model. The proposed method is used to compute the similarity between A and the source models. Model similarities are shown in Table 1.

From Table 1, we can see that the value of $Sim_M(A, A)$ is 1. This is because, of course, A is most similar in shape with itself. $Sim_M(C, A)$ is 0.712, indicating a higher similarity of model C than the remaining models. Model A is a block with a square slot and a cylinder slot. Model C is a block with a square slot. We find that the shape of model C is fairly similar to that of model A. $Sim_M(B, A)$ is 0.633, which also indicates fairly high similarity: model B is a triangular prism with a square slot. There are some shape similarities between model A and model B, and we find that the shape of model B is similar to that of model A to some degree. $Sim_M(D, A)$ is 0.556 and the similarity is low: model D is a block with a semi-cylindrical slot. There are significant shape discrepancies between model A and model D.

From Table 1, we can see that similarity results can efficiently evaluate shape differences between model A and the source models, indicating that the proposed method can be successfully applied to CAD model retrieval.

5 CONCLUSIONS

In this paper, edge number is used to evaluate the shape similarity of two faces. A face similarity matrix is constructed to describe the shape difference between two CAD models. A greedy algorithm is adopted to search for an optimal sequence of face pairs between the source model and the target one. At the same time, a formula for computing model similarity is described. Experimental results show that the method proposed in this paper can evaluate effectively the shape diversity of CAD models.

ACKNOWLEDGEMENTS

This research was supported by the National Natural Science Foundation of China (61502124) and the Natural Science Foundation of Heilongjiang Province, China (F201420).

REFERENCES

Biasotti, S., Cerri, A., Aono, M., Ben Hamza, A., Garro, V., Giachetti, A., …Velasco-Forero, S. (2016). Retrieval and classification methods for textured 3D models: A comparative study. *Visual Computer, 32*(2), 217–241.

Cuillière, J.-C., François, V., Souaissa, K., Benamara, A. & BelHadjSalah, H. (2011). Automatic comparison and remeshing applied to CAD model modification. *CAD Computer-Aided Design, 43*(12), 1545–1560.

Furuya, T. & Ohbuchi, R. (2016). Aggregating sparse binarized local features by summing for efficient 3D model retrieval. In *Proceedings of the 2nd IEEE International Conference on Multimedia Big Data* (pp. 314–321).

Hu, B.-K., Liu, Y.-S. & Gao, S.-M. (2007). Parallel global optimal approach of feedback for 3D CAD model retrieval. In *Proceedings of the 10th IEEE International Conference on Computer Aided Design and Computer Graphics* (pp. 132–137).

Jeon, S.M., Lee, J.H., Hahm, G.J. & Suh, H.W. (2016). Automatic CAD model retrieval based on design documents using semantic processing and rule processing. *Computers in Industry*, 77(1), 29–47.

Liu, A.-A., Nie, W.-Z. Gao, Y. & Su, Y.-T. (2016). Multimodal clique-graph matching for view-based 3D model retrieval. *Proceedings of the IEEE Transactions on Image Processing, 25*(5), 2103–2116.

Sun, W., Ma, T.Q., Li, T. & Su, T.M. (2009). Partial retrieval of CAD models based on topology approximation. *Journal of Computer-Aided Design and Computer Graphics*, 21(12), 1805–1813.

Tao, S.Q., Wang, S.T. & Chen, A. (2017). 3D CAD solid model retrieval based on region segmentation. *Multimedia Tools and Applications*, 76(1), 103–121.

Wang, F., Lin, L.F. & Tang, M. (2014). A new sketch-based 3D model retrieval approach by using global and local features. *Graphical Models*, 76(3), 128–139.

Wang, X.X., Wang, J.L. & Pan, W.F. (2015). A sketch-based query interface for 3D CAD model retrieval. In *Proceedings of the 2nd International Conference on Systems and Informatics* (pp. 881–885).

Electronic Engineering – Wang (ed)
© 2018 Taylor & Francis Group, London, ISBN 978-1-138-60260-1

The research status of direct imaging of exoplanets

Z.J. Han

National Astronomical Observatories/Nanjing Institute of Astronomical Optics & Technology,
Chinese Academy of Sciences, Nanjing, China
Key Laboratory of Astronomical Optics & Technology, Nanjing Institute of Astronomical Optics & Technology,
Chinese Academy of Sciences, Nanjing, China
University of Chinese Academy of Sciences, Beijing, China

D. Ren

Physics & Astronomy Department, California State University, Northridge, USA
National Astronomical Observatories/Nanjing Institute of Astronomical Optics & Technology,
Chinese Academy of Sciences, Nanjing, China
Key Laboratory of Astronomical Optics & Technology, Nanjing Institute of Astronomical Optics & Technology,
Chinese Academy of Sciences, Nanjing, China

J.P. Dou, Y.T. Zhu, X. Zhang, J. Guo & C.C. Liu

National Astronomical Observatories/Nanjing Institute of Astronomical Optics & Technology,
Chinese Academy of Sciences, Nanjing, China
Key Laboratory of Astronomical Optics & Technology, Nanjing Institute of Astronomical Optics & Technology,
Chinese Academy of Sciences, Nanjing, China

ABSTRACT: The detection and research of exoplanets has always been a hot topic in the astronomical field. Most exoplanets have been found by indirect methods, such as the transit and radial velocity methods. The direct imaging of exoplanets can provide a large amount of information, for example, the surface temperature, age, and atmospheric characteristics. The contrast limit of direct imaging in space is estimated to be 10^{-10}, offering reliability in the detection of giant planets and Earth-like planets in the habitable zone of Sun-like stars. This paper reviews the research status of the direct imaging instruments based on space. The instruments' status, objects of observation and the current progress in the field are introduced in respect of coronagraphic and polarimetric imaging. The classification and properties of the polarimetric modulation components are discussed, and the progress of the domestic Chinese research groups is described. Finally, the trends in exoplanet direct imaging are discussed in relation to space observation projects.

Keywords: Exoplanets; direct imaging; coronagraph; polarimetry; space observation project

1 GENERAL OVERVIEW OF EXOPLANET DETECTION

Currently, the Earth is considered to be the only confirmed habitable planet. Finding a habitable planet outside the solar system has always been a desired goal of astronomical observation, which makes it the most favoured topic in the astronomical field. The instrumental and theoretical research of exoplanets has been developing rapidly and significantly in recent years. Mayor and Queloz (1995) found a planet around the 51 Pegasi star using the radial velocity method. This was the first time that mankind had found an exoplanet around a Sun-like star. The period of the planet is only four days, which represented a significant break with the previous knowledge of planets. Since then, the direction of exoplanets detection has been pushing towards low-mass planets with the development of improved instrument and detection technology. Based on the transit method, the launch of the space projects CoRoT (Barge et al., 2006) and KEPLER (Borucki et al., 2008) became a milestone of exoplanet detection, and brought an explosion in growth of the number of exoplanets identified. The transit and radial velocity methods have been the most popular methods in exoplanet detection.

As of May 2017, 3,610 exoplanets had been detected (http://exoplanet.org/). The masses of most certified exoplanets are similar to that of Jupiter. Figure 1 shows the relationship of mass (in units of 1 Earth-mass) to the semi-major axis distance (in the unit of 1 Astronomical Unit, i.e. the distance between the Earth and the Sun, about 1.5×10^{11} m) for the certified exoplanets. The large green circle represents the planets in the solar system, and the blue circle with the letter E indicates the Earth. The red triangle, blue triangle,

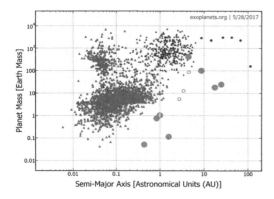

Figure 1. The relationship of the mass and the semi-major axis of the certified exoplanets.

blue hollow circle and blue circle represent the exoplanets found by transit, radial velocity, microlensing and direct imaging methods.

The detection methods of exoplanets are basically classified as indirect and direct methods. Most exoplanets are found by radial velocity and transit methods in the field of indirect imaging, with a small number of exoplanets found with the microlensing method. The radial velocity method measures the shift of the stellar spectrum, which can easily detect the small-track, huge-mass exoplanets. With the help of the spectrometer, the measurement precision reaches 3 m/s (Butler et al., 1996). The precision of the radial velocity method is estimated to be better than 1 m/s, and is aimed at detecting rocky planets around Sun-like and low-mass stars (Kasting et al., 1993). This brings challenges for the imaging precision and quantum efficiency of the instruments. Another major approach in exoplanet detection is the transit method. This measures the light shift of a star when a planet passes in front of it and enables the identification of information such as the track surface of the planet in order to certify candidate exoplanets. The afore-mentioned CoRoT and KEPLER projects use this method to identify exoplanets. The transit method is appropriate for detecting giant planets and some small-scale Earth-like planets.

However, the present methods for detection are all based on stellar light shift signals. The associated rates of misjudgement are unavoidable, which means the methods cannot detect exoplanets with any high degree of precision. The direct imaging method can capture and record the photon signals from exoplanets effectively. Compared with the indirect methods, the direct imaging method is much more reliable in the field of exoplanet information detection.

2 THE CORONAGRAPHIC METHOD OF EXOPLANET DETECTION

There are difficulties in direct imaging exoplanet detection. The results are extremely sensitive to the remote distance of the exoplanets and the ultra-high magnitude of the central star. Direct imaging by use of a coronagraph can suppress the stellar brightness effectively, enabling the detection of dark and faint planets around a bright star.

The principle of the coronagraph comes from the solar coronagraph. The stellar light is suppressed and modulated by special apertures or pupils, which makes it possible to detect the light of the dark and faint planets. Thus, high-contrast imaging is crucial for the coronagraph. The current approaches of modulation consist of a phase-introduced system, a system based on the modulation of aperture shape, and a system based on the stepped-transmission filters. There are advantages and disadvantages to these approaches, which are readily affected by optical aberrations and speckle noise. This makes digital image processing significant in coronagraphic imaging.

The research group of the Nanjing Institute of Astronomical Optics and Technology (NIAOT) at the Chinese Academy of Sciences (CAS) has made a breakthrough in the field of exoplanet direct imaging by coronagraph. In 2015, the group achieved an imaging contrast of 10^{-9} by using a stepped-transmission filter (Liu et al., 2015). Other international research groups have used Deformable Mirrors (DMs) to correct the photon noise caused by the telescopic diffraction and speckle noise caused by the aberrations of the optical components. However, the dark imaging area is not large enough for efficient exoplanet detection. The NIAOT research group proposed a technique to achieve high-contrast imaging in a larger working area. A Spatial Light Modulator (SLM) is applied in the instrument, which creates an equal effect to that of DMs and achieves a contrast of 10^{-10} in the space simulation (Dou & Ren, 2016). The imaging contrast is around 10^{-7} when speckle noise is great, and the red line shows a 10-billion-times imaging contrast in a large working area of 4–45 λ/D after SLM correction.

The coronagraph system is easily built and debugged. It also provides a large working area and an opportunity to apply observations to a wide spectrum range. The observation objects of a coronagraph are classified into three types: direct imaging of certified exoplanets; small-track long-period planets; young giant planets (Guyon et al., 2006). The final observation goal for the coronagraph is to realise the detection of an Earth-mass planetary system in the habitable zone of a Sun-like star.

3 POLARIMETRIC MEASUREMENT OF THE EXOPLANETS

The polarimetric method provides the possibility of direct imaging. The reflected light from the planets provides polarised signals, which can supply a large amount of information about the planet itself. The polarimetry supports the detection of exoplanets, circumstellar discs, debris discs, stellar jets, and other circumstellar media (Schmid et al., 2005). A stellar halo is generated by the telescopic instruments, but

Table 1. Summary of polarimetry instruments.

Research institute	System name	Modulation component
Max Planck Institute	Polarimetric Littrow Spectrograph	1/4 waveplate
National Astronomical Observatory of Japan	High-speed Rotating Waveplate Polarimeter	1/4 waveplate
Paris Observatory	Spectro-Polarimetric Imaging and Characterisation	1/4 waveplate
Zurich Observatory	Zurich Imaging Polarimeter I & II	PEM
NIAOT, CAS	High-sensitivity and High-accuracy Polarimeter	LCVR
ETH, Switzerland	Zurich Imaging Polarimeter III CHEOPS & SPHERE	FLC
William Herschel Telescope	Extreme polarimeter	FLC
E-ELT	EPOL-EPICS	FLC

the polarimetry can obtain useful information from the halo itself. The polarisation components are demodulated from the light intensity, and information such as polarisation degree and surface magnetic field can be deduced. The surface roughness and the atmospheric characteristics can also be deduced from the polarisation components.

Polarimetry for exoplanet detection began with the Zurich Imaging Polarimeter (ZIMPOL) instrument of the Zurich Observatory around 1980 (Roelfsema et al., 2010). This project involved a simple design for the ZIMPOL I instrument with polarisation components. With the development of the polarisation and crystal components, the modulation components of polarimetry shifted rapidly. The current polarimetry instruments and their modulation components are listed in Table 1.

From Table 1, it can be seen that the main modulator components consist of a rotating waveplate, including the Piezo-Elastic Modulator (PEM), a Liquid Crystal Variable Retarder (LCVR) or a Ferroelectric Liquid Crystal (FLC). Their principles and instrumental application will be discussed briefly in the following sections.

3.1 Rotating waveplate

The rotating waveplate method involves the rotation of a waveplate with mechanical structures to achieve the modulation. When the waveplate is rotated at a certain angle, the light intensity will be modulated, and then images can be acquired. The images will be combined and calculated with a Fourier analysis to realise the demodulation process. However, errors are introduced by mechanical error and vibration when rotating the waveplate. The Fourier analysis needs a huge amount of calculation, which adds an overhead to the real-time display of the images and results, and these are key factors in restricting the development of the rotating waveplate method.

Until now, the rotating waveplate method is applied on the Polarimetric Littrow Spectrograph (POLIS) of the Max Planck Institute (Beck et al., 2005), the High-speed Rotating Waveplate Polarimeter (HRWP) of the National Astronomical Observatory of Japan

(Hanaoka, 2012), and the Spectro-Polarimetric Imaging and Characterisation (SPICES) of the Paris Observatory (Maire et al., 2012). The imaging contrast of SPICES is about 10^{-3}, which remains the same level as other instruments. The instruments with rotating waveplate all have a low modulation frequency. The average level of modulation frequency is about 3–4 Hz, except for that of the HRWP, which can reach 30 Hz.

3.2 PEM method

Some polarimetry instruments apply the PEM as their modulator. In the ZIMPOL I and II instruments of ETH Switzerland (Keller et al., 1994), the PEM is applied with a linear polarisation plate and beam splitter as the modulator of incident light.

Two PEMs are situated at angles of 0° and 45°, while the Beam Splitter (BS) is placed after the Linear Polarisation Plate (LIN) to realise the dual-beam imaging. A specially masked high-speed Charge-Coupled Device (CCD) camera is introduced to improve the acquisition speed (Schmid et al., 2012). One-quarter of the CCD surface collects the incident photons, while the other three-quarters transports the electron signal. The specially designed camera makes the imaging contrast stable at 10^{-3}, while the modulation and demodulation speeds are less than 100 Hz.

3.3 LCVR

Besides the rotating waveplate method and the PEM method, the modulation process can be achieved with the characteristics of crystal. With the LCVR, different phase delays are generated with different applied voltages in order to provide different modulation schemes. The polarisation components can be generated by several LCVRs simultaneously. The numerical solution for the polarisation components is accessed easily, without any Fourier analysis, but with different combinations of light intensity. The LCVR reaches its modulation frequency at several tens of Hz. With a fast modulation speed, the imaging contrast can be gradually improved.

The High-contrast High-resolution Imaging Research Group of NIAOT, CAS, proposed a

High-sensitivity and High-accuracy Polarimeter (HHP) instrument with two LCVRs as its modulator (Guo et al., 2017). As well as the LCVR modulation process, a Wollaston Prism (WP) is involved to achieve dual-beam imaging and high-speed modulation. The instrument attains a contrast of 4×10^{-3} and a modulation frequency of 40–50 Hz, which exceeds the performance of other polarimetry instruments using PEM and rotating waveplate methods. The instrument aims to use a camera with a higher frame frequency, which increases the demodulation frequency. The estimated contrast can reach more than 5×10^{-4} and the modulation frequency can reach 200 Hz.

3.4 FLC

Compared to LCVR, FLC has a much faster shifting speed. The crystal cell is disordered when no voltage is applied to the FLC. If a high or low voltage of certain values is applied, the fast-axis direction of the FLC will shift to $0°$ or $45°$. The shifting frequency of the fast axis of the FLC can reach over 1000 Hz, which gives the possibility of fast-speed modulation in polarimetry.

International research groups have been making progress on FLCs. Multiple instruments have applied FLCs as their modulator. The highest imaging contrast is achieved by Characterising Extrasolar Planets by Opto-infrared Polarisation and Spectroscopy (CHEOPS) (Gisler et al., 2004) and Spectro-Polarimetric High-contrast Exoplanet Research (SPHERE) (Thalmann et al., 2008) of ZIMPOL, and the Extreme Polarimeter (ExPo), which is intended for installation on the William Herschel Telescope (WHT) (Rodenhuis et al., 2012). In CHEOPS and SPHERE, FLC and 1/2 waveplate are involved, and a beam splitter is also introduced for dual-beam imaging. ExPo uses a three-FLC achromatic combination with a beam splitter for dual-beam imaging. However, CHEOPS and SPHERE use a specially made mask camera, while ExPo uses a commercial Electron-Multiplying (EM) CCD camera. The former is expensive and involves a long research period, but it has less photon noise. The latter is more replicable as a commercial product. The contrast of these instruments can reach a level of 10^{-5}.

4 OUTLOOK FOR DIRECT IMAGING OF EXOPLANETS

International research groups have been carrying out projects on exoplanet direct imaging based on space. In terms of the observation objects, the planet range for direct imaging starts with young Jupiter-like planets and will expand gradually to Earth-like planets in the habitable zone of Sun-like stars. NASA has proposed a series of space programmes, including ACCESS (A Coronagraph Concept for the Direct Imaging and Spectroscopy of Exoplanetary Systems) and EXCEDE (Exoplanetary Circumstellar Environments and Disc Explorer) (Trauger et al., 2012), which

aim at observing Earth-like and rocky planets in space.

Funded by the Strategic Priority Research Program of the Chinese Academy of Sciences, the High-contrast High-resolution Imaging Research Group of NIAOT, CAS, will continue in their research of the coronagraphic direct imaging of exoplanets (Dou et al., 2014). This will be the preparation for the astronomical satellites JEEEDIS (Jupiter/Earth-twin Exoplanets and Exo-zodiacal Dust Imager and Spectrometer) and ELSS (Exo-Life Search Satellite). As for the polarimetry instruments, the NIAOT research group is considering using FLC as a modulator. The faster modulation speed is estimated to deliver an imaging contrast of 10^{-5}, which brings benefits for exoplanet direct imaging.

In the past few decades, the technology in exoplanet direct imaging has made extraordinary progress. The higher imaging precision has brought convenience for scientific researchers. The knowledge of exoplanets can go further with deeper research into planet formation and deduction. Curiosity about unknown worlds is the motivation for this research and technology. Perhaps, one day, people will find another "Earth" using the direct imaging method, but there is probably still a long way to go.

ACKNOWLEDGEMENTS

This work was supported by the National Natural Science Foundation of China (NSFC) (Grant Nos. 11661161011, 11433007, 11220101001, 11328302 and 11373005), the Strategic Priority Research Program of the Chinese Academy of Sciences (Grant No. XDA04075200), the International Partnership Program of the Chinese Academy of Sciences (Grant Nos. 114A32KYSB20160018 and 114A32KYSB20160057), as well as the special fund for astronomy of CAS (2015–2016). Part of the work described in this paper was carried out at California State University, Northridge, with support from the Mt. Cuba Astronomical Foundation.

REFERENCES

Barge, P., Léger, A., Ollivier, M., Rouan, D., Schneider, J. & Exoplanet CoRoT Team. (2006). Photometric search for transiting planets. In M. Fridlund, A. Baglin, J. Lochard & L. Conroy (Eds.), *Proceedings of the CoRoT Mission Pre-Launch Status – Stellar Seismology and Planet Finding (ESA SP-1306)* (pp. 83–92).

Beck, C., Schmidt, W., Kentischer, T. & Elmore, D. (2005). Polarimetric Littrow Spectrograph - instrument calibration and first measurements. *Astronomy and Astrophysics*, 437(3), 1159–1167.

Borucki, W.J., Koch, D., Batalha, N., Caldwell, D., Christensen-Dalsgaard, J., Cochran, W.D., …Rowe, J. (2008). KEPLER: Search for Earth-size planets in the habitable zone. In F. Pont, D. Sasselov & M. Holman (Eds.), *Transiting Planets: Proceedings IAU Symposium No. IAUS253* (pp. 289–299).

Butler, R.P., Marcy, G.W., Williams, E., McCarthy, C., Dosanjh, P. & Vogt, S.S. (1996). Attaining Doppler precision of 3 m s^{-1}. *Publications of the Astronomical Society of the Pacific*, 108(724), 500–509.

Dou, J.-P. & Ren, D.-Q. (2016). Phase quantization study of spatial light modulator for extreme high-contrast imaging. *The Astrophysical Journal*, 832(1).

Dou, J.-P., Zhu, Y.-T. & Ren, D.-Q. (2014). Current research status of exoplanets. *Chinese Journal of Nature,* 36(2), 124–128.

Gisler, D., Schmid, H.M., Thalmann, C., Povel, H.P., Stenflo, J.O., Joos, F. ...Zinnecker, H. (2004). CHEOPS/ZIMPOL: A VLT instrument study for the polarimetric search of scattered light from extrasolar planets. *Proceedings of SPIE – The International Society for Optical Engineering*, 5492, 463–474.

Guo, J., Ren, D.-Q., Lin, C.-C., Zhu, Y.-T., Dou, J.-R., Zhang, X. & Beck, C. (2017). Design and calibration of a high-sensitivity and high-accuracy polarimeter based on liquid crystal variable retarders. *Research in Astronomy and Astrophysics*, 17(1), 8.

Guyon, O., Pluzhnik, E.A., Kuchner, M.J., Collins, B. & Ridgway, S.T. (2006). Theoretical limits on extrasolar terrestrial planet detection with coronagraphs. *Astrophysical Journal Supplement Series,* 167(1), 81–99.

Hanaoka, Y. (2012). Polarimeter with a high-speed rotating waveplate for the solar observation. *Proceedings of SPIE – The International Society for Optical Engineering*, 8446, 844670.

Kasting, J.F., Whitmire, D.P. & Reynolds, R.T. (1993). Habitable zones around main sequence stars. *Icarus*, 101(1), 108–128.

Keller, C.U., Povel, H.-P. & Stenflo, J.O. (1994). Zurich Imaging Stokes polarimeters I and II. *Proceedings of SPIE – The International Society for Optical Engineering*, 2265(1), 222–230.

Liu, C.C., Ren, D.-Q., Dou, J.-P., Zhu, Y.-T., Zhang, X., Zhao, G., ...Chen, R. (2015). A high-contrast coronagraph for direct imaging of Earth-like exoplanets: Design and test. *Research in Astronomy and Astrophysics*, 15(3), 453–460.

Maire, A.-L., Boccaletti, A., Schneider, J., Galicher, R., Baudoz, P., Stam, D.M., ...Gratton, R. (2012). SPICES: A 1.5-m space coronagraph for spectro-polarimetric characterization of cold exoplanets. *Proceedings of SPIE – The International Society for Optical Engineering*, 8442, 84420I.

Mayor, M. & Queloz, D. (1995). A Jupiter-mass companion to a solar-type star. *Nature*, 378(6555), 355–359.

Rodenhuis, M, Canovas, H., Jeffers, S.V., Ovelar, M.J., Homs, L., Min, M. & Keller, C.U. (2012). The extreme polarimeter: Design, performance, first results and upgrades. *Proceedings of SPIE – The International Society for Optical Engineering*, 8446, 84469I.

Roelfsema, R., Schmid, H.M., Pragt, J., Gisler, D., Waters, R., Bazzon, A., Baruffolo, A. ... Wildi, F. (2010). The ZIMPOL high-contrast imaging polarimeter for SPHERE: Design, manufacturing, and testing. *Proceedings of SPIE – The International Society for Optical Engineering*, 7735, 77354B.

Schmid, H.M., Beuzit, J.-L., Feldt, M., Gisler, D., Gratton, R., Henning, T., ...Waters, R. (2005). Search and investigation of extra-solar planets with polarimetry. *Proceedings of The International Astronomical Union*, 1(C200), 165–170.

Schmid, H.M., Downing, M., Roelfsema, R., Bazzon, A., Gisler, D., Pragt, J., ...Wildi, F. (2012). Tests of the demodulating CCDs for the SPHERE/ZIMPOL imaging polarimeter. *Proceedings of SPIE – The International Society for Optical Engineering*, 8446, 84468Y.

Thalmann, C., Schmid, H.M., Boccaletti, A., Mouillet, D., Dohlen, K., Roelfsema, R. ...Wildi, F. (2008). SPHERE ZIMPOL: Overview and performance simulation. *Proceedings of SPIE – The International Society for Optical Engineering*, 7014, 70143F.

Trauger, J., Moody, D., Gordon, B., Krist, J. & Mawet, D. (2012). Complex apodization Lyot coronagraphy for the direct imaging of exoplanet systems: Design, fabrication, and laboratory demonstration. *Proceedings of SPIE – The International Society for Optical Engineering*, 8442, 84424Q.

Electronic Engineering – Wang (ed)
© 2018 Taylor & Francis Group, London, ISBN 978-1-138-60260-1

Time-resolved spectrum research of laser induced Mg plasma emission

Z.Y. Yang & Q. Li
Lianyungang Normal College, Lianyungang, Jiangsu, China

D.Q. Yuan
Huaihai Institute of Technology, Lianyungang, Jiangsu, China

W.Q. Ni & H.B. Yao
Jiangsu University, Zhenjiang, Jiangsu, China

ABSTRACT: A series of time-resolved spectral lines of Mg alloy was obtained under nanosecond laser shock induced by a pulsed Nd: YAG laser (1,064 nm, maximum energy 500 mJ), which was taken under standard atmospheric pressure and at room temperature. The electron temperature of Mg plasma was calculated by the measured relative emission-line intensity (Mg I 383.2 nm, Mg I 470.3 nm, Mg I 518.4 nm) and rules of the electron temperature were analysed. By contrasting the lines of Mg I 518.4 nm and Mg I 383.2 nm with the biggest relative spectral strength, it was found that relative strength had decreased trend within 100 ns of delay time, although it fell basically in a state of fluctuation. It was believed that in that range of delay time, plasma was at the formation stage, the electron and atomic transition process was in an unstable stage, and the relative strength of plasma did not change in a strict regularity. When the delay time was between 0 ns and 250 ns, electronic temperature reduced rapidly by 7,127 K, and the most obvious and quickly change happened in the delay time of 50 ns. When delay time was between 250 ns and 500 ns, the decreasing speed of electronic temperature decreased quickly. It was believed that because of increased plasma composite radiation, the energy released by composite radiation had a compensating effect for the expansion and cooling of plasma and, simultaneously with the increased extended space of plasma, the collision probability between free electron, atoms, molecule and plasma was reduced, then heat energy could not transform into kinetic energy. By observing spectra of Mg I 518.4 nm which had the biggest relative strength, it was believed that the service life of Mg plasma in this experiment was about 50 µs because the characteristic spectral lines could not be obtained when the delay time was 50 µs.

Keywords: laser plasma, Mg emission spectra, time-resolved, electron temperature

1 INTRODUCTION

Quantitative analysis of chemical elements using a plasma emission spectrum (also known as Laser Induced Breakdown Spectrum (LIBS)) is a very important application (Cristoforetti & Tognoni, 2013; Hahn & Omenetto, 2012; Essington et al., 2009; Bolger, 2000; Tong et al., 2011). Electronic temperature is a very important direction in plasma research. In recent years, attention has been paid to the study of the time distribution and spatial distribution of electron temperature. De Giacomo et al. (2005) studied the parameters of electron temperature and electron density for the temporal evolution of Ti plasma under experimental conditions of standard atmospheric pressure. Zheng Peichao and others (2014) studied the time-resolved spectra of Al plasma, and obtained the time law of electron temperature. The plasma properties of different elements are completely different, but most of the research focuses on elements such as Ti, Al and Cu, and the study of Mg as a future metal is

necessary for the study of its plasma properties. In this paper, the time-resolved spectra of Mg at 500 mJ were obtained by plasma emission spectroscopy. The electron temperature of Mg plasma was calculated and the time evolution of electron temperature was analysed.

2 EXPERIMENTAL DEVICE

The experimental equipment of this experiment mainly by the laser source, light collector, grating spectrometer, DG645 delay and other components, as shown in Figure 1 is the experimental light path.

The experiment was carried out at room temperature and standard atmospheric pressure. The laser uses a Q-switched Nd: YAG solid-state laser. The output wavelength can be switched between 1064 nm and 532 nm. The maximum output energy is 500 mJ, the laser spot diameter is 1 mm, for 12 ns. The output frequency can be set manually and the maximum output frequency is 10 Hz. The plasma radiation signal is

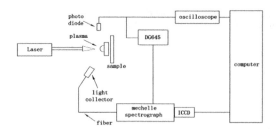

Figure 1. Experimental setup for time evolution of the laser induced plasma.

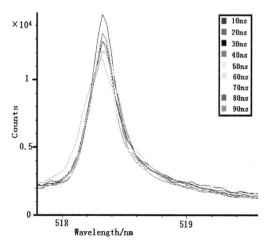

Figure 2. Spectrum of Mg I 518.4 nm when delay time was between 10 ns and 90 ns.

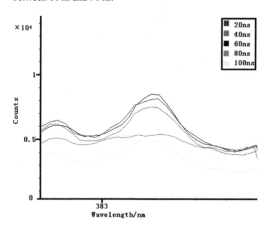

Figure 3. Spectrum of Mg I 383.2 nm when delay time was between 0 ns and 100 ns.

collected by the light collector, and the optical fibre is used as the channel for the signal transmission, connected with the light collector and ME5000 spectrometer. The resolution of the spectrometer is 0.1 nm, the detection range is 200 nm~900 nm, calibrated by a standard mercury argon lamp. The operating temperature of the spectrometer is −20°C. Spectrometer connected with the ICCD camera, DG645 delay control ICCD work, and the photodiode as a trigger signal, control DG645 delay.

3 EXPERIMENT RESULTS AND ANALYSIS

Figures 2 and 3 show the spectra of different delay intervals within 100 ns when the maximum energy of the laser is 500 mJ under the above experimental conditions. Select the 518.4 nm and 383.2 nm two relative intensity of the highest wavelength as the object of comparison.

Figure 2 and Figure 3 show that as the delay time increases, the plasma radiation intensity will decline, indicating that in 100 ns time, the plasma activity has begun to weaken. The collision strength of free electrons, atoms and atomic clusters has a certain slowdown, the brems of radiation reduction. It can be seen that the delay time is not strictly in accordance with the delay time increases, the relative intensity decreases trend, Figure 2, the wavelength of 518.4 nm near the delay time of 50 ns, the relative intensity was significantly greater than the delay time of 40 ns, only slightly lower than the delay time of 10 ns. In Figure 3, the same situation occurs in the delay time of 60 ns, 383.2 nm at the spectral intensity was significantly greater than the other delay time. It can be considered that this is because the plasma is still in the forming stage at 100 ns, and the transition process of electrons and atoms is still in the unstable state, which causes the radiation intensity to fluctuate.

In this study, the Boltzmann slope method is used to calculate the plasma electron temperature, and the time evolution rule is obtained. The formula is as follows:

$$\ln\left(\frac{I_{12}\lambda_{12}}{g_{12}A_{12}}\right) = -\frac{\Delta E}{k_B T} \qquad (1)$$

The characteristic lines with relatively strong relative intensity are selected to minimise the influence of background noise radiation on the calculation results. Mg I 383.2 nm, Mg I 470.3 nm and Mg I 518.4 nm were selected as the data for calculating the plasma temperature. Table 1 lists the calculation of the required,

Table 1. Statistical weight, transition probability and excitation energy.

	Wavelength /nm	Upper level energy /cm^{-1}	Excitation energy /ev	gk	Ak /10^8s^{-1}
Mg I	383.2	47957.027	5.95	15	1.21
Mg I	470.3	56308.381	6.99	5	0.22
Mg I	518.4	41197.403	5.11	3	0.56

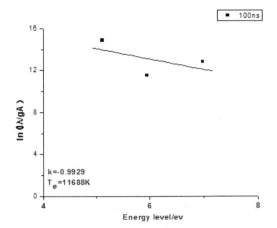

Figure 4. Boltzmann fitting line when delay time was 100 ns.

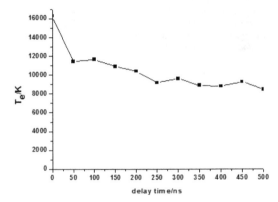

Figure 5. Evolution of the temperature of laser induced Mg plasma with different delay time (under 500 ns).

weight factor and transition probability and other parameters (Griem, 1964; Lochte-Holtgreven, 1968; Ke & Dong, 1998).

Figure 4 is the export of Boltzmann fitting straight line under the delay time of 100 ns, according to Formula 1, and into the parameters of Table 1, and we get the electronic temperature at this time 11,688 K. Figure 5 shows that when the delay time is 500 ns, the electronic temperature changes over time. It can be seen from the figure, in the delay time 500 ns range, the electronic temperature is not strictly in accordance with the law changes, with the increase in delay time decreases, the overall process of change can be divided into two parts: the first part, the delay time at 0–250 ns. At this stage, the electron temperature drops rapidly from 16,300 K to 9,173 K, which reduces to 7,127 K where the delay time is 50 ns, the change is most noticeable and rapid. It is shown that after the increase in the delay time, the plasma is expanded in the direction perpendicular to the Mg alloy, and its thermal energy is rapidly converted into kinetic energy. The expansion space of the plasma increases and the electronic temperature decreases rapidly.

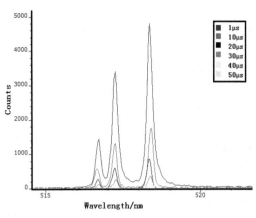

Figure 6. Spectrum of Mg I 518.4 nm when delay time was between 1 μs and 50 μs.

The second part is the delay time at 250–500 ns. At this stage, the electron temperature drop rate decreases rapidly, the change is very obvious, and the electronic temperature has a certain fluctuation, that is the delay time increases and the electron temperature also rises. The second part is smaller than the first part of the slowdown due to the increase in the recombination radiation in the plasma at the second part, and the energy released by the composite radiation compensates for the expansion and cooling of the plasma, while the plasma expansion space increases, free electrons, atoms and molecular plasma collision probability decreases, heat energy cannot be further to the kinetic energy conversion, resulting in the plasma temperature reduction slowdown (Griem, 1964). Figure 6 is the spectrum for the delay time of 1–50 μs, near 518.4 nm. As already shown above, the maximum intensity can be obtained at 518.4 nm, i.e., the maximum intensity of light radiation at 518.4 nm. It can be seen from Figure 6 that it is difficult to observe the characteristic spectrum at 518.4 nm at the delay time of 40 μs. When the delay time is 50 μs, the characteristic spectrum at 518.4 nm disappears completely and cannot be obtained. Only background radiation exists, indicating that the experimental results of the plasma life of about 50 μs.

4 CONCLUSION

In this paper, the relative spectra of Mg I 518.4 nm and Mg I 383.2 nm at the delay time of 100 ns are compared and analysed. It is found that the relative intensity decreases with the delay time of 100 ns, but in the fluctuation state, which is due to the delay period, the plasma is still in the formation stage, and the electronic, atomic transition process is still in a state of instability, resulting in its radiation intensity will be some fluctuations, not strictly in accordance with the law variety.

Then the spectrum of 500 ns delay time is analysed, and the electron temperature with different delay times is calculated. The evolution of electron temperature with time is analysed. When the delay time is 0–250 ns, the electron temperature decreases rapidly and decreases by 7,127 K. When the delay time is within 50 ns, the change is most obvious and rapid. It is believed that this is due to the increase in the delay time, the plasma in the direction perpendicular to the Mg alloy expansion, the heat will be quickly converted into kinetic energy, plasma expansion space increases, the electronic temperature will be rapidly reduced. When the delay time is 250–500 ns, the electron temperature decreases rapidly. The analysis shows that this is due to the increase of the composite radiation in the plasma, and the energy released by the composite radiation has the compensation effect on the expansion and cooling of the plasma. As the expansion space of the plasma increases, the probability of collision between free electrons and atoms and molecules is reduced, and the heat energy cannot be further transformed into kinetic energy, resulting in the slowdown of the temperature of the plasma.

Finally, the spectrum of Mg I 518.4 nm with the strongest intensity in the delay time range of 1–50 μs is analysed. It is found that the characteristic spectrum of the wavelength cannot be obtained at the delay time of 50 μs, and the life of Mg plasma in this experiment is obtained about 50 μs.

REFERENCES

Bolger, J.A. (2000). *Applied Spectroscopy, 54*(2), 181–189.

Cristoforetti, G. & Tognoni, E. (2013). *Spectrochimica Acta Part B: Atomic Spectroscopy*, 90, 1–22.

De Giacomo, A., Dell'Aglio, M., Santagata, A. & Teghil, R. (2005). Early stage emission spectroscopy study of metallic titanium plasma induced in air by femtosecond- and nanosecond-laser pulses. *Spectrochimica Acta Part B: Atomic Spectroscopy, 60*(7–8), 935–947.

Essington, M.E. & Melnichenko, G.V. et al. (2009). *Soil science society of America Journal, 73*(5), 1469–1478.

Hahn, D.W. & Omenetto, N. (2012). *Applied Spectroscopy*, 66(4), 347–419.

Griem, H.R. (1964a). *Plasma Spectroscopy* (pp. 139–140). McGraw-Hill

Griem, H.R. (1964b). *Plasma Spectroscopy* (pp. 303–307). McGraw-Hill:

Ke, Y. & Dong, H. (1998). *Analytical Chemistry Handbook.* Beijing: Chemical Industry Press.

Lochte-Holtgreven, W. (1968). *Evaluation of Plasma Parameter in Plasma Diagnostics* (pp. 156–157). Amsterdam.

Tong, Y.Q., Zhang, Y.K. & Yao H.B. et al. (2011). *Spectroscopy and Spectral Analysis*, 31(9), 2542–2545.

Zheng, P.C., Liu, H.D. & Wang J.M. et al. (2014). *Chinese Journal of Lasers*, 41(10), 260–266.

Electronic Engineering – Wang (ed)
© 2018 Taylor & Francis Group, London, ISBN 978-1-138-60260-1

Analysis of the brightness of molten liquid ejection in a millisecond laser interaction with a silicon plate

L. Zhang & X.W. Ni
Nanjing University of Science and Technology, Nanjing, China

J. Li
Zhoukou Normal University, Zhoukou, China

ABSTRACT: The problem of the brightness of molten liquid ejection is studied in the process of the interaction between a millisecond laser and a silicon plate. First, the process of the millisecond laser interacting with the silicon plate is obtained using the shading method which obtains the sequence of shadows for the process. Among these, the pulse width of the laser is 1 ms, the energy is 7.39 J, and the thickness of the silicon plate is 0.3 mm. Through the sequence of shadow images, observed on the surface of the silicon plate before and after the molten liquid ejection, and ejected in the opposite direction of splashing, the brightness of the molten liquid ejection was significantly higher than the background light brightness. Then, depending on the conditions of the molten liquid ejection, it is presumed that the temperature of the liquid silicon must be greater than its boiling point. Secondly, by using the thermal radiation theory and the emissivity of liquid silicon, the radiation force of 532 ± 15 nm wavelength band about liquid silicon was calculated when the liquid silicon temperature is 3500K. By comparing with the background light intensity, the radiation force of 532 ± 15 nm wavelength band of liquid silicon was 78.7 times the background light intensity. Finally, the radiation force of 532 ± 15 nm wavelength band of liquid silicon at different temperatures is calculated, and the flash phenomenon exists in the process of interaction between millisecond laser and silicon plate.

Keywords: molten liquid ejection, millisecond laser, light intensity

1 INTRODUCTION

Compared with a nanosecond laser, the millisecond laser has higher energy density, longer irradiation and lower power density. Thus, the millisecond laser has the advantage of deep melting depth, high efficiency with material interaction and so on. The process of the millisecond laser interaction with material has aroused widespread concern.

Melting, gasification and the molten liquid ejection phenomena occur in the millisecond laser interaction with the target (Schneider et al., 2007). Among them, the molten liquid ejection is the main form of material removal during laser processing. In general, it is believed that the pressure of gasification causes the molten liquid to eject from the surface of the material (He et al., 2006). In the millisecond laser-material interaction process, the study of molten liquid ejection raises attention. In the process of the millisecond laser interaction with a silicon plate, the melting of the ejection of the brightness problem has not been reported. It is necessary to study this.

In this paper, the process of the millisecond laser interaction with a silicon plate is recorded by the shading method, and the process of the sequence shadow

pictures was obtained. Through these shadow pictures, it can be observed that the molten liquid ejection appeared on the surface of the silicon plate and the brightness of the molten liquid ejection is significantly higher than the background light. According to the mechanism of molten liquid ejection, the temperature range of the surface of the silicon plate is deduced. Then, using the theory of thermal radiation and the emissivity of molten liquid silicon, the radiation force of molten silicon is calculated to be 78.7 times of the background light intensity in the band of 532 ± 15 nm. Finally, the radiation force of molten silicon at 532 ± 15 nm is calculated, and the flash phenomenon exists in the process of the millisecond laser interaction with the silicon plate.

2 EXPERIMENTAL SETUP

The experimental device is shown in Figure 1, the millisecond laser beam through a focusing lens irradiated on a silicon plate of 0.3 mm thickness. On the surface of the silicon plate, the beam radius is about 0.2 mm. The millisecond laser is produced by the BeanTech

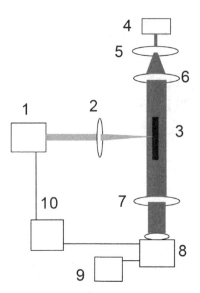

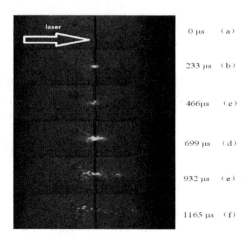

Figure 2. Sequence shadow pictures.

Figure 1. Experimental setup. 1. Millisecond laser, 2. Lens, 3. Target, 4. CW laser, 5. Expansion beam mirror, 6. Collimator lens, 7. Filter, 8. High speed CCD, 9. DG535.

company. The laser model is the Melar50. The wavelength is 1064 nm. The pulse width is 0.5~2.5 ms, and the energy is in the range of 0~50 J. The focal length of the focusing lens is 152 mm in the experiment. The pulse width is 1 ms, and single pulse energy is 7.38 J. A continuous laser beam, in which the power is 500 mW and the wavelength is 532 nm, is passed through the beam expander lens, the collimator lens, the silicon plate, and the bandpass interference filter, with the high speed CCD produced by United States Southern Vision Systems, Inc. recording the whole process. Finally, the image is stored by the computer. The radius of the continuous laser beam is 2 cm; the frame rate of high speed CCD is 4261 fps at a resolution of 1280 × 128; the half band bandwidth of the 532 nm interference filter is 30 nm.

3 EXPERIMENTAL RESULTS

The sequence shadows obtained from the experimental setup of Figure 1 are given in Figure 2. The laser is irradiated on the surface of the silicon plate from left to right. In Figure 2 (b), the brightness of the molten silicon is higher than the background light in the laser irradiation region. In the graphs after Figure 2 (c), molten liquid ejection appears on both the front and rear surfaces of the silicon plate, and the brightness of the melt ejection is higher than that of the background light.

4 ANALYSIS AND DISCUSSION

4.1 *The temperature of the molten liquid ejection*

In general, it is believed that the pressure of gasification causes the molten liquid to eject from the surface of the material. When the gasification pressure is greater than the surface tension, that is $F_r > F_s$, the molten liquid will eject. Therefore, when at least $F_r > 0$, the molten liquid is required to eject. The gasification pressure can be expressed as (He et al., 2003):

$$F_r = 2\pi \int_o^{r_B} r\Delta P(r, T_s) dr \qquad (1)$$

where T_s is the target surface temperature, r_B is the gasification radius, ΔP is the difference between the saturated vapour pressure and the atmospheric pressure at distance r from the laser axis. Therefore, the difference between the saturated vapour pressure and the atmospheric pressure can be expressed as (Jung-Ho & Suck-Joo 2006):

$$\Delta P(T_s) = p_{amb} \exp\left[\frac{L_{ev} m_{Si}}{k_B}(\frac{1}{T_b} - \frac{1}{T_{sur}})\right] - p_{amb} \qquad (2)$$

where k_B is the Boltzmann constant; T_{sur} is surface temperature; T_b is boiling; m_{si} is the quality of silicon atomic; L_{ev} is gasification latent heat; and p_{abm} is saturated vapour pressure.

According to Equation 1 and Equation 2, if $F_r > 0$, then $T_b > T_{sur}$.

4.2 *The brightness of the molten liquid ejection*

4.2.1 *Planck's law*

Thermal radiation is an important way of heat transfer, and heat radiation transfers heat by electromagnetic waves.

In 1901, Planck's quantum theory proved that the blackbody's radiation ability was a function of the blackbody's own temperature T and the wavelength (Li 2003):

$$E_{b,\lambda} = \frac{c_1}{\lambda^5 \left[\exp\left(\frac{c_2}{\lambda T}\right) - 1\right]} \qquad (3)$$

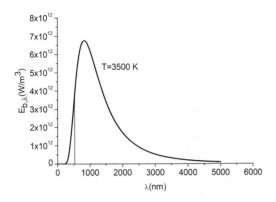

Figure 3. 3500 K blackbody radiation.

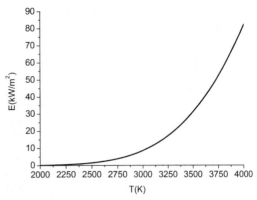

Figure 4. The radiation force of liquid silicon at 517–547 nm at different temperatures.

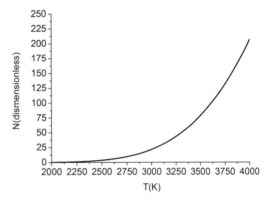

Figure 5. The ratio of the radiation force of liquid silicon at 517–547 nm at different temperatures and background light intensities.

where 1 is the first radiation constant, $c_1 = 3.742 \times 10^{-16}$ W·m². c_2 is the second radiation constant, $c_2 = 1.439 \times 10^{-2}$ W·K.

The band radiation of the blackbody is defined as:

$$E_{b,(\lambda_1-\lambda_2)} = \int_{\lambda_1}^{\lambda_2} \frac{c_1}{\lambda^5 [\exp\left(\frac{c_2}{\lambda T}\right)-1]} d\lambda \quad (4)$$

In the experiment, the bandwidth of the 532 nm interference filter bandwidth is 30 nm and a light range is 517–547 nm wavelength band. Therefore, the high speed CCD records the light intensity about 517–547 nm wavelength band. The 3500K black body of the radiation force is calculated according to Equation 3 in Figure 3, where the shadow part of the area is the radiation force of 517–547 nm wavelength band. It can be seen that at 3500 K, the blackbody can emit light in the 517–547 nm band and the radiation force reaches 3×10^{12} W/m³ or more.

According to the emissivity of liquid silicon at 532 nm is 0.27, and (2), the radiation force of 517–547 nm wavelength band about liquid silicon is (Takasuka et al., 1997):

$$E = \varepsilon E_b = 3.1320 \text{ W/cm}^2$$

In the experiment, the background light source used was a continuous laser, the laser power was 500 mW, the radius was 2 cm, the brightness of the background light was:

$$I_{zm} = \frac{0.5}{\pi \times 2^2} = 0.398 \text{ W/cm}^2$$

The radiation force of the 517–547 nm wavelength band about liquid silicon is 78.7 times the background light source. It can be seen that the radiation brightness of liquid silicon is much higher than that of background light. Therefore, the heat radiation caused by high temperature liquid silicon is the reason why its brightness is higher than background brightness.

According to Equation 3, the radiation force of the 517–547 nm wavelength band was calculated at different temperatures of liquid silicon as shown in Figure 4.

It can be seen that with the increase of the temperature of liquid silicon, the radiation force of the 517–547 nm wavelength band is a non-linear increase.

The ratio of the radiation force of the 517–547 nm wavelength band about liquid silicon to the background light intensity is shown in Figure 5. It can be seen that as the temperature of the liquid silicon increases, the ratio increases. The results show that when the temperature of liquid silicon is 2250 K, the radiation force is basically the same as the background light intensity. When the temperature is 2770 K, the radiation force is 10 times the background light intensity. When its temperature reaches 3500 K, the radiation force is 78.7 times the background light intensity. When the temperature of the liquid silicon is 4000 K, the radiation force is 206.8 times the intensity of the background light. Depending on the conditions produced by the molten liquid ejection, liquid silicon below boiling point does not generate molten liquid ejection, but the radiation intensity is still higher than the background light intensity. Therefore, in Figure 2(b), it can be seen that the brightness of the molten silicon which does not eject is still higher than the brightness of the background light.

5 SUMMARY

The process of millisecond laser-induced molten liquid ejection was obtained using the shadow method. In the sequence shadows, the brightness of the molten liquid ejection was significantly higher than the background light. When a 532 nm bandpass interference filter is installed in front of the high speed camera, the half bandwidth is 30 nm, and the high speed camera records the light intensity at 532 15 nm. Using the theory of thermal radiation, the radiation force of 532 15 nm wavelength band about liquid silicon was calculated. When the liquid silicon temperature is 3500K, the ratio of the radiation force of 532 15 nm wavelength band about liquid silicon to the background light intensity is 78.7. The reason is that heat radiation of liquid silicon phenomenon.

REFERENCES

He, X., Debroy, T. & Fuerschbach, P.W. (2003). Alloying element vaporization during laser spot welding of stainless steel. *Journal of Physics D: Applied Physics*, 36(23), 3079–3088.

He, X., Norris, J.T., Fuerschbach, P.W. & Debroy, T. (2006). Liquid metal expulsion during laser spot welding of 304 stainless steel. *Journal of Physics D: Applied Physics,* 39(3), 525–534.

Jung-Ho, C. & Suck-Joo, N. (2006). Implementation of real-time multiple reflection and Fresnel absorption of laser beam in keyhole. *Journal of Physics D: Applied Physics*, 39, 5372–8.

Li, J. (2003). *Advanced Heat Transfer*. Beijing: Higher Education Press.

Schneider, M., Berthe, L., Fabbro, R., Muller, M. & Nivard, M. (2007). Gas investigation for laser drilling. *Journal of Laser Applications,* 19(3), 165–169.

Takasuka, E., Tokizakim, E., Terashima, K. & Kimura, S. (1997). Emissivity of liquid silicon in visible and infrared regions. *Journal of Applied Physics*, 81(9), 6384–6384.

Design of a high linear VCO

M.Y. Ren, M.Y. Qin & B.Z. Song
Harbin University of Science and Technology, Harbin, Heilongjiang, China

ABSTRACT: With the development of integrated circuit techniques, the design of the clock signal generator VCO is widely used in many fields. A new VCO suitable for the various constraints between area, power consumption and process performance is presented, which has been implemented in a standard 0.18 μm CMOS process. Using an internal hysteresis comparator with high gain and better anti-noise ability, the proposed circuit achieved good linearity, high stability and good temperature effect requirements.

Keywords: VCO; internal hysteresis comparator; high gain

1 INTRODUCTION

Currently, the PLL is widely used in communications, radar, test equipment and other systems, and VCO is the most sensitive block to supply noise and be the performance limiting factor for low-voltage PLL (Kim et al., 2016), which directly generates the output feedback signal. The operating frequency range of the VCO determines the capture range of the charge pump PLL, and its noise suppression ability determines the noise performance of the PLL. Therefore, the performance of the VCO is directly related to the performance of the whole PLL and the ability to work properly.

According to the different output waveforms, the VCO is divided into two kinds of LC oscillator and ring oscillator. Compared with the LC oscillator, the ring oscillators, which have better linearity, lower power consumption and lower cost, have become the key modules of many digital chips and communications systems. Hence, the ring VCO architecture is chosen in this implementation (Tu et al., 2014).

Based on SMIC 0.18 μm CMOS technology, a novel delay cell structure is proposed in this paper. A high performance fully differential ring VCO with a centre frequency of 1 MHz is designed.

2 PRINCIPLES OF A RING OSCILLATOR

The VCO is a circuit that does not need an external excitation signal and output AC voltage by self excited oscillation. In a VCO circuit, the relationship between the frequency of the output signal and the input signal is expressed as:

$$\omega(t) = \omega_0 + K_{VCO} u_c(t) \tag{1}$$

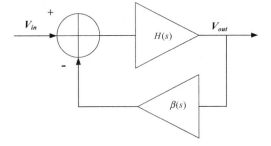

Figure 1. A schematic diagram of a VCO negative feedback system.

In this expression, K_{VCO} is the control sensitivity of the VCO.

A conventional single-ended ring oscillator consists of a series of inverting amplifiers placed in a feedback loop (Wang, 2015). It is a negative feedback system, as shown in Figure 1.

When the loop gain $\beta(s=j\omega_0)=-1$, the closed-loop gain tends to infinity at the frequency ω_0. Therefore, the circuit can amplify its own noise at frequency ω_0 and the amplitude at this frequency can be infinite. But, in fact, if the amplitude increases, the loop gain will be reduced. Eventually, the oscillator establishes a steady state oscillation at one amplitude. The above condition is the "Barkhausen Criterion".

$$|H\beta(j\omega_0)| \geq 1 \tag{2}$$

$$\arg(H\beta(j\omega_0)) = -180° \tag{3}$$

The "Barkhausen Criterion" is a necessary but not sufficient condition for a VCO circuit to oscillate at

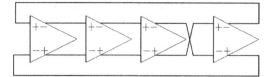

Figure 2. Four-difference ring oscillator sketch map.

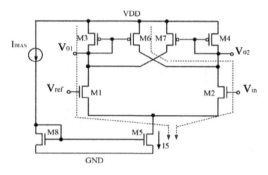

Figure 3. Delay unit of VCO.

frequency ω_0. In the presence of temperature and process variations, the selection of loop gain should be at least twice higher to ensure circuit oscillation.

3 CIRCUIT DESIGN AND ANALYSIS

3.1 Differential VCO structure

Based on the principle of inverter delay, a basic ring oscillator can be formed by an odd number of inverter units connected end to end. The circuit structure of the VCO in this paper is the external connection structure of the four differential ring oscillators shown in Figure 2. It consists of two main inverters and two small auxiliary inverters (Babaie-Fishani and Rombouts, 2016). The first three delay units are connected reversely with each other, and the fourth delay unit and the first delay unit are connected in the forward direction. The period of this VCO is $T = 2NT_{pd}$, where N is the number of inverters in series and T_{pd} is the single inverter delay time.

3.2 Delay unit structure

In this design, the open loop hysteresis comparator is used as a single delay unit in a voltage-controlled oscillator. It can compare the size of the two output analogue signals and controls the level of high and low levels of the output signal. The delay unit is shown in Figure 3.

It is assumed that the power supply voltage is VDD, and the gate of transistor M1 is connected to the reference voltage Vref. When the input voltage value of transistor M2 is much lower than that of Vref, transistor M1 turns on and transistor M2 turns off. Then, transistor M3 and transistor M6 will be on, and transistor M4 and transistor M7 will be off. Resulting in

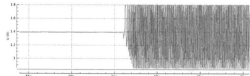

Figure 4. The simulation waveform of the output signal.

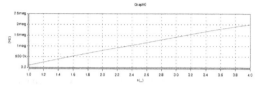

Figure 5. VCO gain linearity simulation diagram.

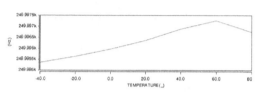

Figure 6. The temperature character of the ring oscillator simulation diagram.

current I5 would all flow through transistor M1 and transistor M2, and output voltage V02 is high level. The current flow in this state is shown in the dotted line in Figure 3. As the input voltage value Vin continuously increases towards the value of the threshold point, some current starts to flow in transistor M2. But the state of the comparator does not change until the current in M2 equals the current in M6.

4 SIMULATION RESULTS AND ANALYSIS

This design uses a Cadence Spectre to simulate the circuit. The oscillator circuit amplifies the tiny noise to start oscillate. The circuit output waveform is shown in Figure 4. From the transient simulation results, the starting time of the oscillator is 55.1 μs, and the oscillation amplitude is about 1.8 V which is approximately the supply voltage. The transient response shows that the oscillation of the oscillator starts well and the start-up time is short.

Figure 5 shows the linear degree within the control range of the voltage-controlled oscillator. It can be seen from this that the linearity of the oscillator is better. Figure 6 shows the temperature characteristics of a phase-locked loop scan which uses the ideal signal source as a reference signal input. It can be seen from the Figure 6, when the ambient temperature changes between $-40°C \sim +80°C$, the frequency jitter of the design is $\Delta F = 3.5$ Hz. It shows that the circuit is not sensitive to temperature, and the temperature characteristic is better.

5 CONCLUSION

Based on the SMIC 0.18 μm CMOS process, a high performance differential ring VCO with a centre frequency of 1 MHz is designed in this paper. The VCO has the characteristics of high linearity, large output swing, insensitivity to temperature and so on. It can operate normally at a 1.8 V low power supply voltage and be widely applied to clock recovery and frequency synthesisers.

ACKNOWLEDGEMENTS

This paper is sponsored by Natural Science Foundation of Heilongjiang Province (Grant No. F2017017).

REFERENCES

Babaie-Fishani, A. & Rombouts, P. (2016). Highly linear VCO for use in VCO-ADCs. *Electronics Letters, 52*(4), 268–270.

Kim, S.G., Rhim, J., Kwon, D.H., Kim, M.H. & Choi, W.Y. (2016). A low-voltage PLL with a supply-noise compensated feedforward ring VCO. *IEEE Transactions on Circuits and Systems II: Express Briefs*, 63(6), 548–552.

Tu, C.C., Wang, Y.K. & Lin, T.H. (2017). A low-noise area-efficient chopped VCO-based CTDSM for sensor applications in 40-nm CMOS. *IEEE Journal of Solid-State Circuits*, 52(10), 2523–2532.

Wang, S.F. (2015). Low-voltage, full-swing voltage-controlled oscillator with symmetrical even-phase outputs based on single-ended delay cells. *IEEE Transactions on Very Large Scale Integration (VLSI) Systems*, 23(9), 1801–1807.

Electronic Engineering – Wang (ed)
© 2018 Taylor & Francis Group, London, ISBN 978-1-138-60260-1

Design a second-order curvature compensation bandgap reference

M.Y. Ren, B.Z. Song & M.Y. Qin
Harbin University of Science and Technology, Harbin, Heilongjiang, China

ABSTRACT: Since entering modern technology, the design of bandgap reference circuits has become an important part of analogue integrated circuit design. With the increasing level of IC technology, the features of CMOS technology such as high power, low power consumption, high-precision and high-level curvature compensation are constantly being realised. Based on a CSMC $0.5\,\mu$m 2P2M CMOS process, a second-order curvature compensated reference circuit is designed.

Keywords: bandgap reference, curvature compensation, operational amplifier

1 INTRODUCTION

Bandgap references are widely used in analogue circuits, digital circuits, and mixed-signal circuits, such as operational amplifiers, linear regulators, DRAM, A/D converters, phase-locked loops, and power converters because of their high accuracy and temperature-independence (Lee et al. 1994). Voltage references are fundamental circuit blocks, ubiquitously used in analogue, mixed-signal, RF and digital systems, including memories. The importance of low power design is self-evident in mobile and energy harvesting applications, whereas resistorless approaches enable the implementation of the circuit in standard digital processes. Basically, the bandgap voltage reference, introduced by Widlar in 1971 (Mattia et al. 2014). Bandgap voltage reference sources are one of the basic circuits, their functions may be single. The bandgap reference supplies a stable voltage to the circuit, compare the output voltage and the results. The data obtained adjust the output voltage, this feature is reflected in the voltage regulator. In voltage detector, bandgap voltage source like a bistable multivibrator gate circuit, In production, the designers must consider the initial voltage accuracy, voltage drift, feed current and the noise.

2 CIRCUIT PRINCIPLES

The normal bandgap reference circuit has no curvature compensation. The principle of curvature compensation is the normal bandgap reference circuit ignores the two order terms of Vbe in the voltage reference. The formula is expressed as:

$$V_{REF} = V_{BE}(T) + K\Delta V_{BE} \tag{1}$$

The following formula shows the expansion formula of the V_{BE}, The highest coefficient of item it is exist, In this case, we cannot ignore the relationship between V_{out} and temperature which, in the waveforms of temperature and voltage, shows a rising parabola. The influence of temperature on the voltage is about 10 degrees. Under special needs, it is important to add the second-order curvature compensation. The formula is expressed as:

$$V_{BE}(T) = V_{G0}\left[1 - \frac{T}{T_0}\right] + V_{BE0}\left[\frac{T}{T_0}\right] + \frac{nkT}{q}\ln\left[\frac{T}{T_0}\right] + \frac{kT}{q}\ln\left[\frac{I_c}{I_k}\right] \tag{2}$$

Using knowledge from advanced mathematics, we can find the following relationship where the formula is expressed as:

$$\frac{\partial V_{BE}}{\partial T} = -\frac{G_0}{T_0} + \frac{V_{BE0}}{T_0} + Z(T) \tag{3}$$

$Z(T)$ is the required part, and will change as the temperature changes. The changes in Vbe is an important part of changing the curvature compensation.

3 CIRCUIT DESIGN AND ANALYSIS

3.1 *Operational amplifier*

In order to increase the primary gain, the experimental input differential amplifier uses a telescope-feed. This kind of structure can effectively reduce the effect of noise on the circuit. At the same time, the circuit is more stable. The output part uses a capacitance add resistance structure, which can increase the gain, as shown in Figure 1.

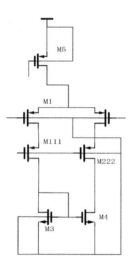

Figure 1. Design of differential input.

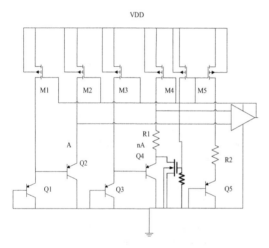

Figure 2. The circuit structure.

It is not only in the design of CMOS integrated circuits. In the signal processing degenerative feedback structure also has the same width range. The wave of the step response may sometimes make the circuit less stable. This reflects the importance of Miller's compensation.

3.2 The curvature compensation

The curvature compensation bandgap reference applies to the relationship between transistors and temperature. In depth understanding of the features of each part of the bandgap circuit. At last we choose a suitable structure to establish the right circuit. The resistance in the circuit can be better matched on the territory. The circuit structure is shown in Figure 2.

The compensation circuit structure consists of three devices, They are fit for the realisation of the circuit structure. When the temperature rises, the resistance current will continue to increase. The V_{gs} of N1 will continue to rise. They final work at the subthreshold area and bear the current of R1, Then the curvature compensation took place.

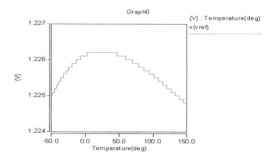

Figure 3. A curve without compensation.

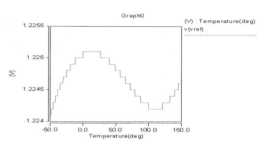

Figure 4. A curve with compensation.

4 SIMULATION RESULTS AND ANALYSIS

The simulate the circuit is use the Capture, using Hspui to simulate the network table, and using Cosmosscope to simulate the waveform. From the two pictures we get, the two bandgap reference circuit at the power supply voltage is 5 V uses the CMSC 0.5 μm technology to simulate the circuit.

The output voltage of the normal bandgap reference circuit should be a standard curve near to 1.2 V as shown in Figure 3. The curve of the curvature compensation bandgap reference circuit is closer to 1.2 V than normal as shown in Figure 4, because the first one ignores the two order terms of V_{BE} in the voltage reference. Then the curvature compensation characteristic is completed.

5 CONCLUSION

In many low supply voltage self-powered systems, such as Wireless Body Area Network (WBAN) nodes, wearable smart devices, and implantable medical devices, voltage references have a significant impact on the overall system performance (Xia et al. 2014). Based on the bandgap reference and operational amplifiers, analysis of the characteristics of the bandgap reference voltage source and the circuit structure are required for curvature compensation. Using the MOS pipe working state, through the calculation and analysis of actual, we can find a suitable amplifier and bias circuit structure, and then calculate the value. The experimental results are changed repeatedly to achieve better experimental characteristics.

ACKNOWLEDGEMENTS

This paper is sponsored by Natural Science Foundation of Heilongjiang Province (Grant No. F2017017).

REFERENCES

Lee, I., Kim, G. & Kim, W. (1994). Exponential curvature-compensated BiCMOS bandgap references. *IEEE Journal of Solid-State Circuits*, 29(11), 1396–1403.

Mattia, O.E., Klimach, H. & Bampi, S. (2014). Resistorless BJT bias and curvature compensation circuit at 3.4 nW for CMOS bandgap voltage references. *Electronics Letters*, 50(12), 863–864.

Xia, L., Cheng, J., Glover, N.E. & Chiang, P. (2014). 0.56 V, −20 dBm RF-powered, multi-node wireless body area network system-on-a-chip with harvesting-efficiency tracking loop. *IEEE Journal of Solid-State Circuits*, 49(6), 1345–1355.

Electronic Engineering – Wang (ed)
© 2018 Taylor & Francis Group, London, ISBN 978-1-138-60260-1

Study on the corona-resistance property of polyimide/alumina nanocomposite films at elevated temperature

S.T. Minh
Ho Chi Minh City University of Technology, Ho Chi Minh, Vietnam

ABSTRACT: A series of polyimide (PI) and polyimide/alumina (Al_2O_3) nanocomposite films were prepared by in-situ polymerization. The incorporated Al_2O_3 nanoparticles resulted in significant improvement on the corona-resistance property of the PI/Al_2O_3 composites. The corona-resistance life was 67 h with the 4 wt% doping concentration at 25°C, which is increased by more than 11 times comparing with pure PI. When the environmental temperature increases to 100°C, the life has a significant decrease. TGA thermograms show the structure of PI matrix is not directly affected by the high temperature of 100°C, which is less likely to reduce the corona-resistance life of PI composites. SEM observations indicated that Al_2O_3 nanoparticles form barrier layer on the surface of composites to bear the impact of electrons form electrode because of the interaction between PI matrix and nanoparticles. Whereas the high temperature impedes the formation of relatively large-sized barrier layer, which leads to accelerated degradation of PI matrix and shorter corona-resistance life. The SAXS data show a negative deviation from Porod's law, which is an evidence of the interaction. Meanwhile the interfacial polarization results in enhancement of relative permittivity of PI nanocomposites. Moreover, FTIR results indicate that the imidization reaction of composite films is complete in this work.

Keywords: Nanocomposite; Polyimide; Alumina; Corona-resistance property; Temperature

1 INTRODUCTION

Polyimide (PI), a kind of condensation polymer derived from dianhydride and diamine, belongs to the important class of organic materials known as "high performance" polymers due to their extraordinary properties of good stability at high temperatures (Ragosta, Abbate, Musto & Scarinzi 2012; Gu & Wang 2015; Ghosh & Mittal 1996). It has been extensively used in different fields, such as the microelectronics, photonics, nuclear, optics and aerospace industry over the decades, owing to its incomparable thermal stability, high mechanical strength, good dielectric properties, high glass transition and high solvent resistance (Yang, Li, Ma, Bu & Zhang 2010; Wang, Zhong, Xu & Do 2005; Yildirim 2014; Wang, Zhou, Dang & Chen 2009; Yang, Su, Guo & Hsiao 2009; Yudin, Otaigbe, Gladchenko, Olson, Nazarenko, Korytkova & Gusarov 2007; Sun, Chang, Tian, Niu & Wu 2016). However, pure PI has its disadvantages inherent in all organic materials that limit its further applications, such as relatively high thermal expansion coefficient, low thermal conductivity, and poor corona-resistance property. To overcome these drawbacks, many PI composite materials were vastly studied especially the inorganic/organic composites, as "synergistic effect" of two components endows composites many new properties. The results show that the addition of nanoparticles in polymers for enhancement of thermal and electrical performance is significantly important (Zhang, Zhao, Zhang, Bai, Wang, Hua, Wu, Liu, Xu & Li 2016; Arai, Noro, Sugimoto & Endo 2008; Feng, Yin, Chen, Song, Su & Lei 2013; Qi & Zhang 2015; Cho, Chen & Daniel 2007; Li, Liu, Liu, Chen & Chen 2007; Weng, Xia, Yan, Liu & Sun 2016).

The addition of Al_2O_3 nanoparticles can effectively improve corona-resistance property and the PI/Al_2O_3 nanocomposite serves as a typical kind of engineering insulating materials in the manufacture of electric motors (Wu, Yang, Gao, Hu, Liu & Fan 2005; Hui et al. 2016; Du, He, Du & Guo 2016; Akram, Gao, Liu, Zhu, Wu & Zhou 2015). It is well known that the operating temperature will increase during the operation of electric motor. While the existing research for corona-resistance property of PI composites mainly aims at the room temperature, which is not enough for manufacture and application of electric motors. Meanwhile, to see the thermal and electrical effects on the insulation life of electric motor, as reported earlier, the accelerated aging test is always carried out at room temperature and high temperature ranging from 100 to 150°C (Haq, Jayaram & Cherney 2008). Therefore, it has a practical significance to investigate the effect of temperature on corona-resistance property of PI/Al_2O_3 nanocomposite film.

In this study, PI/Al_2O_3 nanocomposite films were prepared by in-situ polymerization. The microscopic structure was carefully examined by FTIR spectra and SAXS data. The change of relative permittivity caused by interface, as well as Al_2O_3 nanoparticles is investigated. The effect of temperature on the corona-resistance property of PI/Al_2O_3 nanocomposite films is researched by means of corona-resistance life and comparison of surface morphology images after corona aging at 25°C and 100°C. TGA thermograms are analyzed to help discuss the mechanism of corona aging at elevated temperature.

2 EXPERIMENTAL

2.1 Materials

Pyromellitic dianhydride (PMDA) and 4,4'-diaminodiphenyl ether (ODA) were chemically pure and supplied by Sinopharm Chemical Reagent Co. Ltd. Ethanol absolute and N, N-dimethylacetamide (DMAc) were analytical grade and purchased from Tianjin Fuyu Fine Chemical Co. Ltd. The diameter of α-Al_2O_3 nanoparticles (Beijing DK nano technology Co. Ltd) was about 30 nm and dried at 120°C for 12 h in air oven before use.

2.2 Preparation of PI/Al₂O₃ nanocomposites

The PI and PI/Al_2O_3 composite films were prepared via in-situ polymerization. Pre-weighed N, N-dimethylacetamide, Al_2O_3 nanoparticles and 4,4'-diaminodiphenyl ether were placed into a three-neck round-bottom and were mixed by ultrasonic vibration at 25°C. The ultrasonic wave was used before obtaining the stable suspension. Then pyromellitic dianhydride was added into the suspension at proper ratios and completely dissolved by mechanical stirring for 3 hours at 25°C. The resulting mixture was cast onto a clean glass plate and dried at 80°C for 12 h in an air convection oven. After thermal imidization of mixture was finished in air oven by the method of gradient increased temperature, the PI/Al_2O_3 composite films with thicknesses of 30 μm were obtained.

2.3 Measurements

Scanning electron microscope (SEM) was used to observe the surface morphology of PI/Al_2O_3 nanocomposite films by a FEI Sirion of the Netherlands Philips. Fourier transform infra-red (FTIR) spectra was recorded at 25°C in the range of 500–4000 cm^{-1} on a FT/IR-6100 type-A manufactured by JASCO Spectroscopy and Chromatography Instrumentation. The small angle X-ray scattering (SAXS) experiments were carried out at beam line 4B9A at Beijing Synchrotron Radiation Facility. The storage ring was operated at 2.5 GeV with a current of about 180 mA. The incident X-ray wavelength was selected to be 0.154 nm by a double-crystal Si (111) monochromator. The SAXS data were collected by Mar165 charge-coupled device and the sample-to-detector distance was fixed at 1.52 m. The relative permittivity of PI/Al_2O_3 composite films was measured using the broadband dielectric/impedance spectrometer (Novocontrol Alpha-analyzer) with the frequency range from 10^0 to 10^6 Hz at room temperature. Thermo-gravimetric analysis (TGA) measurements were carried out using a ZRT thermal analysis system in air at a heating rate of 10°C/min and the weight of the sample for each measurement was approximately 8 mg. For the test of corona-resistance property, homemade equipment consisting of a high-voltage ac source, upper electrode and lower electrode was used. The diameter of column upper electrode and circle plate lower electrode were 6 and 60 mm respectively. The distance between upper electrode and lower electrode was 100 μm. All of samples were tested under the electric field intensity of 60 kV/mm in this work.

3 RESULTS AND DISCUSSION

Fig. 1 shows the surface and cross-section SEM image of PI/Al_2O_3 nanocomposite film with 2 wt% doping concentration. It can be clearly seen that there are no obvious clusters of nanoparticles on the surface of tested sample, which indicates that the nanoparticles have good dispersion and minimizes the negative influence of preparation on the properties of PI matrix nanocomposites.

The FTIR spectrum of the PI/Al_2O_3 nanocomposite film with 4 wt% doping concentration is shown in Fig. 2. The appearance of absorption peaks at 1725 and 1780 cm^{-1} are due to both symmetric and asymmetric C=O stretching in imide groups. The peaks at 1380 and 716 cm^{-1} are attributed to C-N stretching and imide ring deformation, respectively. Additionally, 823 cm^{-1} (C-O stretching vibration) corresponding to imide structure is observed and the peak at 1500 cm^{-1} is deemed to the vibration of benzene ring in the polyimide. These results indicate that the imidization reaction of the PI/Al_2O_3 composite films is complete in this work.

Fig. 3a inset shows the two-dimensional SAXS scattering pattern of PI/Al_2O_3 composite film with 4 wt% doping concentration, where pseudo-coloring rings indicate scattering intensity with incidence beam blocked (black rectangle). It can be seen that the SAXS pattern of PI/Al_2O_3 composite film is of rotational symmetry, which implies that the composite material is isotropic and the scatters are spherical in the films. According to

$$q=4\pi\sin\theta/\lambda \qquad (1)$$

where scattering vector q is related to θ (2 scattering angle) and λ (X-ray wavelength), the scattering intensity I(q) versus q plots for PI/Al_2O_3 composite films with different doping concentrations of nanoparticles can be get, as shown in Fig. 3a. It is clear that the composite films have quite similar curves. When q is less than 1.5, the SAXS intensity rapidly decreases. However, at the value of abscissa over 2, the intensity gradually decreases. Furthermore, the microstructure characteristic of PI/Al_2O_3 composite

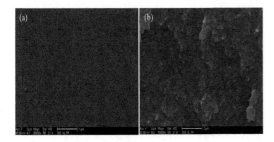

Figure 1. The surface SEM image of PI/Al$_2$O$_3$ nanocomposite film with 2 wt% doping concentration.

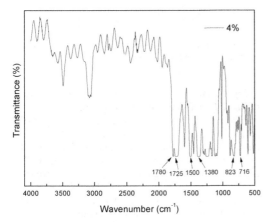

Figure 2. FTIR spectrum of PI/Al$_2$O$_3$ nanocomposite film with 4 wt% doping concentration.

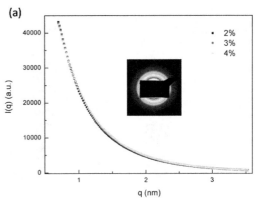

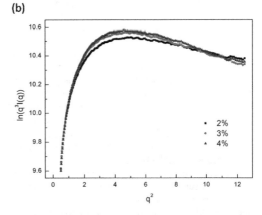

Figure 3. (a) The SAXS scattering pattern of 4 wt% doping sample (inset) and scattering intensity versus scattering vector plot for PI/Al$_2$O$_3$ composite films with different Al$_2$O$_3$ nanoparticle doping concentrations. (b) The plots of ln[q^{3}I(q)] versus q^2 for all 3 doping concentrations.

films can be studied by classic SAXS theory. The plots of ln[q^{3}I(q)] versus q^2 for PI composite films with the same SAXS data set are given in the Fig. 3b. It can be seen that the value on the vertical axis first quickly increases, and then slowly decreases. It is well known that only scattering by ideal two-phase systems with sharp boundaries obeys Porod's law. However, a negative deviation from Porod's law can be observed in the Fig. 3b. The negative deviation can be attributed to the interaction between organic polymer molecular chains and inorganic nanoparticles in the interfaces, which maybe has an effect on the electric properties of PI/Al$_2$O$_3$ composite films.

Fig. 4 shows the relative permittivity of pure PI and PI/Al$_2$O$_3$ composite films as a function of measurement frequency. It can be observed that relative permittivity of pure PI decreases from 3.25 to 3.16 in the frequency range measured, and is lower than that of PI composite films. Meanwhile, the relative permittivity of PI/Al$_2$O$_3$ composite films increases as the doping concentration increases. The increasing of permittivity for PI composite films can be explained by the interaction between organic PI matrix and inorganic Al$_2$O$_3$ fillers. Under the condition of low doping concentration, the more nanoparticles doping means the stronger interfacial polarization. Moreover, the difference of relative permittivity in the relatively low-frequency region (10^0–10^2 Hz) and

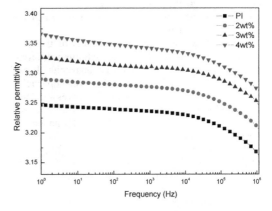

Figure 4. Relative permittivity of pure PI and PI/Al$_2$O$_3$ films as a function of measurement frequency.

high-frequency region (>10^3 Hz) is mainly because the rotation speed of dipole gradually loses the ability to keep up with the increase of measurement frequency. These results suggest that the existence of interfaces can improve the electric properties of

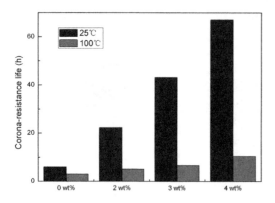

Figure 5. The corona resistant life of Al$_2$O$_3$ doping concentration at 25°C and 100°C.

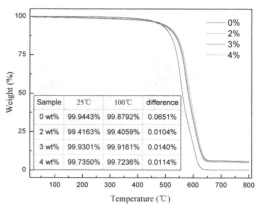

Figure 6. The TGA thermograms of the pure PI and PI/Al$_2$O$_3$ nanocomposite films.

PI/Al$_2$O$_3$ composite films and also can be associated to the change of corona-resistance property.

The corona-resistance life, which is the key parameter of electrical properties of insulation materials, was respectively tested for PI/Al$_2$O$_3$ composite films with different doping concentrations at 25°C and 100°C. As shown in Fig. 5, it is a fact that the corona-resistance life increases with the increase of doping concentration of Al$_2$O$_3$ nanoparticles. At 25°C, the corona-resistance life of the PI/Al$_2$O$_3$ film with 4 wt% doping concentration is 11 times more than that of pure PI (67 h vs. 6 h). The results indicate that the adding of Al$_2$O$_3$ nanoparticles can effectively improve the corona-resistance property of PI matrix nanocomposites, which is beneficial for practical application of PI material. Meanwhile, it is clearly obtained that at the 100°C, both pure PI and PI/Al$_2$O$_3$ nanocomposite films exhibit weak corona-resistance performance comparing with their performance at 25°C, and the corona-resistance life of PI/Al$_2$O$_3$ composite films is still longer than that of pure PI.

Fig. 6 shows the TGA thermograms of the pure PI and its nanocomposite films containing different doping concentrations of Al$_2$O$_3$ nanoparticles. It can be clearly observed that the addition of nanoparticles in PI/Al$_2$O$_3$ nanocomposite films improves the thermal stability, and for the same amount of weight loss, higher temperature is needed for PI/Al$_2$O$_3$ nanocomposite film comparing with pure PI film. The inset is the table that shows the remaining weight percent of pure PI and PI/Al$_2$O$_3$ nanocomposite at 25°C and 100°C as well as their difference. For all of tested samples, the differences of remaining weight percent are less than 0.02%, which indicates that the structure of PI matrix is still stable as the environmental temperature increases from 25°C to 100°C. Therefore, the structure of PI matrix is not directly affected by the high temperature of 100°C, which is less likely to reduce the corona-resistance life of PI composite films. And the effect of temperature on the corona-resistance property of PI/Al$_2$O$_3$ composites will be further discussed by means of the surface morphology images of nanocomposite films.

Fig. 7a and b are the surface morphology images of PI/Al$_2$O$_3$ nanocomposite films with 2 wt% doping concentration after corona aging at 25°C and 100°C respectively. The images can be divided into two parts due to the difference of morphology. Thereinto, the region I is corona aging region and the region II is corona breakdown region. In the region I, comparing with the surface morphology of original PI/Al$_2$O$_3$ nanocomposite film with 2 wt% doping concentration in the Fig. 1, the surface of sample after corona aging is rougher and becomes white. Meanwhile, there are many sunken traces on the surface of samples and the concavity of traces in the Fig. 7b is more obvious. In the region II, the breakdown region of PI/Al$_2$O$_3$ nanocomposite films has a distinct boundary and shows gradient stratified structure. However, the width of breakdown region is smaller and the gradient is greater in Fig. 7b.

The enlarged images of the region I (corona aging region) of Fig. 7a and b are shown in Fig. 7c and d respectively. It is clear that there are many white flocculent clusters on the surface of the two samples and the clusters have a relatively uniform distribution. The diameter of the clusters is mainly in the range of 200–1000 nm. In the initial stage of corona aging, the organic molecular chains of PI matrix on the surface of composite films are continually eroded due to the impact of electrons form electrode. At the same time, the Al$_2$O$_3$ nanoparticles as filler are gradually exposed. Several or dozens of Al$_2$O$_3$ nanoparticles combine PI matrix to form the stable cluster because of the existence of the interaction between PI matrix and nanoparticles. During corona aging, these clusters can bear the impact of electrons form electrode and cover much of surface of composite film. Therefore, they can be used as barrier layer to bear most of the impact of electrons on the surface to reduce the further degradation of PI matrix. Meanwhile, the interface between PI matrix and Al$_2$O$_3$ nanoparticles is also helpful to transport electrons from electrode and reduce the possibility of accumulating electrons in the film, which is beneficial to maintain normal structure of PI matrix.

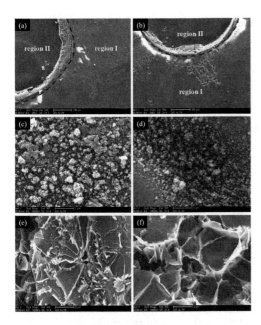

Figure 7. The surface morphology images of PI/Al$_2$O$_3$ nanocomposite films with 2 wt% doping concentration after corona aging at 25°C and 100°C.

This is the reason that the adding of Al$_2$O$_3$ nanoparticles can effectively prolong the corona-resistance life of PI/Al$_2$O$_3$ nanocomposites at 25°C comparing with pure PI films.

However, it can be observed that the size of clusters is smaller and their coverage rate on the surface is lower in Fig. 7d. This is because the increase of temperature enhances the impact of electrons on the surface of composite film so that the clusters lose the ability to continually maintain dimension by means of the interaction between PI matrix and nanoparticles. The original clusters split and become relatively small-sized clusters. One part of them stays put to serve as barrier layer to protect PI matrix; the other part is sputtered out due to the impact of electrons, which may has destruction for the structure of composite films. In this case, the area of exposed PI matrix (the deeper color area) increases and more parts of PI matrix are directly impacted by electrons form the electrode leading to relatively faster degradation of PI matrix. Moreover, it is well known that heat is always produced in the process of electric corona. We believe that the increase of environmental temperature will lead more heat to accumulate in the tested samples, which enhances the destructive effect of corona aging on the PI nanocomposites. Therefore, PI/Al$_2$O$_3$ nanocomposite film has shorter corona-resistance life at 100°C comparing with the sample at 25°C. However, the corona-resistance property of PI/Al$_2$O$_3$ nanocomposite film at 100°C should still be better than that of pure PI film due to the appearance of clusters, which coincides with the results of corona aging test. When the structure of same areas is too weak, the PI/Al$_2$O$_3$ nanocomposite film will has a breakdown because exposed PI matrix (the deeper color area) is still constantly impacted by electrons.

Fig. 7e and f are the enlarged images of the region II (corona breakdown region) of Fig. 7a and b, respectively. It can be seen that both of the two samples show sheet structure. The sample after corona aging at 100°C has bigger structural area, while there are more clusters in the sheet structure of sample at 25°C. The sheet structure can be attributed to the degradation of PI matrix. While at 100°C, the increase of temperature intensifies the impact of electrons and accelerates the PI matrix degradation. Meanwhile, in the corona breakdown region, the interaction between PI matrix and nanoparticles is weakened because the much of structure of PI matrix has been destroyed. Some of them fall into the sheet structure and others are sputtered to the surface of composite films. The sunken traces on the surface of samples in the region II can be attributed to the clusters sputtered. Moreover, the impact of electrons is stronger at high temperature, which will give the clusters greater kinetic energy. Hence, the quantity of clusters falling into the sheet structure is relatively smaller and the clusters sputtered are more destructive. These can be used to explain why the concavity of traces in Fig. 7b is more obvious and there are more clusters in the sheet structure in Fig. 7e.

4 CONCLUSIONS

Polyimide composite films containing different concentrations of Al$_2$O$_3$ nanoparticles were fabricated by in-situ polymerization. The FTIR spectrum indicates that the imidization reaction of nanocomposites is complete in this work. The plots of $\ln[q^3 I(q)]$ versus q^2 with SAXS data show a negative deviation from Porod's law, which is an evidence of the interaction between PI matrix and nanoparticles. The interfacial polarization results in enhancement of relative permittivity of PI nanocomposites. For corona-resistance property, the PI nanocomposite films have an obvious improvement because Al$_2$O$_3$ particles form clusters on the surface of composite films due to the interaction. With a 4 wt% doping concentration, the corona-resistance life is increased by more than 11 times comparing with pure PI at 25°C. When the environmental temperature increases to 100°C, the life has a significant decrease. TGA thermograms show the structure of PI matrix is not directly affected by the high temperature of 100°C, which is less likely to reduce the corona-resistance life of PI composite films. The primary reason is that the high temperature impedes the formation of relatively large-sized clusters, and more parts of PI matrix are directly impacted by electrons, which leads to accelerated degradation of PI matrix and shorter corona-resistance life.

REFERENCES

Arai, M., Noro, Y., Sugimoto, K. I. & Endo, M. (2008). Mode i and mode ii interlaminar fracture toughness of cfrp laminates toughened by carbon nanofiber interlayer. *Composites Science & Technology, 68*(2), 516–525.

Akram, S., Gao, G., Liu, Y., Zhu, J., Wu, G. & Zhou, K. (2015). Degradation mechanism of A1 2 O 3 nano filled polyimide film due to surface discharge under square impulse voltage. *IEEE Transactions on Dielectrics and Electrical Insulation, 22*(6), 3341–3349.

Cho, J., Chen, J. Y. & Daniel, I. M. (2007). Mechanical enhancement of carbon fiber/epoxy composites by graphite nanoplatelet reinforcement. *Scripta Materialia, 56*(8), 685–688.

Du, B. X., He, Z. Y., Du, Q. & Guo, Y. G. (2016). Effects of water absorption on surface charge and dielectric breakdown of polyimide/al2o3 nanocomposite films. *IEEE Transactions on Dielectrics & Electrical Insulation, 23*(1), 134–141.

Feng, Y., Yin, J., Chen, M., Song, M., Su, B. & Lei, Q. (2013). Effect of nano-tio 2, on the polarization process of polyimide/tio 2, composites. *Materials Letters, 96*(4), 113–116.

Ghosh, M.K. & Mittal, K.L. (1996). *Polyimide: fundamentals and applications*, New York: Marcel Dekker.

Haq, S. U., Jayaram, S. H. & Cherney, E. A. (2008). Insulation problems in medium voltage stator coils under fast repetitive voltage pulses. *IEEE Transactions on Industry Applications, 44*(4), 1004–1012.

Hui, S., Lizhu, L., Ling, W., Weiwei, C. & Xingsong, Z. (2016). Preparation and characterization of Polyimide/Al2O3 nanocomposite film with good corona resistance. *Polymer Composites, 37*(3), 763–770.

Li, H., Liu, G., Liu, B., Chen, W. & Chen, S. (2007). Dielectric properties of polyimide/Al2O3 hybrids synthesized by in-situ polymerization. *Materials Letters, 61*(7), 1507–1511.

Qi, K. & Zhang, G. (2015). Effect of organoclay on the morphology, mechanical, and thermal properties of polyimide/organoclay nanocomposite foams. *Polymer Composites, 35*(12), 2311–2317.

Ragosta, G., Abbate, M., Musto, P. & Scarinzi, G. (2012). Effect of the chemical structure of aromatic polyimides on their thermal aging, relaxation behavior and mechanical properties. *Journal of Materials Science, 47*(6), 2637–2647.

Sun, M., Chang, J., Tian, G., Niu, H. & Wu, D. (2016). Preparation of high-performance polyimide fibers containing benzimidazole and benzoxazole units. *Journal of Materials Science, 51*(6), 2830–2840.

Wang, H., Zhong, W., Xu, P. & Du, Q. (2005). Polyimide/silica/titania nanohybrids via a novel non-hydrolytic sol–gel route. *Composites Part A: Applied Science and Manufacturing, 36*(7), 909–914.

Wang, S., Zhou, H., Dang, G. & Chen, C. (2009). Synthesis and characterization of thermally stable, high-modulus polyimides containing benzimidazole moieties. *Journal of Polymer Science Part A Polymer Chemistry, 47*(8), 2024–2031.

Weng, L., Xia, Q. S., Yan, L. W., Liu, L. Z. & Sun, Z. (2016). In situ preparation of polyimide/titanium carbide composites with enhanced dielectric constant. *Polymer Composites, 37*(1), 125–130.

Wu, J., Yang, S., Gao, S., Hu, A., Liu, J. & Fan, L. (2005). European polymer preparation, morphology and properties of nano-sized al 2 o 3/polyimide hybrid films. *European Polymer Journal, 41*(1), 73–81.

Yang, F., Li, Y., Ma, T., Bu, Q. & Zhang, S. (2010). Synthesis and characterization of fluorinated polyimides derived from novel unsymmetrical diamines. *Journal of Fluorine Chemistry, 131*(7), 767–775.

Yang, C. P., Su, Y. Y., Guo, W. & Hsiao, S. H. (2009). Synthesis and properties of novel fluorinated polynaphthalimides derived from 1, 4, 5, 8-naphthalenetetracarboxylic dianhydride and trifluoromethyl-substituted aromatic bis (ether amine) s. *European Polymer Journal, 45*(3), 721–729.

Yao, G., Sun, Z., Gong, S., Zhang, H., Gong, Q. & Liu, L., et al. (2015). Synthesis and characterization of soluble and thermally stable triphenylpyridine-containing aromatic polyimides. *Journal of Materials Science, 50*(20), 6552–6558.

Yildirim, A. (2014). Synthesis and properties of novel high thermally stable polyimide-chrysotile composites as fire retardant materials: journal of polymer engineering. *Journal of Polymer Engineering, 34*(9), 793–802.

Yudin, V. E., Otaigbe, J. U., Gladchenko, S., Olson, B. G., Nazarenko, S. & Korytkova, E. N., et al. (2007). New polyimide nanocomposites based on silicate type nanotubes: dispersion, processing and properties. *Polymer, 48*(5), 1306–1315.

Zhang, P.P. Zhao, J.P. Zhang, K. Bai, R. Wang, Y.M. Hua, C.X. Wu, Y.Y. Liu, X.X. Xu, H.B. & Li, Y. (2016). *Composites, Part A. 84*: 428–434.

Design of CMOS bandgap voltage reference

H. Bao
Mongolian University of Science and Technology, Ulan Bator, Mongolia

ABSTRACT: In this paper, we design a low power, low temperature coefficient, high suppression ratio bandgap reference, first use Capture drawing circuit diagram, export netlist, and then use HSPICE to set simulation parameters and values, and finally use Cosmos Scope for waveform The simulation results show that the temperature coefficient is 2.6 ppm/°C at −50°C to 100°C, consumes 2 mV and has an open loop gain of 70 dB.

Keywords: Bandgap voltage source; high gain; low temperature drift

1 INTRODUCTION

In recent years, a large number of papers on voltage reference sources realized by standard CMOS technology have been reported both at home and abroad. Its technology development is mainly manifested in low power consumption, low temperature coefficient, high PSRR, low noise, and low power supply voltage (Lee et al. 2015, Crovetti 2015, Abbasizadeh et al. 2016, Cai et al. 2017).

The structure of bandgap voltage reference circuit is shown in Figure 1. M1, M2, M3 and M4 constitute the current mirror. The PTAT current is first generated, and the current is added to the resistor to generate the voltage independent of the absolute temperature.

2 STRUCTURE

Amplifier circuit structure shown in Figure 2, by the M1, M2, M3, M4, M5 composed of two stages of the first operational amplifier. M6, M7 form the second stage of the op amp. M8, M9, M10, M11, M12, M13 constitute a bias circuit. The gain of the first stage of the op amp is Av1 and the gain of the second stage of the op amp is Av2. Then the gain of the two stage op amps is Av = Av1 Av2.Av2.

3 SIMULATION

As shown in Figure 3, the abscissa is the frequency and the ordinate is the magnification. Under the normal operating range of low frequency, the open-loop gain of the designed second-stage operational amplifier is about Is 70 dB, the unit gain bandwidth is 52.053 MHz and the phase margin is 64.857°. In theory, it meets the design requirements for the bandgap reference voltage. If you continue to increase the magnification of the op-amp, you need to increase the size of the MOS transistor, The layout design brings difficulties.

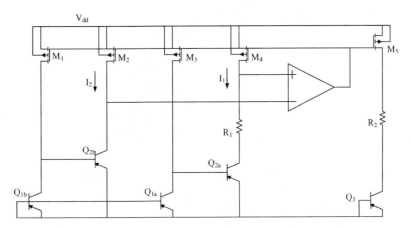

Figure 1. Structure of bandgap voltage reference.

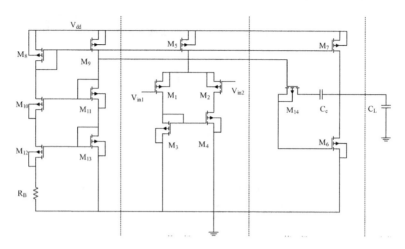

Figure 2. The structure of the two stage CMOS operational amplifier.

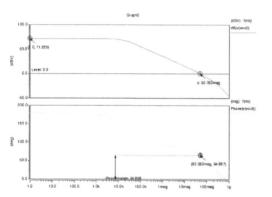

Figure 3. Simulation of DC gain and phase margin.

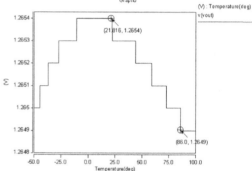

Figure 4. Simulation of temperature coefficient.

A very important parameter of the bandgap reference voltage is the temperature coefficient. The lower the temperature coefficient is, the better the temperature stability of the circuit is. The temperature coefficient of the designed circuit is simulated from −50°C to 100°C. For example, As shown in Figure 4, the maximum and minimum voltages are 1.2654 V and 1.2649 V, so the temperature coefficient can be calculated to be 2.63 ppm. The calculation is as follows:

$$T = \frac{V_{max} - V_{min}}{V_{men}(T_{max} - T_{min})} \times 10^6 (ppm/\circ C)$$
$$= \frac{(1.2654 - 1.2649)}{\frac{1.2654 + 1.2649}{2} \times 150} \times 10^6 = 2.63\, ppm \quad (1)$$

4 CONCLUSIONS

In this paper, a bandgap voltage reference with low power consumption and low temperature coefficient is presented, which is mainly composed of two stages of operational amplifier and basic bandgap structure. The simulation results show that the open loop gain of the designed second-stage operational amplifier is about 70 dB, the unit gain bandwidth is 52.053 MHz and the phase margin is 64.857° under the normal operating range of low frequency. When the temperature coefficient is between −50°C and 100°C. The temperature range is 2.6 ppm/°C, power dissipation is 2 mV, DC voltage rejection ratio is 70 db, meeting design specifications.

REFERENCES

Abbasizadeh, H., Hayder, A. S. & Lee, K. Y. (2016). Highly accurate capacitor-free ldo with sub-1 v −120 db psrr bandgap voltage reference. *Electronics Letters*, 52(15), 1323–1325.

Cai, Z., Luis, E. R. G., Louwerse, A., Suy, H., Veldhoven, R. V. & Makinwa, K., et al. (2017). A cmos readout circuit for resistive transducers based on algorithmic resistance and power measurement. *IEEE Sensors Journal*, PP(99), 1–1.

Crovetti, P. S. (2015). A digital-based virtual voltage reference. *IEEE Transactions on Circuits & Systems I Regular Papers*, 62(5), 1315–1324.

Lee, K. K., Lande, T. S. & Häfliger, P. D. (2015). A sub-μW bandgap reference circuit with an inherent curvature-compensation property. *IEEE Transactions on Circuits & Systems I Regular Papers*, 62(1), 1–9.

Amplifier design in an automatic gain control circuit

H. Bao
Mongolian University of Science and Technology, Ulan Bator, Mongolia

ABSTRACT: The principle and design of an automatic gain control amplifier is analysed in this paper, and introduces that the structure and design of the differential operational amplifier which can realise the function of the amplifier. The parameters of unity gain bandwidth and phase margin, common mode rejection ratio and slew rate are also simulated.

Keywords: CMOS, amplifier, automatic gain control

1 INTRODUCTION

The control circuit which enables the amplifier gain to be automatically adjusted with signal strength is referred to as Automatic Gain Control (AGC). It can keep the output signal amplitude constant or change it within a small range when the input signal amplitude varies greatly. This is not because the input signal is too small to work properly, or because the input signal is too large to make the receiver saturation or blockage (Moller et al., 1994).

This AGC uses the differential operational amplifier and gain negative feedback (Au & Benoit-Bird, 2003). By carefully studying the technical specifications and constructs the structure of the required operational amplifier. Once the structure is determined, it must set the tube size, design the compensation circuit, and select the DC current.

As shown in Figure 1, the circuit is composed of a voltage-controlled gain amplifier, a field effect tube, a rectifying filter circuit and a DC amplifier, and a closed loop control of achieve gain. Output voltage through the rectifier circuit and filter circuit formations voltage-controlled voltage. The output voltage is applied to the gate of the field effect tube. When the voltage-controlled voltage is changed, the resistance between the source and drain is also changed. Therefore, the amplification factor of the amplifier is also changed. When the audio signal is strong, it can automatically reduce the frequency of the amplifier, and when the signal is weak, it can increase the frequency of the amplifier automatically, thus realization volume automatic control and achieving the purpose of automatic gain control (Abdelfattah & Soliman, 2002). This circuit uses a field effect tube as a pressure control component. By changing the voltage of the gate electrode which changes the resistance between the drain electrode and the source electrode, the amplifier gain can be changed to achieve automatic gain control (Frías et al., 2017).

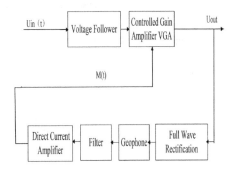

Figure 1. Integral frame structure of an automatic gain amplifier.

It consists of a controlled amplifier and a feedback circuit and the controlled amplifier is composed of a controlled and an uncontrolled, the controlled stage may be a high frequency, intermediate frequency or video amplifier. It can also be a detector or electrically adjustable attenuator. The role of the AGC detector is to properly process the amplitude of the output signal and convert it to a DC voltage that will enable gain control. The role of AGC DC amplifier is to amplify the DC control signal in order to improve the sensitivity of AGC and supply the control power required by the controlled stage. Second, it can also be isolated from the accused and the other levels in order to weaken the influence of each other. When it continuous work, AGC detector demodulation to the high frequency oscillation envelope. The transistor level circuit of an AGC amplifier is shown in Figure 2.

2 DESIGN AND SIMULATION OF A DIFFERENTIAL OPERATIONAL AMPLIFIER

The heart of the AGC circuit is a two-stage operational amplifier. Figure 3 shows a two-stage operational

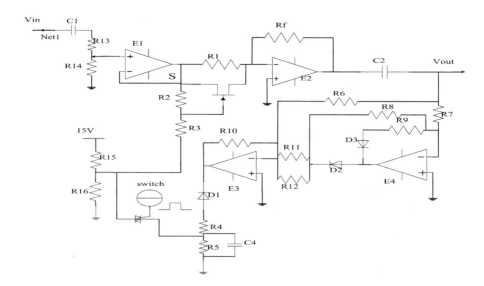

Figure 2. The transistor level circuit of an AGC amplifier.

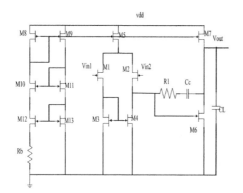

Figure 3. Circuit of a two-stage operational amplifier.

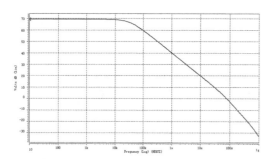

Figure 4. The simulation waveform of open loop gain.

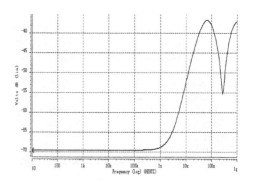

Figure 5. The simulation waveform of the common mode rejection ratio.

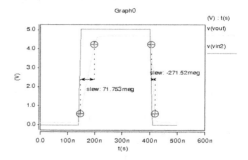

Figure 6. The simulation waveform of slew rate.

amplifier circuit. M1~M4 constitutes a differential amplifier with active load, while M5 provides the operating current of the amplifier. M6, M7 tube constitute a common source amplifier circuit and the output stage of the operational amplifier. M6 provides the operating current to M7. M8 and I1 are components of the bias circuit and provide the operating current of the whole amplifier.

As can be seen in Figure 4, the open loop gain of the amplifier is 70 dB, and the unity gain bandwidth of the amplifier is 110 MHz.

As shown in Figure 5, the common mode rejection ratio is 69.59 dB, which meets the design requirements.

As shown in Figure 6, the op amp's positive conversion rate is 71.753 V/μs and the negative conversion rate is 271.52 V/μs.

3 CONCLUSION

This paper firstly introduces the principle and design of an AGC, then designs and simulates to the differential operational amplifier in AGC amplifier. Finally, the performance of the automatic gain control amplifier is simulated. It meets the design requirements when the unity gain bandwidth is 110 MHz, the common mode rejection ratio is 69.59 dB, the slew rate of positive conversion rate is 71.753 V/us, and the negative conversion rate is 271.52 V/us.

REFERENCES

Abdelfattah, K.M. & Soliman, A.M. (2002). Variable gain amplifiers based on a new approximation method to realize the exponential function. *IEEE Transactions on Circuits and Systems I: Fundamental Theory and Applications, 49*(9), 1348–1354.

Au, W.W. & Benoit-Bird, K.J. (2003). Automatic gain control in the echolocation system of dolphins. *Nature, 423*(6942), 861.

Frías, D.M., Pascual, M.T.S. & Carlos, A. (2017). A CMOS automatic gain control design based on piece-wise linear exponential and logarithm cells. In *2017 European Conference on Circuit Theory and Design (ECCTD)* (pp. 1–4). Piscataway, NJ: IEEE.

Moller, M., Rein, H.M. & Wernz, H. (1994). 13 Gb/s Si-bipolar AGC amplifier IC with high gain and wide dynamic range for optical-fiber receivers. *IEEE Journal of Solid-State Circuits, 29*(7), 815–822.

Design and analysis of CMOS operational amplifier

S. Mallick
The National Institute of Technology Durgapur, Durgapur, India

ABSTRACT: In this paper a RC Miller compensated CMOS op-amp is presented. High gain enables the circuit to operate efficiently in a closed loop feedback system, whereas high bandwidth makes it suitable for high speed applications. A novel RC Miller compensation technique is used to optimize the parameters of gain and bandwidth for high speed applications are illustrated in this research work. The design is also able to address any fluctuation in supply or dc input voltages and stabilizes the operation by nullifying.

Keywords: CMOS; operational amplifiers, common source amplifiers

1 INTRODUCTION

Today, typical supply voltages for analog circuits are around 2.5–3 V, but future trends suggest supply voltages of 1.5 V or even less. With such low values, traditional CMOS circuit solutions can be adopted only if a low threshold process is available. Otherwise, new circuit solutions capable of working with a reduced power supply have to be designed (Lloyd & Lee 1994, Sun & Peng 2002, Yadav 2012).

In this paper, a new operational amplifier that works with a 5-V power supply is presented. It adopts a mirror load in the input stage, which saves input swing, and includes a dynamically biased common source output stage providing both a large output swing and a high output current.

2 STRUCTURE OF AMPLIFIER

Figure 1 shows the structure of operational amplifier, the first stage is differential amplifier, differential amplifier can effectively suppress the environmental noise, which is an important indicator of the amplifier.

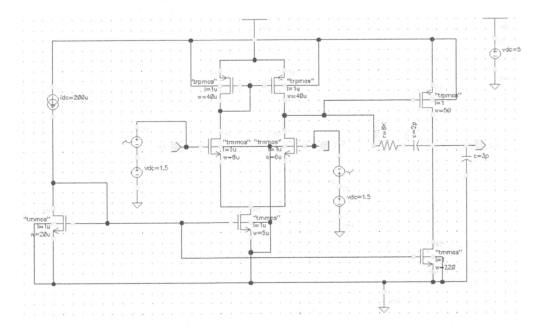

Figure 1. The structure of operational amplifier.

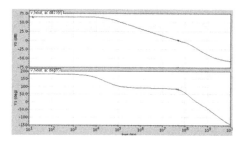

Figure 2. The simulation results of open loop gain and phase margin.

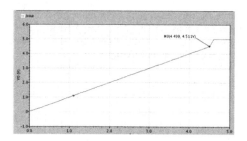

Figure 3. The simulation results of the input voltage range.

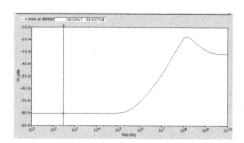

Figure 4. The simulation results of the common mode suppression ratio.

The second stage is common source amplifier, such a common source will not be because of the gain increase caused by the output voltage swing is limited, so that it can be designed to meet the large gain of the second op amp. In addition, a bias circuit is needed to provide a stable current or voltage for all stages of the amplifier.

The simulation results of open loop gain and phase margin can be seen in Figure 2, the open loop gain is 66 dB, the phase margin is 81°.

In order to measure the input voltage range, the output of the operational amplifier and the DC voltage source are connected to two inputs respectively. As shown in Figure 3, the input voltage should be between 0 V and 4.5 V.

In order to measure the common mode rejection ratio, the two ends of an AC voltage source are connected to a forward input and connected to the reverse input and output ends respectively. As shown in Figure 4, the common mode suppression ratio of the circuit is 80 dB.

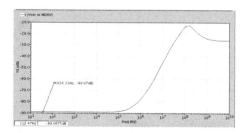

Figure 5. The simulation results of the power suppression ratio.

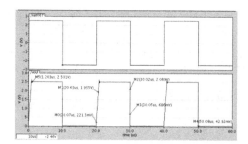

Figure 6. The simulation results of the establishment time.

In order to measure the power supply rejection ratio, the AC voltage source in connected to the supply voltage VDD. As shown in Figure 5, the power suppression ratio is 89 dB.

In order to measure the establishment time, the output of the operational amplifier is connected to the reverse input, the square wave voltage is 20 μs, the delay free signal source is connected to the positive input respectively, then the transient simulation is carried out. As shown in Figure 6, the time of establishment of the amplifier is 1.2 μs.

3 CONCLUSION

A novel RC Miller compensation technique is used to optimize the parameters of gain and bandwidth for high speed applications are illustrated in this research work. The design is also able to address any fluctuation in supply or dc input voltages and stabilizes the operation by nullifying.

REFERENCES

Lloyd, J. & Lee, H. S. (1994). A cmos op amp with fully-differential gain-enhancement. *Circuits & Systems II Analog & Digital Signal Processing IEEE Transactions on*, 41(3), 241–243.

Sun, R. & Peng, L. (2002). A gain-enhanced two-stage fully-differential CMOS op amp with high unity-gain bandwidth. *IEEE International Symposium on Circuits and Systems* (Vol. 2, pp. II-428 – II-431 vol. 2). IEEE.

Yadav, A. (2012). A review paper on design and synthesis of two-stage cmos op-amp. *International Journal of Advances in Engineering & Technology*, 34(22), 2073–2074.

Design and simulation of operational amplifier based on CMOS technology

S. Mallick
The National Institute of Technology Durgapur, Durgapur, India

ABSTRACT: This paper presents a low-power amplifier which is implemented in a commercial 0.35-μm CMOS technology and driving 10 pF capacitive load. The two-stage amplifier achieves over 120 dB gain, 1.515 MHz GBW and 1.3 V/μS average slew rate, while only dissipating 380 μW under 3.3 V supply. In the simulation software Cadence to draw the circuit diagram, using the 0.35 um technology library, the simulation, the final phase margin of 82.61 degrees, gain up to 86 dB, has reached the design requirements.

Keywords: integrated circuit; CMOS multistage operational amplifier; Cadence simulation

1 INTRODUCTION

An amplifier is needed in almost all electronic systems. Its theories and design methodologies have been well developed. However, they have to keep up with the fast advances in present-day technologies. As the channel lengths and supply voltages are further scaling down, single-stage amplifiers based on cascoding transistors are no longer possible. Instead, multistage amplifiers have come into use for low supply voltages. They will prevail especially when high dc gain is compulsory for high-precision purposes (Abou-Allam & El-Masry 1997, Haga et al. 2005, Yu & Mao 2013).

2 STRUCTURE OF AMPLIFIER

The transistor-level implementation of the amplifier is shown in Fig. 1. Transistors M_1-M_5 form the first gain stage. The second stage and the third stage are implemented by transistors M_6 and M_{11}, respectively, to boost the dc gain. C_1 is the Miller capacitor establishing the outer feedback loop. C_2 is the other feedback capacitor which along with the feedback transconductance accomplishing the internal feedback loop.

And then simulate it to get the result shown in figure 2. Its common-mode rejection ratio reaches 87.9 dB, basically meeting the design requirements.

The designed circuit power supply rejection ratio simulation results shown in Figure 3. The power supply rejection ratio of the circuit is 86.3 dB in 100 Hz, which can well suppress the power supply noise and meet the design requirements.

As can be seen from Figure 4, the phase margin of the circuit is 82.61°, the phase margin results greater

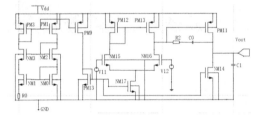

Figure 1. The circuit implementation of the TCFC amplifier.

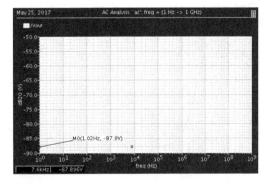

Figure 2. Simulation results of CMRR.

than 60°, indicating that the stability of the circuit designed in this article is still relatively high.

The dc gain, GBW, and phase margin are equal to 83.72 dB, 7.153 MHz, and 82.61°, respectively. The detailed performance of the amplifier is summarized in Table 1.

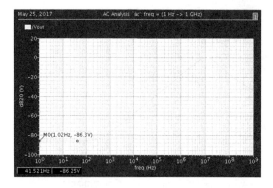

Figure 3. Simulation results of PSRR.

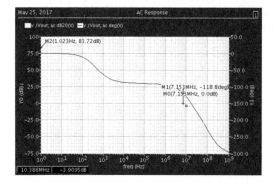

Figure 4. Simulation results of open-loop gain.

Table 1. Performance comparison of amplifiers.

Technology	0.35-μm CMOS
V_{DD}	3.3 V
Total Biasing Current	106 μA
Loading Capacitor	10 pF
DC gain	83.72 dB
GBW	7.153 MHz
Phase Margin	82.61°
Input noise	41.253 nV/$\sqrt{Hz}$ (100 Hz)
SR	1.7 V/μs
Input offset voltage	<60 μV
Output voltage Swing	0–3.3 V
Input common voltage	0.2–2.9 V
PSRR	86.25 dB (100 kHz)
CMRR	87.9 dB (100 kHz)

3 CONCLUSION

The design of a low-power amplifier is presented in this paper. It has been shown that by adequate application of negative feedback, the stability can be well ensured while the high-frequency behavior is not degraded. As demonstrated, the remarkable improvements for both small-signal and large-signal performance have been accomplished in an optimized low-power amplifier powered by low-voltage supplies.

REFERENCES

Abou-Allam, E. & El-Masry, E. I. (1997). A 200 mhz steered current operational amplifier in 1.2-μm cmos technology. *IEEE Journal of Solid-State Circuits*, *32*(2), 245–249.

Haga, Y., Zare-Hoseini, H., Berkovi, L. & Kale, I. (2005). Design of a 0.8 Volt fully differential CMOS OTA using the bulk-driven technique. *IEEE International Symposium on Circuits and Systems* (Vol. 1, pp. 220–223 Vol. 1). IEEE.

Yu, J. & Mao, Z. (2013). A design method in cmos operational amplifier optimization based on adaptive genetic algorithm. *Wseas Transactions on Circuits & Systems*, *8*(7), 548–558.

Author index

An, Y. 45
Ao, B.F. 97

Bao, H. 243, 245

Chen, L. 101
Chen, X.S. 37
Chen, Y.N. 211
Chi, R.H. 85
Cui, Z.J. 9

Ding, D.H. 31
Ding, S.Y. 75
Dong, J.H. 117
Dong, J.W. 117
Dou, J.P. 215

Fan, L.Y. 129
Fan, X.Y. 57
Fan, Y.Q. 139
Feng, Q.X. 27
Feng, X.X. 121, 159, 185

Gao, B.W. 171
Gao, J.F. 113
Gao, Q.K. 69
Gao, X.Y. 207, 211
Gu, X.Y. 1
Guan, X.G. 133
Guo, F.F. 181
Guo, H. 49, 69
Guo, J. 215

Han, D.D. 21
Han, G.H. 175
Han, W.L. 171
Han, Z.J. 215
Hao, G.N. 181
He, L. 13
He, X.F. 125
Hou, J. 121, 159, 185
Hu, S.C. 57
Huang, H. 121, 133, 159, 185
Huang, S.B. 85
Huang, Z.C. 93

Ji, C.Z. 193
Jiang, C.Y. 37
Jiang, J.X. 121, 159, 185
Jiang, S.Y. 201
Jiang, Y.D. 109

Kang, G.F. 93
Kong, W.W. 53
Krasnoproshin, V.V. 37, 109, 113

Lai, Y.Y. 145
Li, J. 225
Li, L.Y. 5
Li, Q. 201, 221
Li, W.T. 69
Li, Y. 85
Li, Y.M. 85
Li, Y.S. 45
Li, Y.W. 13
Liang, F.Y. 155
Liang, H.Y. 31
Liu, C. 89, 117
Liu, C.C. 215
Liu, F. 97
Liu, H. 37, 37, 109, 113
Liu, J.F. 89
Liu, J.H. 17
Liu, X. 81
Liu, X.C. 27
Liu, Y.N. 175
Liu, Y.P. 113
Liu, Y.T. 207
Lu, H.Z. 61
Luo, G. 201

Ma, C.P. 17
Ma, J.J. 53
Mallick, S. 249, 251
Minh, S.T. 237

Ni, W.Q. 221
Ni, X.W. 225

Qiao, Y.J. 139
Qin, M.Y. 229, 233
Qiu, Y.H. 53
Qu, L.W. 37

Ren, D. 215
Ren, M.Y. 41, 229, 233

Shang, Q.H. 155
Shao, J.P. 189
Shen, H.B. 133
Shi, X.J. 69
Shi, Y.C. 175
Song, B.Z. 229, 233
Song, G.M. 201

Song, J.J. 129
Song, M.X. 57
Song, Z.L. 193
Sun, J.D. 5

Tian, H.P. 97

Wang, B. 81
Wang, C.Y. 21
Wang, D.X. 27
Wang, H.Q. 181
Wang, L. 133
Wang, N. 97
Wang, Q. 163
Wang, S.K. 189
Wang, W. 1
Wang, W.X. 61
Wang, X. 133
Wang, X.F. 125
Wang, X.X. 85
Wang, Y. 75
Wang, Y.H. 61
Wang, Y.J. 101
Wang, Y.R. 13
Wei, Y. 193
Wong, K.M. 125
Wu, C. 201
Wu, X.L. 31
Wu, Y.H. 81
Wu, Y.X. 89

Xie, J.X. 61
Xin, X. 61
Xu, J. 53
Xuan, H.W. 109, 113

Yang, B. 125
Yang, C.X. 125
Yang, H.J. 21, 41
Yang, M.Z. 27
Yang, Z.Y. 221
Yao, D.J. 151
Yao, H.B. 221
Yu, B. 109, 113, 193
Yu, G.X. 61
Yuan, D.Q. 221
Yuan, F. 201
Yuan, H. 133

Zang, X.W. 53
Zhan, X.J. 151
Zhang, B. 109, 113

Zhang, C.J. 175
Zhang, C.X. 109, 113, 207, 211
Zhang, G.Z. 163
Zhang, H. 13
Zhang, J.N. 129
Zhang, L. 13, 225
Zhang, R. 45

Zhang, S. 37, 93, 109, 113
Zhang, X. 215
Zhao, C. 181
Zhao, F.Z. 31
Zhao, S. 9
Zhao, Y.Y. 121, 159, 185
Zhou, L.X. 101

Zhou, Y. 101
Zhu, H.L. 89
Zhu, J.L. 133
Zhu, M. 9
Zhu, S.X. 5
Zhu, Y.T. 215
Zuo, Y.Y. 37

PGMO 07/26/2018